The Lyre Book

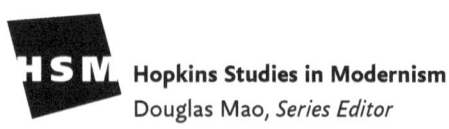

Hopkins Studies in Modernism
Douglas Mao, *Series Editor*

The Lyre Book

Modern Poetic Media

Matthew Kilbane

Johns Hopkins University Press
Baltimore

© 2024 Johns Hopkins University Press
All rights reserved. Published 2024
Printed in the United States of America on acid-free paper
9 8 7 6 5 4 3 2 1

Johns Hopkins University Press
2715 North Charles Street
Baltimore, Maryland 21218
www.press.jhu.edu

Library of Congress Cataloging-in-Publication Data

Names: Kilbane, Matthew, 1988- author.
Title: The lyre book : modern poetic media / Matthew Kilbane.
Description: Baltimore : Johns Hopkins University Press, 2023. | Series: Hopkins studies in modernism | Includes bibliographical references and index.
Identifiers: LCCN 2023011254 | ISBN 9781421448114 (hardcover ; acid-free paper) | ISBN 9781421448121 (paperback ; acid-free paper) | ISBN 9781421448138 (ebook)
Subjects: LCSH: American poetry—20th century—History and criticism. | Lyric poetry—History and criticism. | Poetics. | Intermediality. | Literature and technology.
Classification: LCC PS309.L8 K55 2023 | DDC 811/.509—dc23 /eng/20230802
LC record available at https://lccn.loc.gov/2023011254

A catalog record for this book is available from the British Library.

Special discounts are available for bulk purchases of this book. For more information, please contact Special Sales at specialsales@jh.edu.

For my parents

Contents

List of Figures ix

Introduction 1
Writing Lyric's Media History 1
Writing Media's Lyric History (Lorine Niedecker's
  Bad Reception) 8
Lyres and Scores 18
Records and Transcriptions 29

1 **A Speech-Musical Modernism** 47
   New Arts in New-Old Media 47
   Harry Partch and Speech-Music 2.0 55
   Bitter Music and the Social Indexical 62
   Objectivist Music 71
   Celia Zukofsky, Music's Master 80

2 **Latent Remediation: Radio, Poetry,
   and Social Process** 89
   Poetry's Remedial Fantasy 89
   Verse + Radio = Poetry: John Brooks Wheelwright 93
   Broadcasting Dialect: Sterling Brown 111

3 **Langston Hughes's Songwork** 132
   Lyric Burdens 132
   Lyric/Lyrics, Media/Genre 135
   The Traffic of Songwork 141
   Wartime Kitsch and the Dummy Lyric 152

4　**"In Lieu of the Lyre": Lyric Objects and the Audio Record**　169
　　A Poem's Two Objects　169
　　Marianne Moore, Verbal Pilgrim　174
　　Russell Atkins, Literalist of the Imagination　186

Coda. Helen Adam and the Invention of Poetry　209
　　At "Limbo Gate"　209
　　Lyric Suspension　218

Acknowledgments　231
Notes　237
Bibliography　283
Index　321

Figures

I.1. Jean Elliot, "The Flowers of the Forest," in Sir Walter Scott's *Minstrelsy of the Scottish Border*, 1802, pp. 158-59 34

I.2. Jean Elliot, "The Flowers of the Forest," in *Halfpenny Lyre*, ca. 1840 37

1.1. Harry Partch and his Adapted Viola, 1933 48

1.2. Florence Farr and her psaltery, 1903 53

1.3. Harry Partch, "By the Rivers of Babylon," manuscript, 1932 56

1.4. Florence Farr, "Psalm 137," in *The Music of Speech*, 1909, p. 24 57

1.5. Harry Partch, *Bitter Music*, 1991, p. 128 67

1.6. Harry Partch, *Bitter Music*, 1991, p. 95 68

1.7. Harry Partch, *Bitter Music*, 1991, p. 64 69

1.8. Harry Partch, *Bitter Music*, 1991, p. 26 70

1.9. Celia Zukofsky, *Pericles*, manuscript, 1943 73

1.10. Celia Zukofsky, *Pericles*, in *Bottom*, vol. 2, 1963, pp. 216-17 84

1.11. Cyril Satorsky, frontispiece, in *Bottom*, vol. 2, 1963, p. 7 86

2.1. John Wheelwright, dust jacket, *Political Self-Portrait*, 1940 109

2.2. Norman Corwin, script from *Words without Music*, January 15, 1939, p. 14 117

3.1. Langston Hughes and Herbert Kingsley, "Hard Luck Blues," n.d. 147

3.2. Langston Hughes, worksheet for "Honolulu Yaka-Hula Dixie," 1942 154

3.3. Langston Hughes, prosodic analysis of Edgar De Lange and Sam H. Stept's "This Is Worth Fighting For," n.d. 165

3.4. Langston Hughes, schematic analysis of Edgar De Lange and Sam H. Stept's "This Is Worth Fighting For," n.d. 166

4.1. Transcription of Marianne Moore's "In Lieu of the Lyre," Sanders Theatre, May 11, 1965 180

4.2. Russell Atkins, psychovisual Object-Form, in "A Psychovisual Perspective," 1958, p. 39 194

C.1. Helen Adam and Carl Grundberg, score for "Limbo Gate," in *Songs with Music*, 1982, p. 41 215

The Lyre Book

Introduction

Writing Lyric's Media History

What is "the medium of poetry"? I wrap this phrase in quotation marks to take my distance from a question that is both ill founded and unavoidable. Although a study of poetry's media history surely ought to begin by asking after the medium in which poems are written and read, I argue in this book that poetry's singular relation to the media concept renders the present form of that question bound to misfire. The problem is not simply that the question lacks one definitive answer. Certainly if we desire in a response anything more specific than Ezra Pound's commonsensical assertion, "the medium of poetry is WORDS," then the possibilities do proliferate, from handy abstractions like speech, writing, and print, to innumerable examples thereof: goose quill on vellum, linotyped print on broadsheet newspaper, mimeographed zines, and so on.[1] The contemporary US poet Robert Pinsky, meanwhile, still finds reason to claim that the "medium of poetry is a human body," and this at a moment when some writers are busy stretching "poetry" beyond human language itself—into binary code or bacterial DNA.[2] One is tempted to call the rich confusion a case of practice outstripping theory, but that would imply the prior existence of a theoretical consensus. Has the matter of poetry's medium ever enjoyed such agreement? It seems not.

To complicate the issue further, the phrase "the medium of poetry" can be heard genitively, too, as if poetry itself were a medium in the sense familiar from the visual arts (oil paints, charcoal, marble, etc.), or with reference to a unique expressive capacity tantamount to the poetic as such. Though he doesn't speak of poetry's "medium" per se, it's this latter sense that Roman Jakobson substantiates in his well-known description of the "poetic function." When language users "focus on the message for its own sake," the words and not their meaning, sound rather than sense, they engage the poetic medium, Jakobson argues.³ "Promoting the palpability of signs," a writer is a poet when they arrange words not only with reference to ideational content but also with an ear to how strings of material signifiers—equalized at the level of syllable and stress—combine to form larger schemes and structures.⁴ The medium of poetry is not words, nor writing, then, but a special handling of phonic substances, an orientation toward language from which emerges the possibility of the word patterning we call verse. According to Jakobson, the medium of poetry, its métier, is simply a generative concern for its own linguistic mediation.

John Guillory has noted the striking resonance between the poetic function's concern with "the message for its own sake," introduced by Jakobson in 1958, and that founding salvo of media studies, "the medium is the message," issued by Marshall McLuhan just six years later.⁵ This book exists in part to explore why it's no surprise that media studies and poetry should receive such apparently congruent definitions at precisely *this* postwar moment. For now, we need only spotlight a revealing mistake in Jakobson's analysis. The comparison with McLuhan leads Guillory to observe that Jakobson's exclusive focus on the sound of poetry betrays him into distinguishing, unjustifiably, between the "poetic function" and what he terms the "phatic function," the meta-communicational operation that secures the channel or contact (*Can you hear me now? Check, check! Is this thing on?*).⁶ In Guillory's estimation, a theory of poetic media cannot limit itself to language's phonic substance, or even to language as such, but must convene the poetic and phatic functions together, the "palpability" of the message *and* the physical substrate of the channel, the sound of "WORDS" and their manifestation in written characters, in codex books, in the human bodies they set resonating, or in the digital recording of one such sounding body replayable on a glowing screen. A literary-media theory equal to McLuhan's challenge must encompass poetry *as* and *in* media.

And yet to yoke the phonic and the phatic is hardly a straightforward operation. The fine "palpability" of patterned sound appears categorically different from the artifacts more commonly construed as media by literary historians—from the modernist little magazine, say, a print format that galvanizes social networks, or from Instagram, a digital platform that monetizes them.⁷ To set up a media concept adequate to both a poem's sonic texture and its material instantiation would require traversing scales of analysis and methodological frameworks in ways that threaten to do justice to neither prosody nor history. Here lies the sundering crack in the foundation of our opening question, "What is 'the medium of poetry'?" In strict observance of this fault line, this book contends that poetry on the printed page—what Louis Zukofsky terms "audibility in two-dimensional print"—furnishes a uniquely strange kind of media object, one structured in tension between these two wedded palpabilities.[8] Clearly, Pound's piece of common sense—"the medium of poetry is WORDS"—does not say nearly enough, though attempts to reduce the issue of poetry's media to any contingent substrate (writing, print, book, YouTube video, etc.) must also heed Jakobson's insight that poetry is, by definition, the language practice that reflects a heightened concern for its own status as mediating substance. In short, discussion of poetry's media must involve the material channel of "WORDS," their potential shape in sound, and—what is most challenging—the dynamic relation linking both.

Certainly, one could accuse me of muddling the issue right off the bat by turning up all this complexity. After all, no one is likely to dispute the fact that in the twentieth century, poets published and readers read the vast majority of their poems from *printed media*—from books, yes, but also from pamphlets, magazines, newspapers, and sundry other pulp artifacts of the age of mass print. Poetry has long been at home on the printed and manuscript page, and even today physical poetry collections and their small-press publishers endure as central institutions of US verse culture. Perhaps, then, the print condition of the past century's poetry is a fact both inarguable *and* unremarkable. Perhaps the long cultural processes inaugurated by the invention of alphabetic writing and then by moveable type have so determinedly set the material horizons of poetic practices in the West that all this talk about poetry *as* media is just a confusion of terms, and the question of modern poetry's media really does begin and end with writing and print.

One initial reason for doubting this conclusion is the commonly acknowledged sense that although poems are printed, they're also something more. This intuitive awareness of an incongruity between page and poem is indexed most directly by the word *lyric*, which I have thus far held in abeyance. A deeply contested term, lyric's meaningfulness as the name of a transhistorical object or genre has been the subject of vigorous debate in contemporary criticism. This book will frame lyric not as a genre or mode but as a minimal media condition; it is one good name for the fact that while writing is essentially silent and page bound, poems are (always sometimes) not. One need not hypostasize a "lyric tradition" to agree that the long history of poetry comprises extensive and variegated projects for exploring the relations between word and sound, broadly understood. Whatever else poets, readers, and scholars mean or have meant by the word, *lyric* identifies the kind of writing that harbors the possibility of sound and audition—while also, as we shall see, hailing the interpretive and ideological baggage that this possibility inevitably carries in its wake.[9] First and foremost, however, in this book *to write lyric* will mean to make a notional "music" in language. Notwithstanding innumerable links between poetry and veritably musical practices in the two millennia since Alexandrian scholars first inscribed in writing the so-called lyrics composed and sung by ancient Greeks, the chief means of crafting and sharing lyric sound has been to charge silent writing, and then print, with sonorous potential.

This book studies the tension between sound and writing that has for many centuries characterized the reading and writing of poems. It is also a much narrower inquiry into mid-twentieth-century poetic practice in the United States. My focus on English-speaking American poets and composers (W. B. Yeats and Florence Farr are the lone exceptions) is largely contingent, reflecting the limits of my expertise. As such I hope readers of this book will look eventually beyond the small US cohort I have assembled here and apply its conceptual resources to the media-theoretical engagements of other poets such as Kyn Taniya, Rabindranath Tagore, Gottfried Benn, and Nicolás Guillén, to name at near random just a few who might have featured in different incarnations of the chapters to follow.

My address to the mid-twentieth century, however, is vitally motivated: in the machinic din of the past century, for the first time in a long time, the possibility arises of making poems otherwise. Genuinely new in the period roughly concurrent with Anglophone modernism is poetry's encounter with other, louder, more popular technologies for the recording, reproduction, and transmission of sound and speech, most notably telephony, phonography, and broadcast radio. Between the Second Industrial Revolution and the full emergence of US mass culture at midcentury, machines for technologizing sound denaturalize poetry's relationship to silent writing such that the printed poem, which already bears within it a riddling tension between writing and sound, becomes a still more generatively destabilized media artifact. If our hesitation today before the question of poetry's media is to any extent a legacy of this twentieth-century confrontation with the recording and broadcast of sonorous language, we can answer it only by analyzing, in some historical detail, how these powerful new technologies tested, challenged, expanded, and contoured the media problem at the heart of lyric practice. But to understand how lyric responds to modern media conditions, we must first set into methodological order the longer history from which this twentieth-century episode proceeds. We must ask what lyric is and how to write its media history.

Despite its long dependence on the written page, there are few more enduring commonplaces in literary history than poetry's so-called musicality—the prosodic, ideological, and historical claims to sound that render otherwise silent texts so variously audible. Either as the putative transcriptions of past speech acts or scripts for future ones, the poems often called lyric resist and unsettle their print medium by bringing the silent page into contact with other media, not least the speaking and listening body. The word *lyric* itself alludes to a musical technology—the ancient Greek *lyre*—though lyrics have had little to do with actual instruments since Alexandrian scholars inaugurated a *written* lyric tradition in the second century BC.[10] The etymological persistence of the lyre nevertheless brands lyric as an exemplary media problem, one defined by the historically variable, phenomenally mercurial tension between writing and sound.

I refer to this tension as lyric *intermediality*. N. Katherine Hayles has employed the same term in a posthumanist register to describe the recursive "complex and entangled interactions" between computational machines and human users that sustain our digital environments today.[11] The concept of intermediality developed in this book projects a similar sense of entanglement—as a media condition, lyric is produced through irreducible convolutions of writing and sound. I depart from Hayles, however, by tracing lyric intermediality not to computation but to the basic collocation of visual and sonic systems at the crux of alphabetic writing itself, wherein characters are arranged and combined to represent phonemes.[12] In this sense, all alphabetic writing is intermedial, from newspapers to novels to this very sentence.[13] Unique to poetry is only its supercharged and self-conscious investment in this intermedial condition, an investment complexly overdetermined on the planes of poetic form, cultural history, and aesthetic ideology.

From the quantitative meters of ancient Greek and Latin to Chaucer's iambic pentameter to the ballad stanzas of Emily Dickinson, Charles Olson's projective line, and the page-splay of M. NourbeSe Philip's *Zong!* (2008), the patterns and techniques we fold into the word *prosody* are rooted in alphabetic protocols—in the fact that poetic writing appears to index or prompt sonorous speech. The Greek *prosōdía* (προσῳδία) refers to song with musical accompaniment, and the implication that poetic prosody supplies a superadded, non-semantic accompaniment to verbal meaning will remain with poetry long after poets themselves stop singing to music. Like poetry's conventional associations with lyres, lutes, and sung song, the notion that prosody endows poetic language with an immanent musical accompaniment is an ideological habit buoyed by deep-running nostalgia for poetry's idealized union with song, a nostalgia shared by postclassical humanists, ballad-mongering Romantics, cultural nationalists of many stripes, and latter-day modernist troubadours like Pound, in whose influential estimation "poetry begins to atrophy when it gets too far from music."[14]

If the ubiquitous association of poetry with song is endorsed by the phenomenal experience of lyric sound, it's sealed by the historical actuality of Western poetry's close and generative proximity to real musical practices. A long view of lyric's history, and certainly its *media* history, must look beyond the nostalgic lodestar of poetry's antique union with song to its innumerable relations with cultures of song across time, instanced "on the ground" by entire artistic traditions like opera, troubadour poetry, and hip-hop, as well as by more local, incidental encounters between literature and music of all kinds, religious, folk, commercial, and otherwise. This book's third chapter, on Langston Hughes's songwriting career, explores just such a collision between lyric and song lyrics. For now we need only reflect in passing on the poverty indeed of an English literary history devoid of its ballads, hymns, triolets, villanelles, sapphics, bob and wheels, madrigals, blues stanzas, and other poetic forms appropriated from the realms of song.[15] As a critical heuristic, intermediality attempts to hold together diverse claims to sound and music. The concept yokes the elaborate prosodic structures that alphabetic writing makes possible and the cultures of recitation and performance those sonorous structures encourage alongside literature's busy historical traffic with musical practices and those constant counterfactual avowals by poets, from Propertius to Plath, that lyric just *is* the writing of song.[16] All literatures in alphabetic languages are minimally intermedial, operating between writing and sound. The literature we have consistently called poetry, however—and then lyric poetry especially—is the special kind of writing that takes intermediality as its motor of formal innovation and one chief source of its cultural appeal.

My previous sentence's halting emphasis on lyric sharpens the question of just what I intend by the word. For nearly two decades now, critics skeptical of lyric's elasticity as a generic category have urged important debate in Anglophone poetry studies regarding the meaning of lyric itself. Recognizing that this "archigeneric" designation for short, quasi-musical expressions of a speaker's interior life is a modern invention of the Romantic era, scholars have gone further to maintain that lyric ought to be grasped less as a functioning category for certain poems than as an institutionally codified set of readerly practices.[17] According to this materialist argument, the post-Romantic work of "lyric reading"—a formulation borrowed from Paul de Man and given historicist weight by Virginia Jackson and Yopie Prins—abstracts and subsumes actually existing poetic genres and historically specific scenes and circuits of reception into artifacts of a transcendent lyricism, in which account lyric becomes a context-free ideological container for the imposition of normative subjectivities.[18] In pointed response to such claims, critics like Jonathan Culler have renewed the project of formulating a transhistorical theory of lyric, identifying a set of paradigms and possibilities operating across literary history to produce an identifiable genre of actual poems we can gainfully call lyrics. For Culler, it's on the bases of their constitutive "effects of presentness," their "rich texture of intertextual relations," and "the materiality of lyric language that makes itself felt as something other than signs of a character and plot" that lyric poems can be classed productively as ritual, nonmimetic literary events harnessing the seductive force of patterned language to achieve cognitive and somatic effects—to teach and delight, as Horace says.[19] The disagreement between Culler and those scholars hewing to Jackson and Prins's account revolves essentially around the question of whether *lyric* describes certain poems themselves or rather a historically specific (and specious) way of reading them.

Lyric's media history, by contrast, requires a different sense of its object, one that will lead us to reframe these debates about history, genre, and the sticking point of expressive subjectivity. At several moments throughout this book we will have cause to reflect on how the targets of contemporary critique—namely, lyric's semantic incoherence and ideological inflation—have their roots in what I call *lyric intermediality*. But to avoid unnecessary confusion, it bears repeating at the outset: in the pages to follow, *lyric* refers not to a reading protocol, nor to a transhistorical genre, but to a material condition: the irreducible tension between various types of writing and print, and the claims to sound and audition that materially animate the literary artifacts we have called lyric, as well as a range of similarly intermedial poetic and musical works. As a consequence of this capacious definition, we have occasion to study works that appear no less *lyric* for their overt engagements with a multitude of genres: ballads, blues songs, nursery rhymes, ethnographic notations, closet operas, and musical-theoretical essays. To the charge that this approach indulges an erroneously "super-sized" idea of lyric, I can reply that this book angles merely for a kind of descriptive minimalism.[20] In short, when I use the word *lyric* (or *lyric writing*, *lyric practices*, *lyric materials*, *lyric form*, and *lyric sound*), I am referring to intermedial writing, and not to any generic or historical claims regarding "lyric poetry" as such—claims that, I hasten to add, lyric intermediality underlies, makes possible, and even solicits. By lyric, I mean writing that sounds.[21]

At the same time, this materially minimal account goes hand in hand with a genuinely speculative wager: this book seeks to reactivate lyric's etymological connection to actual lyres. The following chapters will uncover some real-life latter-day lyres (Harry Partch's Adapted Viola, for instance) as well as a few consequential lyre tropes (the radio as a new media lyre or the lyre as a token of poetic autonomy), but lyric's lyre will also come more generally to sign and certify poetry's intermedial condition—the sense of generative material conflict at the heart of lyric practices. To affirm the lyre in lyric writing is to insist on the utility of media and sound studies for accurately describing those material dimensions of poetic practice that everywhere underpin Culler's theory of lyric as a ritual event in language. Meanwhile, though the bold anachronism of recalling the lyre to lyric at first threatens to muddy the waters of an already problematic term, in the final analysis this gesture can in fact enrich a rigorously historical poetics.[22] The media concept of the lyre will permit us to historicize what critical accounts too often abstract as lyric "voice," or reduce and redescribe as prosodic rhythm, or fail to record entirely: the sound of lyric in the twentieth century.[23]

Writing Media's Lyric History
(Lorine Niedecker's Bad Reception)

The sound of lyric attunes us to other, extraliterary frequencies. Detailing the media history of midcentury poetry involves, first and foremost, situating that poetry as an intermedial practice of sonorous writing. And yet, simultaneously, this book also enlists this poetry in an effort to reconstruct the history of *other* sound technologies in the twentieth century: that is, to write media history *from the perspective of lyric*. It will be one major argument of this book that intermedial writing renders the poetic page uniquely sensitive to the media-infrastructural developments and technological upheavals giving texture and shape to everyday life in the first half of the twentieth century in the United States. As a consequence, lyric forms reflect and document the wider media ecologies in which they are produced and circulated.

To the degree that written poems are concerned with their own claims to sound, they tend to bear the impress of a culture's other means for reproducing speech. If this seems especially true of what David Trotter calls the "first media age," the midcentury moment beginning with the widespread adoption of broadcast radio in the 1920s, it's because in this period technologized mass media, previously the province of spectacular marvels and newfangled oddities, becomes for the first time a daily, omnipresent, and indeed "overwhelming" fact of modern life.[24] The poetic and musical works studied in the following pages were all fashioned in a media context of "ubiquitous communication," wherein telephony, phonography, radio, and cinema had fully emerged to join photography and mass print as legible and increasingly familiar components of the cultural-industrial ecosystem.[25] The pages to follow will probe some of the unpredictable ways this media context reset the terms for modern poetry.

It's lyric writing's apparent *resistance* to the shattering influence of new machines—the key respects in which its materials do *not* keep up with the times—that makes poetry acutely relevant to the writing of media history. And so before rushing on to examine the influence of unprecedented technological advancements, we'd be right to expend a brief spell wondering at just how *little* poetry was transformed by developments that would otherwise so momentously alter the patterns of everyday life and the fabric of social relations, shaping everything from habits of sensory perception and psychic desires to the very spatial and temporal coordinates that subjects use to imagine a common world, leaving Americans indelibly changed as consumers, citizens, and social beings. Though media theorists like Friedrich Kittler have asserted a causal link between technologized sound and the modernist poem's newfound self-consciousness as a material object—whether as the sonorous "word as such" or as printed "black riders" careening across the white page—in 1935 W. H. Auden found ready cause to reaffirm on behalf of all poets in English: "Of the many definitions of poetry, the simplest is still the best: memorable speech."[26] In the first media age, even after machines like the phonograph had usurped poetry's ancient mnemotechnological function, the end of poetic writing remains the inscription of sound on the page with an ear to iteration.[27] Intermedial writing, in the twentieth century, is an increasingly "residual" practice, to borrow Raymond Williams's term, but the residual position is by no means an irrelevant one. It may even prove a condition of aesthetic innovation and critical purchase—for all its outmodedness, "an effective element of the present."[28]

In tracing how printed poems respond to new technical formations like phonography and broadcast radio, this study will show that these negotiations are important not only for what they reveal about the intermedial substance of poetry—"memorable speech," now and then, as ever—but also for what they disclose about phonography and radio themselves, technologies still dynamically under construction in the first half of the twentieth century. The relative familiarity of these machines by 1935, for instance, must not obscure the fact that their cultural and political meanings were by no means settled or fully institutionalized. Decades of intellectual work on the social construction of technology have taught us to regard media technologies as the processually realized expression of social needs—as complex assemblages of machines and protocols organized and developed at the unruly intersection of users, markets, and institutions.[29] When it comes to the study of modern sound media, poetry can be a research partner in efforts to fill out the social history of technologized sound.

Consider the case of radio. By the late 1930s, radio broadcasting in the United States had graduated into what would later be christened its golden age, the oversimplified outlines of which remain familiar to us today: three major networks (NBC, CBS, and the Mutual Broadcasting System, or MBS) exercised market dominance through affiliations with just under half of all stations nationwide, political figures like Franklin D. Roosevelt and Father Coughlin (and from overseas, Chamberlain, Mussolini, and Hitler) were testing radio's powers to convene and mobilize "intimate publics," and innovative radio dramas by Orson Welles and Norman Corwin were bending the ear of a nation tuning in, on average, five hours each day.[30] While nothing in this recognizable profile of the medium can be accounted wrong, per se, in seeking to recover a more particularized and variegated sense of how radio was experienced by listeners in the 1930s and how it came to possess these cultural meanings so retrospectively ready to hand, we must also consider aspects of broadcasting history that unsettle and contravene this picture. It would be a mistake, for instance, to associate the radiophonic voice too essentially with fascist propaganda, or with the charismatic paternalism of the New Deal state, since—as the radio radical John Wheelwright illustrates in chapter 3—the political horizons of long- and shortwave radio were much more undecided, even as late as the 1930s. Likewise, we would be wrong to render 1930s radio synonymous with the prestige programming of major networks when, in fact, three-fourths of all news programs were produced by local stations across the country.[31] Radio didn't subsume the regional so much as shoot the local through with cosmopolitan difference, as a case study of the Wisconsin poet Lorine Niedecker can help us to appreciate.

I include a discussion of Niedecker in this introduction because radio's golden age was also the period of the poet's "folk-tongue," the hybrid mode of transcriptional composition that issued in the short, numinous poems of "anonymous authorship, of proletarian origin, and of subtly subversive intent" collected in her first book, 1946's *New Goose*.[32] "Looking around in America, working I hope with a more direct consciousness than in the past," as she wrote to Harriet Monroe in 1936, the Niedecker of the *New Goose* period relinquished the aggressively surrealist strategies of her earliest poems for modes of social commentary befitting the poetic fashions and political exigencies of the moment.[33] As Niedecker began formulating her new speech-based poetics, it was the satirical, apotropaic dimension of Mother Goose rhymes—their "carnivalesque puncturing of authority"—that afforded her the "more direct," more critical purchase she'd been looking for.[34] *New Goose* "speaks and sings against all that's predatory in 'Mother Goose,'" ran Louis Zukofsky's jacket copy for the book, though it levels its satire not at early modern nobility but at the rapacities of industrial capitalism, the violent legacies of settler colonialism, the straits of contemporary gender relations, and the social turmoil of the Depression and World War II.[35]

In the first *New Directions* annual of 1936, Niedecker published the following poem in a group she titled "Mother Geese":

> In speaking spokes the mighty
> come down from welding wires
> to light up the farmers
> with electricity.
>
> For sun and moon and radio
> farmers pay dearly;
> their natural resource: turn
> the world off early.[36]

When she published *New Goose* ten years later, Niedecker left off the first of these two quatrains, but both stanzas are fair examples of the chiming, riddling density she could achieve by marrying "everyday speech" to the balladic and epigrammatic textures of nursery rhymes.[37] Trademark products of her poetic "condensery," each stanza presents readers with a knot of semantic ambiguities drawn tight around several figural indecisions.[38] For instance, we might visualize the "speaking spokes" of "mighty" voices broadcasting radially out from some inferred transmitter to the "welding wires" of receiving antennae on farmers' radio sets. Because "electricity" suggests another infrastructural system altogether, however, we can also refer those "wires" to electric lines "welding" or joining isolated communities. One could resolve this infrastructural confusion, to be sure, by taking "electricity" figuratively, as that affective charge the farmers receive from the "mighty" voices in their living rooms, or by recalling that radio waves are indeed properly electromagnetic. But if we are tuned in to the satirical mood Niedecker is channeling from the original Mother Goose, another more persuasive reading emerges: the poem slyly implies how its farmers might *assume*—confusing the networked, invisible phenomena of radio waves with electrical currents—that electricity *is brought to them by*, courtesy of, the "mighty" voices coming through on the plugged-in radio. The social medium and the technical apparatus are ideologically fused in a feint of the powerful, and Niedecker offers us something of a media-minded Marxist critique in four lines and two rhymes.

The second stanza ranks radio among the "sun and moon," a crossing of nature and technology that recalls futurist techno-fantasies (Velimir Khlebnikov's "radio of the future" as "the spiritual sun of the country," for example, or F. T. Marinetti and Pino Masnata's "la radia" that will "destroy the hours of the day and night") while also registering, more practically, that radio was "the first electronic medium to program day and night," as Kittler reminds us.[39] But if this first line naturalizes radio, distantly indexing broadcast's etymological roots in agricultural practice (the "broadcasting" of seed), it also technologizes nature: the "sun and moon" become the ironic light bulbs that render these natural light sources obsolete and that "farmers pay dearly" to keep lit. The adverb nods at monetary straits and—more gravely—at the costs in human toil, these last invoked by the connotation of Edison's extended working day. But Niedecker observes a compensatory "natural resource" in the farmers' ability to hit the lights and silence the "mighty" voices of radio, to "turn/the world off early." The discourse of radio's celebrated connectivity in this case feeds an ironic fantasy: if, as Hadley Cantril and Gordon Allport claim in *The Psychology of Radio* (1935), "a turn of the wrist immeasurably expands [the] personal world" as the "poor man escapes the confines of his poverty," "the country dweller finds refuge from local gossip," and "the villager acquires cosmopolitan interests," then a flick in the other direction figures the revenge of the non-hegemonic periphery.[40]

We can also sense in the poem's wry punchline what Marshall Berman has described as the dialectic of "thrill and dread" native to the "maelstrom of modern life": those swept up in the rapids of modernization "are moved at once by a will to change—to transform both themselves and their world—and by a terror of disorientation and disintegration, of life falling apart."[41] The possibility of "turn[ing]/the world off early," of making all that is solid melt *on* the air (to adjust Berman's Marx), is both a thrill to imagine and a disaster to fear. In miniature, this dialectic informs the *jouissance*-tinged sensation of flipping off the radio dial and silencing a technological medium of such wide horizons. The action engenders a fantasy of enormous agency, but of course your own private apocalypse will mean nothing come tomorrow; yours is not one of those "mighty" voices on high.

However unassuming, "In speaking spokes . . ." resonates with a strain of critical-philosophical inquiry similarly concerned with the experience of switching off, with the "natural resource" of concerted refusal. In the early decades of radio, as Kate Lacey recounts, the possibility of shutting down the apparatus was "a constant refrain for critics of the numbing and isolating effects of broadcasting."[42] For Marxist critics in particular, recourse to the dial—that least surface of contact between the individual and the alienating technical system—signals an ironic and ineffectual protest against capitalist society's misuse of radio technology. In Bertolt Brecht's well-known formulation, the commercial and governmental application of radio as "an apparatus for distribution" squanders its potential as "the finest possible communication apparatus in public life," the "vast network of pipes" it might afford "if it knew how to receive as well as to transmit, how to let the listener speak as well as hear, how to bring him into relationship instead of isolating him."[43] This "critical error," according to Walter Benjamin, perpetuates a "fundamental separation" between speaker and listener, such that "the audience, thoroughly abandoned, remains inexpert and more or less reliant on sabotage in its critical reactions (switching off)."[44] Unhappily, the medium's sorry state collapses the political distance between the fully exercised critical subject and the desperate saboteur of last resort.

In his writings for the Princeton Radio Research Project, Theodor Adorno elaborates Benjamin's position by identifying "switching off" as one of the few identifiable "countertendencies" to network radio's "ubiquity-standardization," by which he means not the standardization of programming but the structural logic of broadcast itself, the "more or less authoritarian offer of identical material to a great number of people."[45] Unsurprisingly, these countertendencies—twirling the dial to select different stations, tinkering fetishistically to improve reception, writing fan letters, and so on—prove limited as modes of resistance and agential expression, but only "switch[ing] off the radio" "hints at an irrationality beyond the concrete act" itself.[46] The listener "derives a certain amount of pleasure from this gesture," Adorno surmises, as if reveling in their apparent ability to "condemn even the most powerful dictator"—the "mightiest" of voices—"to silence." With "a sort of wild joy," the listener "who cannot possibly alter the ubiquity-standardization of the radio phenomenon transforms it and every pleasure he might get from it into the pleasure of destruction," though "this gesture of opposition is the most fruitless of all." Twenty years later, Jean-Paul Sartre will concur by pushing the analysis even further. Concerned to describe that species of collectivity he calls an "indirect gathering," Sartre notes that "the mere fact of *listening to the radio* . . . establishes a serial relation of *absence* between the different listeners."[47] This gathering is thrown up before the "impotent listener" precisely when they attempt to resist the inexorable passivity of their position vis-à-vis the objectifying radio voice by flicking off the set, "for this purely individual activity changes absolutely nothing in the real work of this voice," which "will continue to echo through millions of rooms and to be heard by millions of listeners; I will merely have rushed into the ineffective, abstract isolation of my private life."

But for all the ways Niedecker's brief media critique echoes the discourse around "switching off"—with Benjamin's recourse to sabotage, Adorno's fruitless annihilation, and Sartre's serial impotence—her poem also quite evidently *privileges* the farmers' "natural resource" in a manner our theorists do not. Niedecker's sketch of radio listening supplies a useful analogy for the material operation of her speech-based writing more broadly, a media-critical poetics of "bad reception" that responds to the social fact of radio not only at the level of content but equally at the level of form.

On the page, the poetics of bad reception involve cutting off one's lyric material at the source. When Kittler notes that "radio runs as easily as water from a tap," he is channeling anxious critics of early broadcasting who described the distracted, indiscriminate consumption of programmed content as "tap listening."[48] Niedecker was fifty-nine when she finally installed running water in her cabin. That same year, 1962, also saw the long-delayed publication of her second collection of poems. Interlacing these events, Niedecker's link between the water tap and her own poetic economy effects a contrast with the flowing taps of radio:

> Now in one year
>   a book published
>   and plumbing—
> took a lifetime
>   to weep
>   a deep
>   trickle[49]

Against Sartre's gendered concerns regarding an "impotent" listening subject, Niedecker configures her poetic practice as a strategic disruption of flows, as a partial shutting down or straitening of the normal channels by which messages move through media, through the ether to the ear, and from the ear to the page. The critical purchase, the "natural resource," which the farmers discover in tuning out, Niedecker exercises in a speech-based writing practice that enacts medial difference in a similarly dreadful, similarly thrilling manner. Her "bad reception" is an aesthetic strategy made possible by the dynamic I term lyric intermediality—the tense encounter between writing and sound that produces poetic writing's dual existence as transcription and score, index of past and future sound. Niedecker writes a speech-based poem, but speech is not simply set down on the page. As I describe in more detail below, what might appear as straightforward transcription actually belies a complex re-sourcing, a destabilizing of the prepositional relation of poetry to speech predicated on the strategically poor reception of the latter. Like her farmers, Niedecker troubles normative media channels with a recourse to silence, though she effects this silence not with an apocalyptic flick of the dial but with a self-reflexive regard for the affordances of the textual page. Put more directly, Niedecker's tool of sabotage is silent print.

As a study of modern poems like Niedecker's, this book summons a finer-grained view of radio's history not merely because the unfixed meanings of radio so intensely excited poets and readers of the period, many of whom were busy fantasizing a radiophonic future for verse. I argue that poems can also help *reveal* this history. Once a radio writer herself, Niedecker was interested in the phenomenal experience of radio listening.[50] In a 1952 letter to Louis Zukofsky, she casts a glance back at the exuberant prospects for poetry's radiophonic future that characterized the late 1930s: "Radio *should* be a good medium for poetry—speech without practical locale. Stage with all its costumes and place and humans tripping about too distracting sometimes. Poetry and poetic drama—suggestion—the private printed page plus sound and silence. I remember when [Archibald] MacLeish and Norman Corwin were so excited about radio. It doesn't pay, that's all!"[51] It's unclear whether the phrase "speech without practical locale" refers primarily to poetry or radio, though that's much to the point: radio suits a speech-based poetry exactly because broadcasting reinforces the former's already well-attested capacity for setting speech suggestively adrift via a process linguistic anthropologists term "entextualization," thereby complementing with effects of "sound and silence" poetry's intrinsic properties on "the private printed page."[52] I conclude this introduction by noting how Niedecker's radio-like querying of sources relies on the protocols of lyric intermediality. The question before us now is whether Niedecker's poetic "speech without practical locale" reveals anything further about the particular social history of radio itself.

I suggest that to the extent radio's impress is visibly traceable on the textual surface of poems like "In speaking spokes . . ." it appears via the unlikely proxy of nursery rhyme prosody. Despite their professed sources in "everyday speech," the *New Goose* poems resist direct attribution; entextualized on the page, these utterances turn away from their practical sources only to re-source themselves in a multitude of possible speech acts. Niedecker manages this effect by leveraging the silence of print, but the effect is notably amplified by her employment of the loosely measured, trochaic, often jauntily rhymed prosodic structures of the ballad and epigram. These deeply ingrained "mnemonics of childhood" recall us not so much to any putative "folk" experience as to a sense of collective oral utterance, without definable origin and long mediated by print tradition.[53] The sheer oddity of the *New Goose* procedure should not escape us: Niedecker takes the transcribed speech of twentieth-century individuals and works it into the sonorous patterns of printed nursery rhymes. In this way, an intimate address becomes, by virtue of its prosodic structure, also the anonymous, multiply sourced, collective, and palpably remediated language of lore:

Remember my little granite pail?
The handle of it was blue.

> Think what's got away in my life—
> Was enough to carry me thru.[54]

To propose that poetic engagement with Mother Goose should somehow necessarily educe the experience of radio broadcasting may strike one as implausible, but in fact nursery rhymes are closely woven into the history of sound technologies. The association stretches back even to the phonograph's famous first words, "Mary had a little lamb," which Edison shouted into his brand-new machine in 1877.[55] One might also adduce the vigorous debates in popular culture and scholarship during the 1930s regarding the developmental effects on children of radio listening, an anxious complement to the vast amount of programming directed at young audiences.[56] Even Norman Corwin's *Words without Music*, a program discussed in chapter 2 that one radio historian refers to suggestively if not quite accurately as a "series of socially conscious nursery rhymes," "vitalized" Mother Goose verses in one of its earliest episodes.[57] Refracted onto Niedecker's page, these prosodic schemes represent a self-conscious engagement with residual cultural practices that are recontoured and reinvested with meaning by modern media conditions.

Provided we learn to read them as intermedial technologies in their own right, poetic forms are extraordinary media-historical archives. But accessing these archives will require avoiding too-simple models of influence or reflection. It demands instead that we commit ourselves to the deep social contexts of radio and poetic practice both. In this respect, another poet offers a cautionary tale. The immodest Ezra Pound professed to have "anticipated" radio in the early *Cantos* because the experience of reading his collaged, interrupting voices plausibly evoked the experience of twirling the dial on a receiver set.[58] Critics concerned to track the influence of new technologies can occasionally fall into the trap set by Pound's brag and purport to read texts *analogically* for the presence of, say, radio in a work's formal structures. McLuhan makes a similar claim for T. S. Eliot's poetics of the "radio-tube grid circuit," while commentators today repeat the error when they draw too straight and simple a line between the internet and contemporary aesthetic programs for literary practice.[59] Whether in the spirit of analysis or prophecy, such interpretations run the risk of forgetting that media as much as poems are themselves socially mediated, and that the former's most salient ramifications may have little to do with their most distinctive technical features. In Pound's case, it would be better to say—as

Daniel Tiffany has—that as far back perhaps as the early Imagist experiments, his poetry anticipates not so much the phenomenal experience of radio listening as the historically specific articulation of radio to fascism emblematized by the debacle of his own Rome Radio broadcasts.[60] The principal link between Pound's poetic trajectory and the medium of radio is nothing essentially technological or poetic but rather the long historical processes that eventuate in European fascism. Adopting a similar methodological attitude, the chapters to follow begin with the supposition that because modern poetry is structured in tension between sound and print, its reaction at the level of form to the emergent protocols of phonography and radio hands down a complex reflection—and often a critical refraction—of the social processes by which these technologies are themselves constructed. Only because intermedial forms are peculiarly responsive to technology's social history is lyric writing an exquisitely sensitive barometer of technological change.

Lyres and Scores

The act of convening modern poetry and media history raises a number of challenging questions, the furthest reaching of which are taken up in turn by this book's chapters. First and foremost, if lyric is a media problem, what kind of problem is it, and in what sense does the methodological purchase offered by media and sound studies remodel our sense of lyric writing as material practice (chapters 1 and 2)? To what extent are poems shaped by their cultural proximity to other sound technologies like phonography and broadcast radio, and what can we learn about their social history by reading poetic forms for the mediated traces of their influence (chapter 3)? What is the relationship between modern poetry and its mass-mediated sister art, the song lyric (chapter 4)? How does lyric writing's media condition inform our sense of the kind of durable objects poems are (chapter 5)? How might a media theory of lyric help us account for hypostasizing claims made on behalf of poems, like that of their vaunted "autonomy," and what sight lines does it cast into the future for a lyric studies committed both to poetic sound and literary history (coda)?

Answering these questions requires an expansion of the lyric archive. The present study arranges printed poems alongside such artifacts as pop songs, art music, radio broadcasts, and ethnographic recordings. Poetry's encounters with musical practice are indispensable sites for historicizing lyric sound because they dramatize and amplify lyric writing's signature contest between sound and silent print. To this end, chapter 1 turns to the

American "hobo" composer Harry Partch and his brief collaboration with W. B. Yeats. Rekindling Yeats's turn-of-the-century dream of a "new art" uniting word and music using the antique psaltery, Partch's experiments in the 1930s setting poetry to microtonal music involved notating the subtle melodies of speech with new scales and instruments—homemade lyres, in fact. Built to compete with the phonograph, these new-old media pressed lyric to its absolute limit as a symbolic practice, clarifying both lyric's intermediality and its sensitivity to sonic-industrial developments. When economic necessity forced the composer to transplant Yeats's dream of a regenerative oral poetry to the transient shelters of the Depression-era West and he began notating the voices of fellow migrants, his compositional practice as a lay ethnographer entailed unprecedented possibilities for the literary inscription of speech and those social stratifications speech can be heard to disclose. The chapter then pursues the late modernist afterlife of speech-music in the Objectivist program of Louis Zukofsky. Whereas Zukofsky's poetics rest on an idealized aestheticization of lyric intermediality, we witness in Celia Zukofsky's speech-musical setting of Shakespeare's *Pericles*—supposedly the crowning achievement of her husband's theoretical armature—a critical deconstruction of just those ambitions.

Chapter 2 considers how the printed pages of late modernist poems receive the impress of broadcast radio through effects of *latent remediation*, my term for medial influence as it is routed through the more total social processes forming the traditional object of Marxist cultural analyses. Whereas existing studies of broadcast and poetry often account for radio's influence by way of formal analogy—Pound's poetics of scrambled radio dials and disseminated voices, for example—this chapter explains poetry's messier, more contingent, less direct negotiations with a new cultural apparatus whose social meanings are yet unfixed. Centered on the techno-utopian expectation, circulating headily in the 1930s, that poetry would soon leave its medium of print for radio, "an instrument much more amazing than the lyre ever was," this chapter describes how the poetries of John Wheelwright and Sterling Brown were all in a precise sense *latently remediated* by radio—that is, materially transformed by the protocols of a technologized soundscape—prior to or despite any actual appearance on the air.[61] Ministering to these lyric negotiations with radio's golden age generates critical strategies for making sense of the literary repercussions of technological media more broadly.

Any media history of lyric must sooner or later address what Langston Hughes identifies as its singular twentieth-century "burden": poetry's relationship to popular song. Chapter 3 approaches lyric writing's vexed proximity to the pop song by redressing one specific literary-critical symptom of this larger problematic: the relative neglect by Hughes scholars of the poet's exceptionally prodigious career as a collaborating songwriter in the 1940s. This chapter focuses especially on Hughes's quest to write a patriotic hit during World War II. Reading the voluminous song drafts collected in the poet's archive, I show how Hughes accommodated the standardizing demands of the culture industry by sensitizing his flexible compositional process to the affordances of kitsch diction and the generic expectations of multiple, incompatible publics. Daniel Tiffany's renovated notion of poetic *kitsch* and the nonce media-theoretical heuristic of the "dummy lyric" light up Hughes's songwriting practice as a historic confrontation between lyric and commodity forms. In lending these lyrics a critical ear, this chapter proposes the category of *songwork*, inclusive of Hughes's song lyrics *and* poems, as a ruined or riven concept capable of reflecting back the social contradictions fracturing an impossible lyric whole to which, repurposing Theodor Adorno's quip regarding low and high art, the torn halves of pop songs and poems simply do not add up.

Chapter 4 looks into the kind of *things* poems are. More specifically, it asks how midcentury cultures of ubiquitous recording informed the degrees of objecthood ascribed to discrete poems. The chapter opens with discussion of Marianne Moore's late literary celebrity, locating in her extensive audio archive a vivid emblem of the *displaced objecthood* reshaping lyric practices at midcentury. It then follows Moore to her midcentury encounters with a poet and composer whose ingenious efforts to objectify lyric point up the modes of latent remediation motivating his own historically specific desire for a "genuine" lyric objecthood. Russell Atkins responds to midcentury displacements in lyric materiality, exemplified by Moore's recording career, with elaborate ideological defenses of poetic objecthood, providing eloquent testimony to the connection between modern attempts to objectify poetic practice and lyric writing's transforming media condition. I demonstrate how Atkins's "psychovisual object-forms" respond to a culture of ubiquitous recording by finding license therein for a fully democratized lyric practice.

The book concludes in league with the Scottish-American poet Helen Adam, whose time-lapsing ballads provide an occasion for surveying how poems move through history—a consideration perhaps more urgent now than ever, as we ask in our age of rapidly expanding digital platforms what it would mean for poetry if the book were eventually to share the fate of the lyre. A reading of Adam also occasions one final consideration of lyric writing as an interdisciplinary meeting place for media studies and literary history, and one final opportunity to convince theorists and archaeologists of media that lyric writing demands and repays their specialized forms of attention. This book's first task, however, and the goal of this chapter's remaining pages, will be to convince students of *any* disciplinary object, but perhaps scholars of poetry most of all, that something called lyric intermediality is veritably at the animating heart of poetic history.

I can lend some specificity to the concept of lyric intermediality by spotlighting a few moments in Western poetic history when this peculiar media condition is brought by various circumstances into high relief. While the episodes to follow seem to me especially illustrative slices of lyric history, this whistle-stop tour makes no claims to comprehensiveness, and those with other priorities would surely adduce different examples. Rather, I have assembled these particular poems because they preview the arguments of later chapters and because they invite me to establish the two broad protocols that together compose the medial dynamic inherited by poets in the twentieth century: the poem as a score, script, or prompt for sound, and the poem as a transcription or sound recording. The score and the transcription model two ways of figuring a printed poem's relationship to sound. Both arise, I suggest, as material responses to the fact that lyric writing is writing that misses its lyre.

For ancient Greek poets, the lyre accompanied sung poetry. By offering the singer "certain stable pitches . . . to use in centering and articulating a song," the instrument coordinated speech with music and composition with performance.[62] In this way, the lyre may seem, in hindsight, like the perfectly ordained solution to the printed lyric's richly problematic intermedial condition: the historically variable tension between writing and sound. As a consequence, wherever they surface in the archives and fantasies of print literary history, the lyre and its avatars—the harp and the lute, or later the radio, the record, and even Bob Dylan's guitar—can be taken as tokens of poetry's conspicuous overinvestment in a media condition that all forms of alphabetic writing share, but none so intensively and creatively as lyric writing. Lyric intermediality is synonymous with the lyre's constitutive absence on the page. Poems with pretensions to sound always get written, to repurpose a phrase from Marianne Moore, in lieu of the lyre.

But we needn't rush to embrace the absent lyre just yet. Though the genealogy to follow may appear an exceedingly partial and eclectic one, any story about Western lyric writing requires an early stopover in Lesbos, where we find Sappho, in Fragment 118, addressing a lyre that was quite real indeed:

> yes! radiant lyre speak to me
> become a voice[63]

The rhetorician Hermogenes recorded this fragment for posterity in his treatise *On Types of Style*, where the cited lines exemplify the "sweetness" conveyed when speakers ascribe "rational qualities to things that are irrational."[64] In other words, Fragment 118 evokes the imaginative pleasure of personification, the giving of human speech to an inanimate object. Insofar as Sappho apostrophizes her lyre in expectation of a reply ("speak to me"), the poem more specifically instances the mode of personification known as prosopopoeia, the implicit animation of the artifactual, natural, or otherwise nonhuman.

Literary theorists in the poststructuralist tradition have identified apostrophe, prosopopoeia, and anthropomorphism more generally as foundational poetic tropes. Several decades ago, Jonathan Culler elaborated the significance of lyric's vatic and sometimes embarrassing gestures of apostrophic address, whereby poets—the poets of "O wild West Wind" or "Death, be not proud," for example—triangulate their speech by hailing an object or absent entity in pretended ignorance of their legitimating audience, the reader. As for J. S. Mill and Northrop Frye, in Culler's view apostrophe is definitional for lyric poetry because it's the trope that tropes "the circuit of communication itself," endowing poetic language with the "ritualistic, hortatory" dimension that removes the utterance from ordinary language use and remarks it as "a special sort of linguistic event in a lyric present."[65] Apostrophe, in this manner, institutes the poetic occasion; it convokes at once the poem's object *and* its expressive subject. When poets address themselves to inanimate or absent objects, they render that addressee "potentially responsive," and this act "implies a certain type of *you* in its turn."[66] Indirect address constitutes and authorizes the poetic speaker as a speaker of lyric as such.

Fragment 118 presents us with a rarer specimen of lyric apostrophe, for here Sappho's addressee is not an absent lover, natural object, or supplicated deity, but the lyre itself, the instrument with which she and other ancient Greek lyric poets composed and performed. If for some lyric theorists triangulated address is fundamental to lyric poetry—if to different extremes all lyrics trope their rhetorical situation—then addressing the radiant lyre renders this tropological operation not merely explicit but explicitly material. Indeed the fragment self-references not only its context of communication but just as pointedly its "materialities of communication," to borrow a banner phrase of the so-called material turn in the humanities.[67] In this case Sappho exercises a mode of address we might term *phatic* apostrophe, recalling Jakobson's identification of "phatic" messages as those that refer to the communication channel itself. Invoking her lyre, Sappho performs a kind of elegant mic check.

The phatic address of Fragment 118 differs from paradigmatic instances of lyric apostrophe in one further respect. The poem literalizes the animating work of prosopopoeia. Much of apostrophe's rhetorical magic derives from the fact that when poets address dead lovers, nightingales, and nations, they impute to these objects a capacity for response, lending them voice and face (*prosopo* + *poeia* means "making face").[68] Sappho enjoins her lyre to speak outright—to "become a voice"—and of course it *does*, ringing out in accompaniment to the performance of her own sung poem. Keats's nightingale, Shelley's west wind, and Whitman's America do not actually respond to these poets' invocations, but Sappho's lyre makes veritable sound. The ambiguity seeded by the translator's choice of the word "become" (does the lyre become a voice itself, perhaps supplanting Sappho's own, or does it merely ornament the latter, as a becoming addition?) articulates a logic of supplementarity that will be central to the media history of modern lyric writing. Throughout this book we encounter cases in which medial artifacts that first appear utterly incidental to poetic practice in print (Yeats's antique psaltery, Hughes's interest in songwriting, or Celia Zukofsky's musical settings) are revealed as crucially foundational, possessing a voice all their own. And the instances of lyric accompaniment, actual and imaginary, that I submit to scrutiny will bear out the truth of Sappho's observation that lyric poems depend on the sharing of poetic agency between poets and their lyres—by which I mean their media conditions.

Since Sappho, the image of the lyre has continued to serve poets as a token of lyric phaticity and a sign of poetry's claims to sound and music. The lyre figure offers literary historians a recurrent index of the intermedial ambitions attending Western poetry throughout its long histories in manuscript and print. The example of Sappho reminds us that before it became a symbol or trope, the lyre, in its several forms and alongside other musical-poetic instruments of the ancient world, was a historical medium for composition and performance.[69] Though by the time Sappho was canonized by Alexandrian scholars, her poems and those of other Lyric Canon poets like Pindar, Alcaeus, and Anacreon were preserved, edited, and studied as *texts*, not as musical works.[70] It's not too much to say that lyric has *always* been a matter of writing, since the word does not even appear until the second century BC, several hundred years after the waning of those widespread performance cultures in which the archaic and classical versions of so-called lyric once flourished.[71]

The corollary fact, that lyric writing's "lyre" has always been a trope, may explain why to poets like Nathaniel Mackey it resembles a confounding figment of the imagination: while lyric's etymology is "something I've been unable to forget," Mackey writes, "at times I've wondered if it were something I'd made up or been misinformed about."[72] The lyre's tropological dimension is also one reason for its cultural longevity—for why it has been difficult for other poets and readers of written and printed poems to disabuse themselves of the lyric's musical metaphorics. From practically its first moment of literary-historical institution, lyric poetry has been missing its lyre, its musical-medial supplement, and it misses it still. Wherever the lyre crops up in the poetic tradition, readers are reminded of poetry's intermedial ambition to score sonorous phenomena, whether as sung song or prosodically shaped speech.

One could organize a chronicle of poetic history around the lyre image, trailing the gesture of phatic apostrophe as it appears, time and again, in mutating forms. A first exhibit in the lyre's afterlife might feature Horace, who—it bears emphasizing—neither sang his poems nor played the lyre. Though written some five centuries after Sappho, the Roman poet's own address to the instrument, poem 32 in his first book of *Odes*, begins by acknowledging the same shared agency for which Sappho pled when she asked her lyre to "become a voice." Appropriately enough, Horace follows Sappho in the stanza shape that bears her name:

> We've been commissioned. If ever we've played
> in an idle hour under the shade some piece
> that lives on, sing me a Latin song now, O
>   lyre from Lesbos[73]

Horace's next verses make clear that his model lyre belongs to Alcaeus, Sappho's countryman, who

>   fought
> furiously on land and sea, yet, when he
> laid his arms down . . .
>
> would always sing Wine, Music, and Venus

This swapping of weaponry for poetic accoutrement echoes a theme common in antiquity: the lyre's intimate relationship with the bow, Apollo's *other* signature instrument and the lyre's weaponized twin. The twentieth-century poet Octavio Paz will exalt this pair into a magisterial allegory for poesis itself in his 1956 treatise *The Bow and the Lyre* (*El arco y la lira*), and we shall find the slippage between music and martial conduct reprised, at the end of this book, in the treacherous fairy realms of Helen Adam.[74] For now it's sufficient to observe how suggestively the poem's concluding invocation of the lyre resounds alongside Sappho's fragment.

> O glory of Phoebus, lyre welcome at feasts
> of Jupiter most high, O sweet solace
> of all my trials, hear me whenever I
>   duly invoke you.

If the lyre possesses a voice, it has an ear, too, and the poem binds the poet and his fictional lyre in a recursive circuit that in truth returns us to the posthuman dynamics of intermediation described by Hayles: here the poet "invokes" in song his lyre, which hears that song, and "sing[s]" back in echo "a Latin song now." Two interanimating poles, poet and lyre link up in a single cybernetic feedback loop.

Sixteen centuries later, a not dissimilar scene of poet-lyre recursion is given early modern staging in Thomas Campion's "When to her lute Corrina sings" (1601). In its first line, as John Hollander notes, the poem boldly confuses accompaniment with apostrophe, since "to her lute" can mean both *accompanied by* and *addressing* the instrument.[75] The poem then evolves the conceit of an affective correlation between Corrina's voice and the gently personified lute:

> When to her lute Corrina sings,
> Her voice revives the leaden stringes,
> And doth in highest noates appeare

> As any challeng'd eccho cleere;
> But when she doth of mourning speake,
> Ev'n with her sighes the strings do breake.[76]

Corrina's implied song derives its emotional force from an identity of words and music—Corrina's "voice" and its "challeng'd eccho" in "highest notes"—but underlying this aesthetic union is a suspicion that Corrina and her lute are materially united in a manner deeper and more unusual than is metaphorically accounted for by lyrically expressive music. The hunch is confirmed in the second and final stanza, when the speaker strides into the poem to intercede between Corrina and her instrument.

> And, as her lute doth live or die,
> Led by her passion, so must I:
> For when of pleasure she doth sing,
> My thoughts enjoy a sodaine spring;
> But if she doth of sorrow speake,
> Ev'n from my hart the strings doe breake.

The poem's conventional conclusion purports to tell us something about the speaker, who now reveals themself as the real instrument of Corrina's affections, "[l]ed by her passion" back and forth from "sodaine spring" to the brink of heartbreak or death. And yet the poem's amatory turn ("so must I") also retrospectively clarifies Corrina's relationship with her lute, and the reader/listener's relationship to Campion's poem, initiating an unwieldy chain of musical-poetic chiasmi. If the speaker is like the lute, then surely the lute is like the speaker—erotically (over)involved with its object of affection, that is, with Corrina herself. At this point, we must remember that like Sappho but unlike Horace, Campion's lute—his latter-day lyre—is both figure *and* instrument. The poem appeared in Campion's 1601 *Book of Ayres* with music for lute and viola da gamba, composed with the aid of future "king's lutenist" Philip Rosseter.[77] Imagining a scene of performance, then, we discover Corrina's relationship with her lute consequently doubled by the performer's relationship to the lute they (or their accompanist) are presently strumming, placing the putative audience in the position of the poem's speaker, listening and so thoroughly seduced, enthralled, and supersensitively played upon as to resemble, in fact, a merely responsive instrument. The *mereness* or apparent triviality of the instrument is only an illusion, of course, for as we have seen, this entire series of enfolded relations has been set into motion by the mediating lyre, as thematic figure and tool of performance, consubstantial with the poem's evocation of feeling and mode of address.

The entangled identity of poets and their lyres we have remarked in Campion, Horace, and Sappho becomes, in the Romantic period, a full-blown theory of phenomenal experience, though in this case the model medium is neither the lyre nor lute but the aeolian harp, that player-less, open-air instrument "whose strings / The genii of the breezes sweep."[78] For news of the harp we can look to such Romantic landmarks as Samuel Taylor Coleridge's "The Eolian Harp" (1796) and Percy Shelley's "Alastor, or The Spirit of Solitude" (1816), but it receives perhaps its most ingenious treatment in the latter's "A Defence of Poetry" (1821/1840). There, Shelley has recourse to this musical toy in his description of what it's like for a subjectivity to maintain intercourse with the world: "Man is an instrument over which a series of external and internal impressions are driven, like the alterations of an ever-changing wind over an Æolian lyre, which move it by their motion to ever-changing melody."[79] Much can be made of this evocative figure, which presciently forecasts materialist explanations for human consciousness. Timothy Morton, for one, has erected an "object-oriented defense of poetry" upon the poet's intuitive grasp of poetry and intellection itself as material processes; in his view the "Aeolian lyre image provides all the tools we need for including thinking in a physicalist realism."[80]

For our purposes, however, most crucial is what this figuration of consciousness discloses about poetic media. Nearly in the same breath that Shelley introduces the metaphor of the aeolian lyre, he feels the need to qualify it substantially, for "there is a principle within the human being, and perhaps within all sentient beings, which acts otherwise than in the lyre, and produces not melody alone, but harmony, by an internal adjustment of the sounds or motions thus excited to the impressions which excite them."[81] The metaphorical value of the harp is its sensitivity to stimuli, and its capacity to imitate and be in-formed by thought and sensation. But lest he insinuate too static and simple a sense of the material and spiritual processes at issue, Shelley must dynamize the relation between self and world by according the former further degrees of agency. As it happens, the need to rejigger his analogy to make room for the mind's "internal adjustments" forces Shelley to ditch the aeolian harp altogether for an older model: "It is as if the lyre could accommodate its chords to the motions of that which strikes them, in a determined proportion of sound; even as the musician can accommodate his voice to the sound of the lyre." Shelley's rhetorical efforts to reground lyric poetry—"the expression of the imagination"—on a new, cutting-edge theory of how the imagination materially works ends up reaffirming lyric writing in its most persistent aspect, as an intermedial relation between sound and poetry. Shelley requires for his Romantic account of poetic cognition exactly that responsively supple sharing of agency between singer and lyre—that flexible relation of accompaniment—announced by Sappho two millennia prior.

Though it can be counted on to mutate and evolve, accruing diverse meanings across distant periods and among far-flung communities of readers, lyric intermediality is an enduring constant of poetic history in the West. The lyres, lutes, and harps that so insistently populate poems in silent print are reminders that poets have often responded to this media condition with the concerted ambition to script sonorous phenomena. Lyres on the page communicate the need to promise print readers a future recuperation of sound.

If these transhistorical claims strike one as overlarge, it's worth pausing to recognize that media theorists have made still bolder claims for the ancient lyre as a piece of cultural-technical equipment. In his idiosyncratic investigations of the Greek alphabet, Kittler identifies the lyre as an indispensable tool for elucidating the "essential unity of writing, number, image and tone," a momentous cultural insight distinguishing pre-Socratic Greece as the originary home of a knowledge of Being subsequently foreclosed by Western thought.[82] Since the Greeks used their alphabet both for writing and counting, language and number, a Pythagorean thinker like Philolaus could exhibit music's physical bases in whole-number ratios by directing his students to shorten their lyre strings by ratios expressed in letters like δ καὶ γ, or 4:3, a perfect fourth. In this way, the lyre came to mediate not only the lyric song and poetry of Philolaus's contemporaries, but also number, sound, and language. Cooperationalizing the linguistic, mathematical, and musical affordances of the first vowel alphabet, the "lyre is not only a musical instrument such as exists in any culture," writes Kittler, "but also a magical thing that connects mathematics to the domain of the senses."[83] In the words of John Durham Peters, for Kittler "the Greek alphabet liberated and recorded the harmonies within poetry, numbers and music, but it also embodied the harmony between them," and only the lyre—as an instrument of poetry, music, and mathematical research and a vital supplement to the alphabet—could elaborate this Pythagorean unity.[84]

Kittler calls the lyre an "epistemic thing," borrowing a concept developed by the historian of science Hans-Jörg Rheinberger to delineate objects of scientific inquiry that are constrained and constructed by the experimental systems in which they are embedded, objects whose nature and meaning are determined only by the contingent unfolding of research itself. When abstracted from the research process, epistemic things "present themselves in a characteristic, irreducible vagueness," which is "inevitable because, paradoxically, epistemic things embody what one does not yet know. Scientific objects have the precarious status of being absent in their experimental presence."[85] Though lyres may seem a different class of entity from the biosynthesized proteins that are Rheinberger's focus, Kittler designates the lyre an epistemic thing to underline just this generative "vagueness." It may be odd to speak seriously of lyres in the context of printed poems or odd to speak of them as anything other than the nostalgic trope they also absolutely are. But this strangeness issues partly from the fact that lyres, like epistemic things, embody what we do not yet know about lyric writing's intermedial condition. In a manner analogous to the "vagueness" of epistemic things, the lyre's half-articulated, "magical" quality may indicate its very centrality to the media-historical research this book aspires to undertake. As a sign of poetry's medial unconscious, the lyre invites us to reconstruct the historically specific conditions for the operation of lyric sound in social space.

But before we begin exercising this mode of material analysis on twentieth-century lyric writing, we must first make room in our brief genealogy for a second medialogical current in poetic history, one concerned less with the ambition to *sound out* poems with lyres and more with the possibilities of the page itself as a *recording* technology.

Records and Transcriptions

If Sappho and her fellow singers initiate lyric practice in the West, one credible starting point for English lyric writing is "Caedmon's Hymn." The story of the hymn's miraculous composition by the seventh-century Northumbrian cowherd—"the single most sustained discussion of vernacular poetic production and performance to survive the Anglo-Saxon period"—comes to us by way of the Venerable Bede's *Ecclesiastical History of the English People* (AD 731), and it commences with the silence of a refused lyre.[86] On occasions when custom dictated the passing of the harp from reveler to reveler around the fire or feasting table, Bede tells how Caedmon, illiterate and uninitiated in the arts of song, would unceremoniously disappear from the gathering before his turn to sing. Asleep in the stables one night after just such a vanishing act, the herdsman dreams the appearance of an unknown figure who duly commands the former to "sing me something."[87] After some initial demurrals, much to his own surprise Caedmon answers the visitor's request with a song "about the beginning of created things" in nine alliterative distichs addressed to God the Creator of the universe. Upon waking, Caedmon remembers the dream-verse, and so do we; it's the earliest English poem by a poet whose name we have not forgotten:

> Now we must praise / heaven-kingdom's Guardian,
> the Measurer's might / and his mind-plans,
> the work of the Glory-Father, / when he of wonders of every one,
> eternal Lord, / the beginning established.
> He first created / for men's sons
> heaven as a roof, / holy Creator;
> then middle-earth / mankind's Guardian,
> eternal Lord, / afterwards made—
> for men earth, / Master almighty.[88]

The modern English quoted here from the first page of *The Norton Anthology of Poetry*, that totem of late print culture, participates in a constellation of diversely mediated translations. The text of Bede's *Ecclesiastical History* gives the poem only in a Latin gloss, and the version above translates the Old English poem copied by scribes into the *History*'s margins. Meanwhile, this original poem, we recall, itself represents the transduction into writing of an oral composition, one that—if Bede is to be believed—entered the sphere of human relations in the language of dream.

Adducing these circumstances, the poet-critic Allen Grossman, in his inimitably grand style, has dubbed the hymn a "precisely impossible poem": "an account of an action that cannot be witnessed, creation," by means of "an exploit"—the skillful application of Anglo-Saxon prosody to Christian themes by an illiterate, unmusical dreamer—"that cannot in any practical sense occur."[89] Caedmon "knew no songs" and yet created ex nihilo the song of the ex nihilo creation. Its patent impossibility accounts for the abiding mystery the hymn installs at the very origin of Christian poetry in English. Bede goes on to relate how Caedmon reveals his dream-poem to the abbess and brethren of the nearby monastery in Whitby and then, much to their astonishment, repeats the poetic feat, again and again, begetting spontaneous oral verses of remarkable wit and skill. Caedmon spends the remainder of his life fashioning into accomplished poetry the scripture and doctrine conveyed to him by literate monks, into whose monastery this "most religious man" was welcomed.[90] Caedmon's illiteracy, the very condition of his poetic powers, was taken by the brothers, by Bede, and by English posterity, as a sign of grace. As Katherine O'Brien O'Keeffe explains, the poet's "continuing formal illiteracy suited Bede's narrative, not because it necessarily portrayed him as an oddity within the monastery, but because it enabled Bede to claim him as a pure vessel for receiving the matter of his poetry through the learned instructions of the brothers of Whitby."[91]

Yet underlying the specifically Christian meaning of this mystery is a more basic medialogical conundrum. What is indelibly miraculous in this story stems from a confrontation between the oral culture to which Caedmon belongs and the situation of "mixed orality and literacy" obtaining in the emergently scriptural economy of the monastery.[92] As Grossman and others have emphasized, if a historical Caedmon *did* exist, his genius would have consisted in applying an Anglo-Saxon prosody he would have already surely known—the technics of composition that were the common possession of scops and bards—to the subject matter of a lettered Latin Christendom.[93] But this is no simple wedding of form to content. Bede narrates, instead, a dramatically uneven encounter between orality and literacy; the mystery of ex nihilo creation can be understood as a redolent symptom of—and ideological solution to—halting medial change. In other words, Bede portrays Caedmon accomplishing a paradox that is indeed "strictly impossible": the first English poet *writes his poem orally*.

To apprehend this paradox, it's necessary to grasp spontaneous creation itself as an ideological figment of literate cultures, where the fixity and permanence of writing endows the process of composition—oral or not—with a contrasting and newfound pretense of ephemeral immateriality.[94] In Caedmon's case, that immateriality shades into sheer illiteracy. Bede shows Caedmon composing a poem like we imagine (mistakenly) any literate poet would—alone, absconded from the social round, working up a sui generis artifact from the whole cloth of his own native language and inspired sensibility, and then returning to the public realm to inscribe that poem, to "publish" it, for posterity on minds and pages. The hymn's composition has all the trappings of writing—save a literate poet. But Caedmon's inability to *actually* write hardly matters in this rehearsal of literary composition. Or it does but only—per O'Keeffe—as a signature of God's grace and as an index of medial transformation in the *longue durée*.

According to John D. Niles, what the Caedmon myth "chiefly affirms is not the status or power of the bard" but rather "the power of writing, in a monastic setting, to absorb and subsume all things."[95] Caedmon's oral efforts are bent ultimately to the task of versifying scripture, and Bede's *History* itself serves to capture and contain in the Latin alphabet the oral poem and the oral myth of its miraculous origin. The mystery of the *written oral* poem is therefore an *effect* of the conquest of literacy, though it also surely *resists* this conquest, since the wonder it provokes indicates just how finally inassimilable oral composition remains to the order of writing. The poem's miraculous production can be read as a narrative device for papering over the yawning gap between "oral and literate modes of thought," and while no such gap exists today for the vast majority of audiences, it has been the prerogative of lyric writing to tarry with this medial mystery.[96] Put otherwise, poets and their readers have a hard time awakening themselves, if not from Caedmon's dream, then most certainly from Bede's bardic dream of Caedmon, in which orality and literacy spontaneously cooperate. It's altogether fitting, then, that when Barbara Holdridge and Marianne Mantell decided in 1952 to start a record label dedicated to "spoken word" recordings for a mass audience, jump-starting the age of the commercial audiobook, they called it Caedmon Audio.

There is one moment in Bede's narrative that pierces the semblance of miracle to show up Caedmon's hymn for the media-historical problem it is. The Caedmon chapter's decorous Latin prose features exactly one metaphor. This figure appears not during the relation of Caedmon's dream, where we might expect such an indulgence, but later, in the short description Bede furnishes of what we would today identify as Caedmon's "poetic practice": the literate monks at Whitby first introduce "a passage of sacred history or doctrine" to Caedmon, who then, "memorizing it and ruminating over it, *like some clean animal chewing the cud*," composes the scripture "into the most melodious verse: and it sound[s] so sweet as he recite[s] it that his teachers bec[o]me in turn his audience."[97] Bede illustrates Caedmon's marvelous, paradoxical ability to write an oral poem with appeal to the double meaning of *ruminando* (thinking and chewing cud), celebrating the upstart poet's spellbinding and utterly otherworldly production of "melodious verses," while at the same time linking the poet to those rough beasts he tends in the stables. If it cannot be spirited away to the realm of divinely inspired dream, the act of poetic mediation—the mysterious confrontation between literate and oral protocols by which Caedmon gives prosodic shape to scriptural material, applying the common possession of bardic *technē* to the alien task of solitary composition—can be spoken of only from behind a tropological screen of the nonhuman.

Kittler has argued that Romantic literature is predicated on a thorough *forgetting* of poetic materiality. Discourse network 1800—the closed, gendered, socially instituted feedback loop of readers and writers, imagined voices and manuscript hands—ensures a spiritualized self-presence that makes of writing the direct representation of inner speech, an "auditory hallucination."[98] Arguably, this phonocentric forgetting of inscriptive media—a forgetting of inscription *as* mediation—survives today in tenacious notions of lyric voice, transcendent lyric presence, and above all the capacity for the written page to record speech.[99] Bede's tale gestures at the longer history of this medial erasure by making of Caedmon a black box, if you will, for *poiesis*: in goes scripture, out comes oral poetry. At the same time, the story unwittingly prompts us to recall that the real history of poetic writing is materially quite messy, though the mess may be submerged in figurations of the nonhuman or smoothed over by religious miracles.[100] As if to complement the long, lyre-laden tradition that would conceive written poems as *scripts* or *scores* for musical performance, the example of Caedmon initiates for us a parallel, more characteristically vernacular strain of poetic practice wherein the lyric ideal and ambition of highest note is the *transcription* or *recording* of sonorous events. In this way, the entextualization of poetic speech joins the "'lyre-harp-lute' constellation" as another timeworn guise of lyric intermediality.[101]

The ideal of lyric transcription was fully artifactualized in the English eighteenth century's widespread fascination with medieval and early modern ballads, as well as in the process of this cultural activity's uptake in the nineteenth century, where we find the ballad ideologically overcoded in the interlocking discourses of literary history, cultural nationalism, antiquarianism, and proto-ethnography. Together, the cadres of ballad collectors, scholars, broadside printers, imitators, and forgers who undertook over the long Romantic period to inscribe in print an oral poetic past for England and Scotland generated what Maureen N. McLane has termed a "romance of orality," an ideological concern for sound that brands a wide range of poetries in English, from Wordsworth's "The Solitary Reaper" to contemporary slam.[102]

THE FLOWERS OF THE FOREST.

I've heard them lilting, at the ewe milking,
    Lasses a' lilting, before dawn of day ;
But now they are moaning, on ilka green loaning ;
    The flowers of the forest are a' wede away.

At bughts in the morning, nae blithe lads are scorning ;
    Lasses are lonely, and dowie and wae ;
Nae daffing, nae gabbing, but sighing and sabbing ;
    Ilk ane lifts her leglin, and hies her awae.

In har'st at the shearing, nae youths now are jearing,
    Bandsters are runkled, and lyart or gray ;
At fair, or at preaching, nae wooing, nae fleeching ;
    The flowers of the forest are a' wede awae.

Figure I.1. Jean Elliot, "The Flowers of the Forest," in Sir Walter Scott's *Minstrelsy of the Scottish Border*, 1802, pp. 158-59. Courtesy of the Hathi Trust

With regard to the history of the ballad, this romance points both toward and away from the actual terms of lyric intermediality at the turn of the nineteenth century. The cherished notion of balladry as an antique, collective, anonymous, oral practice prevailed in Romantic circles *despite* the fact that this craze for old ballads was fueled by a multi-mediated ecology of new and cheap print. To the extent that, in Meredith L. McGill's words, "the history of ballad collecting and ballad scholarship is one of aggressively weeding out cheap print so as to imagine that the central drama of the ballad is an encounter between orality and literacy," we ought to take careful

At e'en, in the gloaming, nae younkers are roaming
  'Bout stacks, with the lasses at bogle to play;
But ilk maid sits dreary, lamenting her deary—
  The flowers of the forest are weded awae.

Dool and wae for the order, sent our lads to the border!
  The English, for ance, by guile wan the day;
The flowers of the forest, that fought aye the foremost,
  The prime of our land are cauld in the clay.

We'll hear nae mair lilting at the ewe milking;
  Women and bairns are heartless and wae:
Sighing and moaning on ilka green loaning—
  The flowers of the forest are a' wede awae.

The following explanation of provincial terms may be found useful.

Lilting—Singing chearfully. *Loaning*—A broad lane.—*Wede away*—weeded out. *Scorning*—Rallying. *Dowie*—Drearie. *Daffing and Gabbing*—Joking and Chatting. *Leglin*—Milk-pail. *Ha'rst*—Harvest. *Shearing*—Reaping. *Bandsters*—Sheat-binders. *Runkled*—Wrinkled. *Lyart*—Inclining to grey. *Fleeching*—Coaxing. *Gloaming*—Twilight.

distance from the methodological assumptions according to which that history has been told.[103] Brief consideration of one ballad across two different species of print will allow us to appraise ballad culture's contribution to the development of lyric intermediality without rehearsing facile contests between orality and writing.

When Sir Walter Scott included "The Flowers of the Forest" in his epochal anthology *Minstrelsy of the Scottish Border* (1802), he appended a headnote describing the popular ballad's rather complicated provenance (fig. I.1). These "well known and beautiful stanzas" elegizing young Scots cut down in battle were once thought authentically antique, before Robert Burns, Scott, and Scott's source—a "Dr. Somerville"—set the record straight: here "the manner of the ancient minstrels" is only "happily imitated."[104] Though largely "successful" in effect, the poem is evidently a replica; "the manners indeed are old, but the language is of yesterday."[105] Composed scarcely fifty years prior, "The Flowers of the Forest" is the work of one Jean Elliot, a "lady of family in Roxburghshire" who was challenged by her father to write an old-style ballad on the theme of the Battle of Flodden (1513), commemorating not the fateful death of James IV but instead the many, though less memorable, young Scots of Ettrick Forest who "fought aye the foremost" and also lost their lives.[106] Yet these facts do not necessarily determine the ballad as fraudulent. Through the testimony of his oral informants, Scott establishes that the initial and refraining fourth line of the first stanza derive from a genuinely antique ballad. This now-lost ballad is also the source of the air to which Elliot's poem was written and to which it's commonly sung, making the music itself several centuries old. Elliot borrows her title from this popular tune, and the melody known as "The Flowers of the Forest," which elicited two other sets of lyrics in the mid-eighteenth century, would have been well known to her.[107]

If Elliot's work is not an authentic early modern ballad, then, it's also patently more than a mere imitation, since she adopts its title, refrain, subject matter, and music straight from this missing original. Corey Gibson maintains that while it may seem as if Elliot has "extrapolated or intuited her way (from two surviving lines)" to a hypothetical sixteenth-century poem composed shortly after the battle, by putting the original refrain "in the mouths of the 'lasses a-lilting'" where it "appears in implied quotation marks," Elliot in fact devises a powerfully self-reflexive artifact.[108] Effectively citing itself— or the lost, eponymous ballad it pretends to be—the poem evokes the very oral tradition it seeks to imitate. Even more strikingly, though, the poem documents its own specific passage through that tradition—its own oral currency prior to the fixity of Elliot's page. In short, the ballad recorded in Scott's *Minstrelsy of the Scottish Border* pretends simultaneously both to *be* and to be *about* the lost song called "The Flowers of the Forest."

Here's the crucial point to register: the romance with orality we discover figuratively folded into this ballad is predicated on its intermedial nature— on the fact that in the print ecosystem of ballad culture, "tunes and texts travel semi-independently, converging and diverging."[109] As exemplified by an Edinburgh broadside of the ballad (fig. I.2), by the mid-nineteenth century,

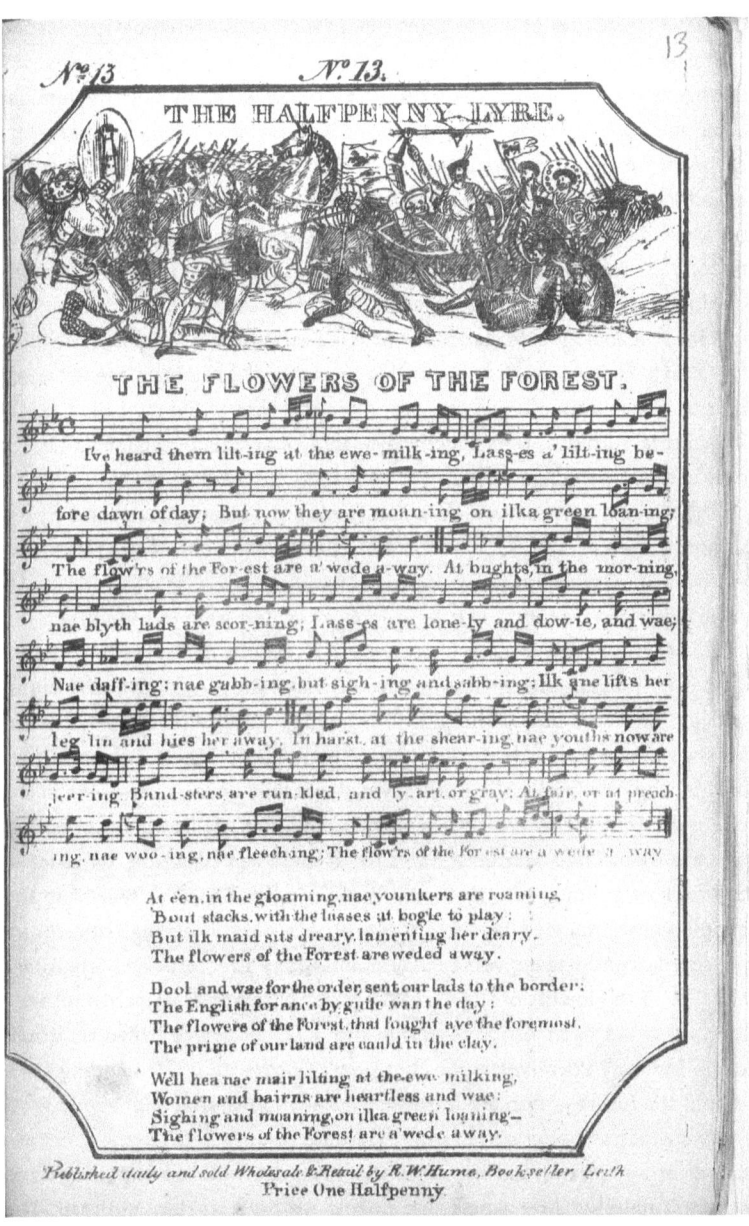

Figure I.2. Jean Elliot, "The Flowers of the Forest," in *The Halfpenny Lyre*, c. 1840. Inglis Collection of Printed Music, National Library of Scotland

the poetic medium of cheap print (here felicitously emblematized by the *Halfpenny Lyre*) had brought Elliot's ballad and the old air into close association.[110] But one hundred years earlier, Elliot herself knew only the familiar tune, and not the words. Penning a new set of the latter *for and on the subject of* the former, Elliot crafted a poem that seems positively to digest the entire ballad tradition as a cultural problematic of "double-temporality," in Gibson's phrase.[111] Chapter 3 and this book's coda will return to the ballad's proximity to song, its suspensive temporalities, and its central role in a media theory of modern lyric writing. In a larger sense, though, all the chapters to follow will involve analogous instances in which efforts to cross media thresholds by entextualizing speech or recording music on the printed page leave sonorous remainders of various kinds. As in the case of Elliot's ballad, these intermedial surpluses then become available for aesthetic resignification, for new cultural meanings reflective of lyric writing's media history.

In the twentieth century, the poetic ambition to transduce sound on the printed page can be discovered across the cultural field. Brent Hayes Edwards, for instance, locates a "poetics of transcription" at the heart of African American literary practices, and Charles Bernstein has identified the "scoring" of sonorous speech in poetry and song lyrics as "one of the primary sites of . . . invention" in late or "second wave" modernism—in the blues- and jazz-inflected poems that Edwards highlights, certainly, but also in Zukofsky's Lower East Side English and in the lyrics of Cole Porter and Charley Patton.[112] Bernstein's use of the word "scoring" is apposite, since in the twentieth century, as ever, poems both *score* or transcribe sonorous events and are themselves veritable *scores* for sound's reproduction, whether in the form of an imaginary and mellifluous song on Sappho's lyre or in the supple vernacular of Claude McKay's dialect poetry. Though these two possibilities endorse a diverse range of imaginary structures and temporary effects—"The Flowers of the Forest" figuratively exploits its fictive reliance on past sound to incorporate within the poem an entire ballad tradition, while Sappho's vocatives secure her poems' posterity by demanding voicings in the future—both are intrinsic to the lyric material (alphabetic writing) on which any poet sets to work. They are two moments in the sonorous career of any poem, and to notice that texts are both transcriptions and scores is merely to recognize that poems are both written and read. The sonorous events to which instances of lyric writing refer—the sounds they might once have been and the sounds they might yet be—never coincide or exactly correspond, of course. These sounds and the subjectivities they index are "immiscible"—they don't mix.¹¹³ Of interest to the study of poetics are the formal structures and social activities that mediate these two moments. Lyric writing's ability to communicate itself, to find readers in the world, is in the care of a human-sized gap between the poem-as-transcription and the poem-as-score. Our exceedingly swift tour through the history of lyric intermediality has highlighted that any discrete literary practice may emphasize either possibility—some poems resemble transcriptions, others are more akin to scores for performance—but of course both protocols inhere in the practice of lyric writing, which makes conspicuous and critical issue of its intermedial condition.

In Niedecker's transcriptive poems, the tension between script and record comes to a decisive head. The result is a speech-based poetics that overturns its own bases by undermining accustomed procedures of transcription: in "recording" speech on the silent page, Niedecker neither attributes these voices to specific persons, nor does she completely abstract them from their originating milieu. According to Jenny Penberthy, insofar as "the majority of the *New Goose* poems are built almost entirely out of overheard local speech," Niedecker's "method depended on an opportunistic ear, ready for the irregular sounds of living speech, for any undiluted linguistic possibility."¹¹⁴ In a 1966 letter to local poet Ron Ellis, Niedecker scribbled the following explanatory note to accompany one such transcriptional poem: "My mother who was a kind of Mother Goose, straight out of the people etc. told me this just as it is. You see at this time—1930s—there was this rage to get poetry into direct, simple speech i.e. this was William Carlos Williams' cry and having always lived among the folk and feeling anyhow that the source of all literature is *there*—folk, Nature . . . and could we say Yeats with his kind of music."¹¹⁵ Niedecker's invocation of Depression-era commitments to "direct, simple speech" accords well with our sense of the decade's proletarian and populist urgencies. All the same, her formulation announces a slight departure from Williams's demand for the American idiom, just as it differs, with a telling prepositional adjustment, from even that wider modernist (but also late Wordsworthian) "search for a . . . modern colloquial idiom," and indeed from the bardic orality renewed by Yeats and discussed in chapter 1.¹¹⁶ The "rage to get poetry into direct, simple speech" does not specify a desire to get everyday speech *into poetry* by the approved means of quotation or direct transcription, severing speech from context to secure a sort of "good reception" in print, as it were. The odder, literal meaning packed into the phrase "poetry *into . . . speech*" avows the primacy of speech and the aspirational retention of its context. In other words, Niedecker limns the possibility of a lyric practice openly mediated by its sources in the social world. Here is a transcriptional poetics that strains against its own tendencies to recontextualize an utterance, that strains—we should say "rages"—against its own media conditions in print.

The poem that Niedecker appended to this revealing note—*New Goose*'s "The museum man!"—allegorizes her devotion to bad reception as artistic practice:

> The museum man!
> I wish he'd taken Pa's spitbox!
> I'm going to take that spitbox out
> and bury it in the ground
> and put a stone on top.
> Because without that stone on top
> it would come back.[117]

This comic vignette has mysteries to spill about the object world, about archival habits across cultural-institutional and domestic spaces both, about the fond grievances of marriage, and about grief's displacing specters. But to attend to the meta-commentary on offer in this poem that arrived, we remember, "just as it is" from Niedecker's mother, we must be willing to read the "spitbox" as a glancing figure for the speaking mouth. The poem's central interpretive quandary is how closely the speech-based poet's activity corresponds to that of her mother, the speaker, and also to that of the interloping museum man whose absence spurs Daisy Niedecker to a frustrated, perverse archival practice all her own: rather than removing the artifact from its original context for preservation and display, she hopes to consign the spitbox to decay (if not to oblivion necessarily, since the stone is also a marker) by burying it deep as she can in its native soil. But where, presently, *is* this spitbox, against which both the museum man and Niedecker's mother are conspiring with their own agendas for fixity? To where does it so infuriatingly "come back"? If it has evaded the cool white walls of the gallery, and as yet escaped the speaker's future plans to "bury it in the ground / and put a stone on top," between this alternative past and potential future, we cannot be sure of its location, our only coordinate the vague deictic "that"— "I'm going to take that spitbox out." This poem of unsettling prepositions and toppled hierarchies leaves us with two rather improper locations for a spitbox (museum and grave) and the spitbox itself just a ghostly streak, stratified across the past and the future, "speech without practical locale."

In Peter Middleton's view, the archival alternatives posed by the museum man and Niedecker's mother throw up a "parable of the problem faced by the poem itself," which purports to display folk speech for a metropolitan audience but is wary of modernism's universalizing gaze: the poem is destined for "either burial in the locality of Niedecker's world, or the museum of international poetry publishing."[118] Steering focus to the material medium itself, we see how Niedecker not only outwits the decontextualizing fixity of her own transcriptional poetics by "using folk speech as both medium and theme" and thus "anticipat[ing] such recontextualization" (Middleton's point); we also recognize her challenge to the normative logic of a speech-mimetic poetry in print. "The museum man!" is allegedly sourced in her mother's speech, to be sure—but since readers are privileged to suffer that speech's silent, bad reception in print, they cannot locate this source any more definitively than they can the spitbox. The poem sends its sonorous source into a kind of flux or flight, such that it seems to always "come back" to a place that we as readers of the poem cannot quite pin down. More plainly, speech is not simply *before* or prior to this nonetheless speech-based poem. Because Niedecker supplies no obvious frame for this putative transcription, when a reader goes in interpretive search of its source, the best they can do is reproduce it phonetically by reading the poem aloud—or silently aloud, as Niedecker would have it—consigning it thereby to an afterlife both ghostly and mobile, a source we can wade back to only insofar as it waits ahead of us.[119]

My treatment of Niedecker has focused on her proximity to broadcast radio. But in fact it is the phonograph, a machine capable of materially "scoring" sound in a manner printed poetry simply is not, that is primarily responsible for driving poetry's protocols for scoring and transcription into unprecedentedly close relation. In just this particular light, the phonograph seems a miracle worthy of Bede's Caedmon. While the cultural impacts of recorded sound are wide and various, phonography's most drastic implication for lyric media in particular stems from its mode of inscription.[120] Because the machine shares the same mechanism for recording and playback (the stylus in a groove), it materially links these two sonic events. Listening to a poem on a phonograph record, we hear a "recording" that is also a "score," the past in the present. The phonograph realizes a dream that the poems we call lyric have entertained for a long time, and its rivalrous example drives lyric practices into something of a media-historical crisis.

Lyric intermediality's encounter with phonography will take center stage in the next chapter, on Harry Partch, whose speech-musical medium pushes the symbolic lyric as close to phonographic protocols as it can possibly go, and then once again in chapter 4, when we examine the effects of ubiquitous recording on the kinds of objects that poets and their readers make of poems. Perhaps somewhat surprisingly, however, this book is not about poetry's audio archive per se. Literary-critical addresses to recorded poetry—what Jason Camlot has dubbed "phonopoetics"—are teaching us how to appreciate and historicize "audiotexts" increasingly available on indispensable digital archives like PennSound and UbuWeb.[121] This book, by contrast, takes aim at less direct routes of medial influence. Excepting chapter 4, my study concludes just shy of the turn to performance in verse cultures of the 1960s, a development that, in tandem with the widespread adoption of magnetic tape, exploded the size of the recorded literary archive. Poetry on record is not our analytical starting point but rather one development in modern lyric writing's long encounter with technologized sound. To put this another way, this book presents one slice of the audio archive's richly contentious prehistory. I do engage with a number of extant sound recordings, to which I refer readers throughout the chapters to follow. But most of the listening this text requires takes the form of an invitation to hear behind and beyond these express engagements with technology a more complexly historical kind of music, a socially mediated confrontation between lyric intermediality and newer technologies for the communication of the sonorous speaking voice.

One final example will substantiate the point. Founded in 1952, Caedmon Audio was the first US commercial venture dedicated to recording spoken word. I remarked a moment ago on the fitness of the label's name, but what is it exactly that makes the reference to Bede's cowherd so apposite? One answer lies in Caedmon's imprimatur as oral bard par excellence. Prior to founding Caedmon with her college friend Barbara Holdridge, Marianne Mantell studied for a Columbia PhD in medieval literature. The pair were thus well positioned to pitch their literary LPs as new-old media: "a revival of ancient traditions of reading aloud" by way of "the twentieth-century's technological breakthroughs."[122] Such an account tallies with the story of Caedmon's first smash hit, Dylan Thomas's *A Child's Christmas in Wales and*

Five Poems (1952).[123] When Holdridge and Mantell approached Thomas with the idea of recording his voice, the rhapsode was already in the middle of his second American tour, playing the bard to rapt audiences around the country. Caedmon's founders captured on record what seemed an essentially oral poetry, bypassing print to access and preserve, in Mantell's words, "the sounds Thomas heard in his head when he wrote."[124] In a manner we'll see repeated throughout this book, recorded sound at once rescues a poetic tradition long mired on the printed page and solicits from future listeners brand-new forms of poetic attention, from the clamorous middlebrow fandom surrounding poetic celebrities like Thomas, to the strategies of audio-textual "close listening" promulgated by scholars in recent decades.[125] In any version of this story, print and writing appear only as problems to be overcome.

But there is another story one might tell about Caedmon, one that heeds the "messiness" of media history while clarifying the material dynamics of printed poetry itself.[126] This is the story about the cultural preconditions that make the naming of Caedmon Audio seem, with hindsight, so inevitable. It's also a story about how these preconditions come to shape poetries on the page. Indeed, just as Caedmon's oral poem illuminates the scriptural economy and its emergent forms of life, the cultural dream of Caedmon Audio says as much about print as it does about phonography. At stake is the possibility of demonstrating that before they ever make it to acetate disc, Thomas's poems already bear the influence of twentieth-century sound technology. Such a proof would require study of Thomas's own extensive engagements with radio, beginning in 1933, when one of the nineteen-year-old's earliest "published" poems was broadcast on the BBC, running to the posthumous radio masterwork *Under Milk Wood* (1954).[127] But the completest account would need to make room for wider and potentially unconscious determinations as well, including the role played by technologized sound in the modernist tendency to materialize the word as such—that desire, in Thomas's phrase, "to treat words as a craftsman does his wood or stone or what-have-you, to hew, carve, mould, coil, polish and plane them into patterns, sequences, sculptures, [or] fugues of sound."[128]

Is it defensible to posit that before Thomas recorded "Fern Hill" for Holdridge and Mantell in 1952, the poem's "tuneful turning" had something materially to do with the tuneful turnings of the phonographic disc?[129] McLuhan seemed to think so. He trumpeted the media-theoretical insight he claimed to discover in Thomas, whose "romantic preference" for a "pre-literate world," like that of Yeats and Joyce, was bound up with "advancing or unfolding technology."[130] "The ease with which he took to microphone and phonograph," writes McLuhan, "was equal only to his joyous storming over his audience with an eloquence which owed more to the bardic than the literary tradition. But it is precisely the microphone, phonograph and radio that have readied our perception again for enjoyment of poetry as speech and song."[131] It's unclear in these remarks whether sound technology sets in motion the desire for bardic speech and song or whether this desire itself fed the widespread adoption of microphone, phonograph, and radio. By refusing to unmuddle the lines of causation, whether he meant to or not, McLuhan hails a dialectical complexity—the complexity of social process—that this book aims to dwell with and to map.

Modern poetry's media history is not a long, inexorable march away from print or an agonistic struggle between the possibilities of the page and the affordances of sound technology. The history worth writing will assume the form instead of a complex negotiation, one keyed at all points to the mediations of technology *and* of social needs and desires. Poetry is an easy mark for these needs and desires, and thus a receptive instrument for their registration, because it misses its lyre. In the story of Bede's Caedmon, that absence is smoothed over. In the twentieth century, sound technologies promised to fill it. But to the extent that poetry remains an art of the page, we can trace much of what is distinctive about the sociality and formal power of lyric writing to this extraordinarily generative lack.

To argue for connections between twentieth-century poetry in print and the radio or phonograph or between the lyre itself, so long largely silent in literary history, dormant beneath layers of cultural ideology, and modern poetry—these are tall scholarly orders. They are nonetheless exactly the kind of speculative connections the present study aims to substantiate. That these connections should *appear* so speculative is an epiphenomenal result of disciplinary limitations that beset the study of literary-media history. To put this difficulty most plainly: it's hard to know how media interact in history. The finer terms of this challenge have been laid out by thinkers like John Guillory and Richard Grusin, but it implicitly motivates twenty-first-century scholarship across the heterogenous terrain of media studies, from Bernhard Siegert's theory of "cultural techniques" to Wendy Hui Kyong Chun's critical work on "habitual new media."[132] One central intellectual task of media thought in the twenty-first century is to theorize the occluded relationship between material *media* and larger processes of social *mediation*. This book pursues that project by direct appeal to the poetic archive.

I have only gestured in this introduction toward the analytical routes by which one might travel from Dylan Thomas's printed poems, for example, to their technologized soundscapes—namely, through modernist ideas about the possibility of printing sound and the investment of writing with a dynamic sensuousness, ideas themselves maturing in dialectical response to technologically enforced transformations in perception and competitive negotiations with new media. Subsequent chapters will expound in greater detail how, for instance, the consequential relations between radio and the poetic page take hold only via the informing context of the US social field in the 1930s, how the pop lyric refracts modern poetry through the prism of the commodity form, and how elaborate ideas about lyric writing's objecthood evolve in anxious relation to a literary culture of ubiquitous recording. My next chapter will address this book's paradigmatic case of rerouted or unexpected medial influence: how a modernist desire to return speech to music with ancient lyres is above all a matter of the phonograph. In every case, modern lyric practice supplies a conceptual space and a material archive for exploring how media are themselves historically mediated.

Lyric is an ideal venue for this task because it is, itself, a media problem. Perhaps it's surprising that we should locate such a rich nexus between literary and media studies not in the usual suspects of digital or computational literature but in that most familiar object of humanistic inquiry: the printed poem, close read for all its sensuous artifactuality. But the critical orbit of lyric writing's own conflicted mediality renders poetry unusually expressive of the contingent, contested, socially embedded dynamics of technological transformation. Adorno calls the modern poem "a philosophical sundial telling the time of history."[133] This book ventures to read lyric writing as a media sundial telling the history of technology. But even here Adorno anticipates us. Some twenty years before branding the poem a philosophical timepiece in "On Lyric Poetry and Society," he tested the metaphor on another technology of speech, the phonograph, envisaging "some later point" in the future when "instead of doing 'history of ideas' [*Geistesgeschichte*], one were to read the state of the cultural spirit [*Geist*] off of the sundial of human technology," tracing historical processes as they are momentarily configured in the technical construction and social application of cultural technologies like the phonograph.[134] The pages to follow will place side by side these two "sundials," modern poetry and sound media, to see what they measure in each other. But another ambition of this study—simpler, though perhaps more profound—is just fully to recognize, with the combined resources of literary and media studies, that these two sonorous technologies catch and reflect the very same sunlight.

1 A Speech-Musical Modernism

New Arts in New-Old Media

In 1934 the young composer Harry Partch traveled from California to Dublin to meet W. B. Yeats, and an old dream of modern poetry's flared to life again. For several years Partch had been experimenting with the means of recalling an enervated, abstract European musical tradition to the fine inflections and "spoken vitality" of the human voice.[1] Judging himself ill suited to the conservatory and classical training on the piano, Partch "came to the realization," he would later write, "that the spoken word was the distinctive expression my constitutional makeup was best fitted for, and that I needed other scales and other instruments."[2] Inspired by Hermann von Helmholtz's *On the Sensations of Tone* (1863), he had abandoned equal temperament—the conventional twelve-tone division of the octave—for microtonal scales in "just intonation," a tuning system based on small whole-number ratios. Because these new scales replaced the fraudulent intervals of equal temperament with intervals derived from a single note and its overtones, Partch dubbed his new system "Monophony," and its hairsplitting microtones equipped him with pitch resources sufficient for recording and harmonizing the subtleties of speech. Ditching the piano and its "twelve black and white bars in front of musical freedom" for a Monophonic medium of his own design, the Adapted Viola, Partch resolved henceforth "to allow the spoken words of lyrics to govern the melody and rhythm of the music" (*BM* 12) (fig. 1.1).[3] "Speech Music," he called it—"a modern renascence of the most ancient musical ideals."[4]

Partch had already begun exercising his Monophonic system in settings of Li Po lyrics, bits of Shakespeare, and the Psalms, but for his first full-scale dramatic application of speech-music he settled upon Yeats's 1928 transla-

Figure 1.1. Photograph of Harry Partch, his Adapted Viola, and experimental keyboard model, ca. 1933. Box 9, Folder 27, Harry Partch Estate Archive, 1918-1991, The Sousa Archives and Center for American Music. Courtesy of the Harry Partch Estate

tion of *King Oedipus*, a text that neatly articulated his interest in ancient musical ideals to his admiration for those ideals' most eloquent modern proponent. To Partch, Yeats had long been a "voice in the wilderness" for "the inherent musical beauty of spoken words," for language "unburdened by symphony orchestras and the Abstract beauty of sustained 'word' tones" (*G* 38). And when his hero in their bardic common cause responded encouragingly to a letter from the fawning composer, granting permission to set

King Oedipus and expressing "complete sympathy" with his Monophonic principles, Partch leveraged Yeats's endorsement to secure a research grant from the Carnegie Corporation to fund his trip abroad (*BM* 23). Physical proximity to Yeats was crucial. Partch's compositional practice often involved transcribing the minute melodic contours of texts-in-recitation, and he aimed to derive his setting of *King Oedipus* from the authorizing aural imprint of Yeats's actual voice. Like the other spoken songs crafted in the "flexible medium" of Monophony, this composition would satisfy the two distinguishing requisites of Partchian speech-music: (1) the principle of thoroughgoing harmonization, such that vocal parts are "recited in tones that are physically part of the instrumental accompaniment"; and (2) what I will term the *speech-music interdiction*, according to which "the line of the spoken inflection determines the melody."[5] "Words *are* music," Partch avowed in support of this latter principle, echoing Yeats's own virulent objections to the "*bel canto* mockery of words" effected by the distortions of opera and art song (*BM* 12, 167). "No word of mine must ever change into a mere musical note," the poet had proclaimed, "no singer of my words must ever cease to be a man and become an instrument," and in interdictions like these Partch recognized the potential shape and force of his own intuitive ambitions.[6] "I had been drawn to Yeats because of that marvelous experience of seeing eye to eye with him through his writings over a period of years—writing in which he expounded, and hoped for, a union of words and music in which 'no word shall have an intonation or an accentuation it could not have in passionate speech'" (*G* 333).[7]

When Partch approached Yeats's Rathfarnham home with his Adapted Viola in tow, it had been a long quarter century since the poet had written those words in spirited defense of his "new art" of speaking verse to musical accompaniment.[8] As Partch later lamented, ultimately "Yeats was too early for me, and I was too late for Yeats" (*BM* 167). Nevertheless, this ill-starred encounter is an instructive episode in the media history of modern poetry, one all the more generative for its seemingly oblique relation to the literary historical record. As scholars of modernism attend ever more resourcefully to poetry's catalyzing negotiations with sound technologies like phonography and broadcast radio, poetic writing has come increasingly into view as a sonic media practice in its own right, fully enmeshed in the period's rapidly transforming media ecologies.[9] But just what *kind* of media practice modern poetry is, and how printed poems are shaped by the technological and social protocols of other nonprint media, remains under-theorized. As we shall see, in their quixotic attempts to reintroduce *actual* lyres into the twentieth century, Yeats and Partch underline the peculiar mediality of that writing called lyric—or better, its *inter*mediality. More radically even, their experiments in the technological conditions of writing sound invite us to grasp lyric itself, fundamentally, *as* a media problem.

The speech-musical ambition to realize a hypostatized union of music and language occasioned the elaborate construction of "new-old media": material practices for the pursuit of bardic ideals under modern media conditions. In Partch's case especially, the Monophonic system bears the conspicuous impress of contemporary acoustic science and a technologizing soundscape. It is precisely speech-music's vexed, "new-old" position astride different media systems that recommends it as an illuminating analytical surrogate for the similarly conflicted, intermedial conditions of modern poetry—which also aspires to negotiate sonorous speech and symbolic notations. Yeats's interest in the antique psaltery and Partch's construction of his own lyre-like instruments suggest that modern lyric writing, crucified as it is between writing and sound, is always missing its lyre (medialogically as much as etymologically). These endeavors also suggest that much of what is most distinctive about lyricized twentieth-century poetry can be traced to the ways a regressive or fantasized ambition toward the union of words and music defies, disturbs, or appropriates its own irresistibly modern conditions—to the ways lyric minds the gap between our technological means for representing speech and sound. This chapter thus proposes to frame Partch as a lyric theorist of sorts and his speech-musical practice as an unlikely artifact of the still-to-be-written media archaeologies of those printed things many persist today in calling lyrics. After introducing speech-music in its Yeatsian and Partchian modes, I take up one specific literary application of the latter, Partch's "hobo" journal *Bitter Music*, which in transplanting Yeats's dream of a regenerative oral poetry to the transient shelters of the Depression-era American West, exploits those intermedial dynamics intrinsic to lyric practice, entailing unprecedented possibilities for the narrative inscription of sonorous speech.

Harry Partch ratchets lyric intermediality to a certain material limit. But to write or read an avowedly lyric poetry, as we know, is often to blithely disrespect that limit—to make a counterfactual music of speech. In the second part of this chapter, I follow speech-music into its late modernist afterlife to examine one especially inventive attempt to reconcile the tension between sound and print. At the same moment that Partch was formulating his early speech-musical experiments, the poet Louis Zukofsky documented his own wrestling match with print and sound in the founding documents of the so-called Objectivist movement. Whereas Yeats and Partch built new instruments to negotiate lyric intermediality, Zukofsky idealizes the tension with a lyric-ideological sleight of hand that comes in for illustrative critique twenty years later in his wife Celia Zukofsky's 1963 musical setting of Shakespeare's *Pericles, Prince of Tyre*. This piece of music has never, to my knowledge, been performed in public; I ask us to hear it as it was performed on a spring evening in 1952, in the Zukofskys' home, for a private audience that included Marianne Moore. From Yeats and his collaborator Florence Farr, to Partch, to Celia and Louis Zukofsky, this chapter traces three manifestations of the speech-musical impulse across the first half of the twentieth century. In each case, intimate speech-musical sounds resonate with the wider currents of media history precisely by evading mechanical inscription.

"I would of course see you with pleasure," writes Yeats in his letter inviting Partch to Dublin, "but doubt if my unmusical mind would be of much help to you. Your work interests me very much but I have no knowledge of music . . . I made no attempts to carry out my theories (setting spoken words to music), nor have I done so, since Florence Farr died."[10] Ronald Schuchard has indispensably documented these efforts of Yeats and Farr, the actress, singer, and prime mover in the poet's campaign for musically regulated spoken verse, a campaign that, despite his letter's demurral, "consume[d] more of [Yeats's] energy and interest than any single aspect of his poetic and dramatic art" in the century's first decade.[11] Their goal was to develop and promote a modern mode of accompanied declamation that would answer Yeats's long-held desire "to hear poems spoken to a harp, as [he] imagined Homer to have spoken his."[12] This "new art" represented a veritable artistic *movement*, Schuchard advises—by 1908 "the most visible poetic movement" in Great Britain—involving various theatrical institutions, international touring circuits, and a crush of acolytes readily testifying to the ostensible resurrection of ancient melic and troubadour song.[13]

As one major front in the ethno-nationalist program of the Celtic Renaissance, Yeats's pursuit of the bardic arts expressly staged in the cultural arena a heated contest that he saw playing out in the "Irish imagination and intellect," a contest between the world that "sang and listened" and the world, the English world, that "reads and writes."[14] We might say that by marrying a nativist cultural politics with a crude media history, Yeats enlisted spoken verse in an effort to decolonize the Irish sensorium. A restored oral culture, he argued, would herald a new "spiritual democracy" in which "the common man has some share in imaginative art"—a vision dependent on art finding its "way to men's minds without the mediation of print and paper": "I myself cannot be convinced that the printing press will be always victor . . . The world soon tires of its toys and our exaggerated love of print and paper seems to me to come out of passing conditions."[15] Yeats aimed to flee the "Gutenberg galaxy," but unlike later theorists of a "secondary orality," it wasn't on the wings of any *new* media.[16] Though Yeats would indeed "re-mediate" an oral tradition for BBC Radio in the 1930s, these engagements were a long way off in 1906.[17] Rather, the poet's familiar brand of modernist primitivism dictated that any truly experimental, "really new" artist "goes backwards till one lights upon a time when [the art] was nearer to human life and instinct."[18] Yeats is an innovator in what we might call, after Partch, the field of "new-old media"; and the ambition of Yeats and Farr's "new art" to obviate the mediation of print necessitated the (re)appearance of a "new-old instrument," one that vividly marks, in the archaic shape of a lyre, poetry's intermedial problem (*BM* 160-61).[19]

I am speaking of Farr's psaltery, that performatively antique, ur-object of the "new art" (fig. 1.2). The harp-like instrument was custom-commissioned by Yeats from the early modern music authority Arnold Dolmetsch, who tuned the satinwood instrument to the unique range of Farr's voice—from the "G below middle C . . . to the G above," each chromatic semitone doubled with its octave for a total of twenty-six strings.[20] In light of Partch's later rejection of equal temperament, the psaltery's semitonic, or twelve-tone, design is a surprising solution to the problem of setting speech to music. Yeats recounts that before Dolmetsch "taught us to regulate our speech by the ordinary musical notes," he and Farr *did* experiment with other notational methods, including the use of quarter tones.[21] They also tried indicating pitch contours analogically by means of "wavy lines" below the text, a kind of faux-phonographic inscription. But the construction of the psaltery put an end to these experiments. It is possible, then, to see in the conventionally semitonic psaltery a retrograde operation, a betrayal of the media conditions distinguishing Friedrich Kittler's discourse network 1900, that moment when the phonographic possibility of sound's physical capture rendered obsolescent such forms of symbolic notation as the alphabet and

A Speech-Musical Modernism 53

Figure 1.2. Photograph of Florence Farr and her psaltery, 1903. The Picture Art Collection, Alamy Stock Photo

twelve-tone scale. But this easy dismissal mistakes the psaltery's function, which is not one of reproduction but one of mediating accompaniment: the lyre precariously coordinates sound and writing.

Here is how Farr explains her method for affixing to the page the "inherent melody" of spoken poetry: "[I] spoke the first line of a poem in the most impressive way that occurred to me, and immediately after sang and wrote down the notes I found I had used as starting points for the spoken words."[22] She provides the important caveat that "it was impossible to capture the inherent melody of each spoken word," only the broader melodic contours of entire phrases. The result was a highly limited use of pitch material, but it was precisely the instrument's *limitations*—its "cunning simplicity"—that supposedly freed the performing voice to pursue the finer modulations that were its esteemed aesthetic payoff.[23] "If one is sufficiently practiced to speak on [the psaltery] without thinking about it one can get an endless variety of expression," observes Yeats.[24] "All art is, indeed, a monotony in external things for the sake of interior variety, a sacrifice of gross effects to subtle effects, an asceticism of the imagination."[25] But here is the speech-musical (also the media-technological) rub: though the psaltery's "ascetic" capacity to minimally regulate pitch opens the way for extemporaneous vocal elaboration, these "subtle effects" are strangely unaccountable in the specific economy of accompaniment avowed by Farr, Yeats, and, in due time, Partch. They transpire as a kind of excess via the mediating operations of the psaltery. According to the principles of speech-music, "words *are* music" prior to any formalization; "the little tunes contained in every word" necessarily *precede* their own regulation on the psaltery, and no word set to music "shall have an intonation or accentuation it could not have in passionate speech."[26] "I simply speak as I would without music," writes Farr, summarizing her composition process.[27] Notice the two contradictory directives composing the source code of the "new art." In performance, *speech should follow the psaltery's lead*—this, after all, is the reason for plucking the psaltery—but in composition, *the psaltery must strictly follow speech*. We can observe a supplemental logic distinguishing the psaltery's medial operations: the instrument *merely* accompanies, and must never interfere with, the sonorous character of "passionate speech," but it is nevertheless absolutely constitutive of speech-music. The new-old medium of the psaltery—and the medium includes, in this case, the satinwood instrument, its associated techniques, and the cultural-political investments of the Celtic Renaissance—ultimately effects, creates, props up, or naturalizes the "natural music" of poetic speech, that which it was supposed simply to support, record, or convey.[28]

And so the "new art" hinges on the mediating capacity of the psaltery, which, by processing the concerted sonorousness of an intoning voice, offers up not only the hypostatized union of speech and music but also a semblance of that union's raw materials: a "passionate speech" that is also already a music of words. Then, like any good medium or information channel, the psaltery disappears from the artistic situation it established in the first place, now merely ancillary to the ideal possibility it leaves in its wake: the possibility of poetry and song's lyric reconciliation in the twentieth century.²⁹ In the subtitle to her short book *The Music of Speech*, Farr describes her speech-music in ouroborian terms, as "verse set to its own melody"; what makes this recursive definition possible is the lyric medium that it also excludes. Quite improbably then, it would seem that a real-life lyre returns to poetry in the first years of the twentieth century only to immediately vanish.

Harry Partch and Speech-Music 2.0

By the time Partch took up the gauntlet of the speech-music interdiction (Yeats had set it down after Farr's 1912 departure for Ceylon as his focus shifted to Noh-inspired dance plays), the conditions of possibility for intermedial speech-music had been fundamentally altered by Partch's familiarity with industrial and scientific advances bearing upon the reproduction of sound.³⁰ Their initial interview only barely came off, as Partch recounts:

> When I arrive I feel that his cordiality is perfunctory, and I sense worry and uninterest in his voice. But I insist on playing "By the Rivers of Babylon" on my viola, intoning the words to its music. I feel fairly wretched.
>
> I draw my last bow, and there is silence. But I am not uneasy for long. All the things that I have been trying to get across to Yeats in letters and in talking suddenly become significant to him—and, in a way, the flood of comment that finally comes, in his deep measured voice, epitomizes the total comprehension that I have striven for so long. (*BM* 25)

"By the Rivers of Babylon," or "Psalm 137," had been Farr's "*pièce de résistance*," and so Yeats must have heard in the young composer's rendition an uncanny reprise of his "new art," though quick comparison of the scores' first lines reveals the substantial differences between these two bids for spoken song (figs. 1.3 and 1.4).³¹ Partch's Monophonic system, represented here in his "ratio notation," allows him to set every syllable and also to harmonize these finer inflections with two strings of Adapted Viola accompaniment, while in Farr's setting the psaltery notes serve only as approximate guides for the voice.³² The more "significant" ground of Yeats's recognition almost certainly concerned the speech-music interdiction itself—that is, Partch's ability to sing without singing, to the accompaniment of a lyre-like instrument only deceptively dispensable to the overall effect. Nevertheless, in its passage from Yeats and Farr to Partch this interdiction appears to change sign, as it were. No longer a negative proscription only, the interdiction against song begins with Partch to entail a positive, constructivist freedom: thanks to

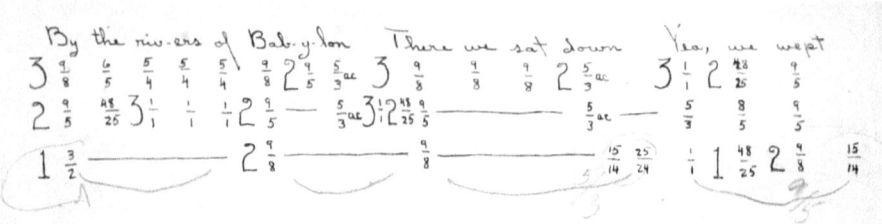

Figure 1.3. Opening of Harry Partch's "By the Rivers of Babylon" (Psalm 137) in 1932 ratio notation manuscript. Version A. Box 15, Folder 7, Harry Partch Estate Archive, 1918-1991, The Sousa Archives and Center for American Music. Courtesy of the Harry Partch Estate

certain technologically afforded transformations in the cultural understanding of sound, by 1934 musical procedures can record, harmonize, and systematize the musical-material stuff of speech to an unprecedented extent.

After meeting Partch, Yeats dashed off an eager note to Edmund Dulac, another musical collaborator. "I want you to hear him": "His whole system is based upon a series of notes within the range of the speaking voice . . . He never sings or even chants to his viola but always speaks, and every inflexion and tone of his voice is recorded in the score. He is, in fact, working out what Florence Farr and I attempted but with a science and a knowledge of music beyond Florence Farr's reach."[33] Partch's "scientific" and musical advance on Yeats and Farr began with his conviction that speech-music on the psaltery had been compromised from the first by the instrument's semitonic design. Because the score could only approximately regulate the tones of a speaking voice, "in the actual rendition the spoken tones and the music were generally not integrated, except by accident" (G 39). This lack of harmonic integration is exactly what Partch was bent on redressing when he abandoned equal temperament for forty-three microtones in just intonation and a Monophonic system "physically" integrating every note of the voice and its accompaniment. Partch constructs this new-old medium—and manages to honor the interdiction subordinating music to speech—by seizing inventively upon the technological and scientific resources available to him for making empirical sense of oral and aural phenomena.

Given Partch's ambition to work upon ever finer records of inflected speech, it is no surprise that phonography played a conspicuously rivalrous role in the development of the Monophonic medium. In January 1934, angling for a research grant, Partch described to Guggenheim Foundation ex-

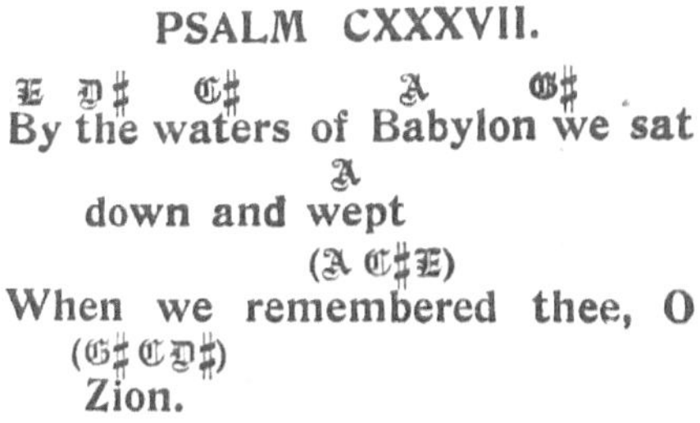

PSALM CXXXVII.

E D♯ C♯ A G♯
By the waters of Babylon we sat
        A
down and wept
  (A C♯E)
When we remembered thee, O
(G♯ CD♯)
Zion.

Figure 1.4. Opening of Florence Farr's "Psalm 137," as printed in her *Music of Speech* (1909), p. 24.

ecutive Henry Allen Moe his plans for setting *King Oedipus*: "The procedure I would follow, I think, would be to have phonograph records made of certain lines, as spoken by the players. These, notated, would then become the basis for the musical setting."[34] Though this transcriptional approach had been part of the composer's method for years, in all previous cases he had worked from a live voice in performance; the expedient of the phonograph was a new addition. He wrote again two days later, evidently to convince his funder of Yeats's "unprecedented" speech-musical credentials and to emphasize the feasibility of his own enterprise. After quoting Yeats's version of the speech-music interdiction ("no word shall have an intonation . . . ," etc.), he returned to the subject of phonography: "I spoke of recording interpretations on phonograph records which, notated, would become the basis for a setting. For several months last spring I was engaged in notating phonograph records of California Indian songs for the Southwest Museum, Los Angeles. Many of the songs bordered on a spoken rendition. After many repetitions of a part of a record I duplicated its tones on my viola and notated the result. Therefore, I know the procedure is practicable."[35] Though Partch finally did not use a phonograph to record the voices of Yeats and his Abbey Theatre actors, relying instead on paper and pencil to sketch "rough graph[s] of their spoken inflections," the fact that he considered it plausible suggests his assumption of a certain mediological parity between the phonograph and the Monophonically adapted viola, both capable of recording for reproduction.[36] The experience Partch adduces for the practicability of his method took place in the spring of 1933 when folklorist Eleanor Hague contracted the composer to transcribe a collection of Native American songs from deteriorating wax cylinders. In hiring Partch, Hague cited his research on "musical intervals of smaller scope than the ordinary diatonic intervals"—"Indian tunes are replete with such intervals," she supposed—and the entire process furnished, in turn, a kind of aural whetstone for Partch's own musical practice.[37] As Richard Kassel explains, the work of notating "pitch and rhythmic nuances," work made difficult by "the age and limited fidelity of these cylinder recordings," helped Partch "tune his ear" in unwitting preparation for *Bitter Music*, discussed below, in which he would transcribe "the speech-music of his fellow hoboes."[38]

Of course, to spend one's time painstakingly converting the signals of wax cylinders into the conventional symbols of musical notation, however radically refined, is to swim brazenly against the teleological currents of media history. But this was scarcely apparent on the ground in 1933. Archivists could better preserve paper than wax (the songs were copied eventually onto durable aluminum discs), and musicologists needed ways of visually materializing sound for analysis. We might compare Partch's work to Hungarian composer Béla Bartók's transcriptions of the Serbo-Croatian folk songs famously recorded on aluminum discs by classicist Milman Parry and his student Albert Lord in their efforts to infer the compositional procedures of Homeric poetry by analogical appeal to "the living epic tradition of the Yugoslavs."[39] While it is certainly true, as Wolfgang Ernst notes, that Parry and Lord captured on their discs "not just oral poetry . . . but noise as well"—the stochastic surplus of the "real audio signal"—and that "transcribing a sonic event into musical notation obliterates such acoustic events," neither Bartók nor Partch was under any illusions in this regard.[40] Both composers comprehended the limits of symbolic notation vis-à-vis the phonograph's ability to record and reproduce sonic signals. On the first page of *Serbo-Croatian Folk Songs* (1951), Bartók admits that "an absolutely true notation of music (as well as of spoken words) is impossible," that the "only really true notations are the sound-tracks on the record itself."[41] In Partch's case, acknowledging these limits meant testing his Monophonic system against the phonograph and schooling his method in the values of preservation and fidelity cultivated by sound reproduction technologies.[42] In other words, placed alongside the phonograph, the Adapted Viola's microtonal scales no longer produce only superfine melodies but also less-than-perfect records. At the Southwest Museum, the emergent medium of Monophony grew *conversant* with the protocols of phonography, even as—Ernst's reminder—they remained two essentially different media for sonic registration.

A Speech-Musical Modernism

According to Partch's biographer Bob Gilmore, we ought to insist upon this distinction between the composer's microtonal transcriptions and mechanical recordings, "not because [the former] were so much less accurate . . . but because they were never intended as anything but a means to an end."[43] That "end" is of course the properly Monophonic systemization of speech patterns, whereby each note coheres harmonically with every other. But since these "means" appear indispensable to speech-music—how else honor the speech-music interdiction?—Partch's compositional procedure lodges itself undecidably *between* phonographic and musical protocols. Here, then, we locate another salient instance of what we have identified as lyric intermediality. Partch's works exhibit something of a split personality, simultaneously transcriptions *of* past sonic events and scores *for* future ones, thereby rehearsing that same supplemental logic we observed in the case of Farr and Yeats, where the psaltery serves both merely to "record" or "fix" the music of speech and also verily to *produce* that same music. Similarly, the Adapted Viola can be said to *accompany* speech, rather than simply reproduce it, insofar as the instrument unstably mediates the two axes of reproduction: the preservation of sound in transcription and the provision or scoring of accurate playback. Since sound recording collapses these two axes in one mechanical inscription, Monophony's proximity to the phonograph—a function of both Partch's brief gig as an ethnomusicologist and the media situation circa 1930 more generally—ratchets this tension to the extreme, driving transcription and score into abrasive nonrelation. As Theodor Adorno opined the same year Partch met Yeats, the phonograph record "reestablishes by the very means of reification an age-old, submerged and yet warranted relationship: that between music and *writing*."[44] Though Partchian speech-music represents an attempt to drag that "age-old" relationship into the twentieth century *without* reifying it in the commodity form of the phonograph record, Monophony registers the reverberations of phonography all the same.

The Monophonic system also makes innovative use of developments in acoustic science, and Partch's orientation toward the physical features of sound and hearing was catalyzed as early as 1923 when the composer stumbled upon Helmholtz's *On the Sensations of Tone* in the Sacramento Public Library.[45] This landmark achievement of nineteenth-century empiricism aspired to "connect the boundaries . . . of *physical and physiological acoustics* on the one side, and of *musical science and esthetics* on the other," by founding a "theory of music" on a "theory of the sensations of hearing."[46] Specifically, it was Helmholtz's experimental description of the harmonic series of upper partials or overtones (the mathematically deducible series of resonant frequencies that accompany any pitched note, which we hear as one tone and designate by its lowest or fundamental frequency) as the basis for phenomena of consonance that spurred Partch to abandon equal temperament for just intonation using simple, precise ratios. The composer assembled his microtonal scales with intervals derived from relationships among positions in this harmonic series.[47] Though Partch stopped at forty-three, theoretically one could knit an infinitely thick "fabric" of pitches by extrapolating ratios from the series, filling the space of an octave with limitless gradations of ever-less-consonant intervals.[48] We can trace a strikingly direct line, then, from Helmholtz to the satisfaction of speech-music's twin principles: microtonal Monophony (1) enables Partch to improve upon Yeats and Farr's practice by fully and "physically"—that is, harmonically, as a matter of vibrational ratios—integrating accompaniment and voice, while also (2) permitting the rigorous subjection of song to speech's subtler music.

As a twentieth-century beneficiary of Helmholtz, speech-music keeps suggestive company with other speaking media. Jonathan Sterne points out, for instance, that the "theory of upper partials or overtones" is at work whenever we pick up the phone.[49] Helmholtz's discovery that "sounds could be best distinguished from one another by upper partials" has meant that telephone systems do not have to carry "the entire range of audible sound"; relying on "a little psychoacoustic magic," receivers instead trick the brain into reconstructing full sounds from just these easily distinguishable frequencies. But telephony supplies only one instance of the momentous shift in the analytical regard for sound occasioned by Helmholtz and the acoustic science pursued in his wake. When it becomes possible to demonstrate that an aural quality as elusive as instrumental timbre or vowel sound "is not due to anything more mysterious than its upper partials," then "all sounds become, in principle, synthesizable."[50] Inasmuch as Partch's intonational innovations rely upon a similar embrace of sound's synthetic possibilities, Monophony represents a strange, bardic repurposing of the scientific and technological conditions of modern communication systems.[51] This was a *conscious* repurposing, one might add; to drum up support for a Guggenheim application in 1933, Partch won himself a letter of recommendation from Harvey Fletcher, the "father of stereophonic sound" and director of physical research at Bell Labs, that seedbed of techno-sonic innovation.[52] Perhaps more significant than even this explicit vote of confidence was the confirmation Partch lit upon in Fletcher's 1929 *Speech and Hearing*, which declares it "infeasible to set a boundary" between speech and vocal music and which testifies to the clear psychoacoustic perceptibility of forty-three tones to the octave.[53]

When Partch wrote his magisterial treatise *Genesis of a Music* (1949), he girded its theoretical expositions with citations of acoustic research in order to insist upon Monophony's physical, physiological, and psychological bases. But as befits a new-old medium, these efforts to authorize speech-music's *modernity* indissociably involved a return to *antiquity*, to unities of speech and song, science and art. For Partch, the surest means of historical regress were those proposed by empiricists like Helmholtz, whose own tome, after all, supplied "the physiological reasons," deferred a few millennia, "for that enigmatic numerical relation announced by Pythagoras."[54] The mission to return musical practice to perfect ratios represents only one facet of the composer's career-long identification with ancient artifacts and forms of life. Though this animating obsession would reach its apogee with Partch's integrated music and dance dramas of the 1950s and 1960s, it was already visible in his early-sworn fidelity to the speech-music interdiction and in the sheer contempt for artistic specialization and disciplinary "purity" that led him to build his own instruments and study acoustic science (*BM* 191).[55] As *Genesis* illustrates, Partch brewed a uniquely severe variety of cultural pessimism, fusing D. H. Lawrence's disgust with bourgeois mores to the typological historiography of Oswald Spengler.[56] The only appropriate response to modernity was time travel: "Let not one year pass," he writes in 1960, "when I do not step one significant century backward" (*BM* 188).[57] But as we have seen, this regression can take place only via a musical apparatus that has been considerably abetted and recast by modern media conditions, one that has internalized the rivalrous impact of phonography and the knowledge of acoustic and aural materialities. As a new-old medium, speech-music's regressive program for reuniting word and song manifests the shaping influence of a media-technologically transformed understanding of sonorous speech as a matter of signals everywhere transduced by records, radios, and telephones. "Words further musicalized by accompaniment of instruments have been a constant source of joy to man thruout history"; by the time Partch writes this sentence, to "further musicalize" means also to sonically engineer.[58] The very terms of lyric's constitutive intermediality—that enduring tension between sound and printed notations—have been immanently retooled by their technological conditions.

Bitter Music and the Social Indexical

Partch spent just a few days with Yeats in November 1934, exhibiting the Monophonic system and raptly noting in "rough graphs" the "vibrant tones" of the poet's voice in renditions of *King Oedipus* (*BM* 167). With a promise to meet Yeats again in London in April, Partch then resumed his research into the history of intonation at the British Museum, enjoying while in England his new Yeats-backed passage into a "world of London intellectuals" like Dulac and Dolmetsch and pursuing his plans to construct a new microtonal organ.[59] In December, Partch moved on to Rapallo, likely at Yeats's suggestion, "to organize [his] research materials and to compose."[60] After a sojourn in Malta, he returned to London only to discover that a convalescing Yeats would be unable to meet again after all, and on March 30, 1935, the composer set sail for Portland, Maine. Though he would never see Yeats or Europe again, by all accounts his trip had been a success: *Oedipus* was underway, and he had uncovered ample historical documentation—"a solid rock of apostasy"—for speech-music and just intonation.[61]

In a jarring reversal of fortunes, Partch's return to an economically ravaged United States almost immediately inaugurated the eight-year period he would later term his "own personal Great Depression"—the "hobo years" that so indelibly characterize his reputation still (*G* 323). He explained his embrace of an itinerant lifestyle as a wholesale refusal of the musical establishment and as a necessary consequence of his unwillingness to rely any longer on the good graces of patrons. "My friends are very kind," he writes in 1935, "but there are only two things to do—return to Los Angeles and resume begging under the apology of my music, or, wearying of begging, with a whoop and a holler go on the road" (*BM* 35). Looking back in 1974, he recalls this moment a bit more soberly:

> For several years . . . 1930–1934 . . . I had entrée into the best places—well, some of the best places, in San Francisco and Los Angeles and New York . . . and I knew a lot of people. But after I came back from England, I'd had it; I just was not going to be a sycophant for these wealthy people . . . which I could have been, very easily, after a trip to Europe . . . And you see, it was also a Depression . . . I went from that world to a world of no jobs. Total poverty, really total. But Roosevelt had been in office by that time two years, and so he'd established these transient shelters, they were called, for homeless guys; and they did very well, they really did very well.[62]

S. Andrew Granade has persuasively detailed the extent to which Partch's decision to "go on the road" was accommodated by the robustness of the New Deal's Federal Transient Bureau, which at its peak oversaw the care of a half-million people nationwide, an outsized portion of which had migrated to Partch's California.[63] In any case, by 1949, the composer's recourse to vagrancy had become the stuff of myth: "My return was to a jobless America," he recollects in *Genesis*, "and I took my blankets out under the stars beside the American River (the river of gold!), carried my notebook, kept a journal, and made sketches. I called the journal *Bitter Music*" (*G* 323).

Inspired perhaps by Yeats's praise for his writerly skill, Partch eventually fashioned this eight-month journal into a recognizably literary achievement, albeit a deeply unconventional one.[64] *Bitter Music* marries speech-music to autobiographical journaling to effect a sui generis literary form, a kind of illustrated operatic memoir for page and piano. The novella-length narrative offers a tender, often funny, passionately discerning, and only at odd moments condescending portrait of the Depression-era artist among other young men who had likewise "found [their] place in the American economic system, as the learned Dr. President advised [them] to do—in a camp for transient bums" (*BM* 36). As Partch shuttles between shelters and work camps in California, Oregon, and Washington, we find the composer struggling to come to terms with his precipitous change in fortune, with the fact that his "own efforts to bring beauty into the world have come to aimless wandering" (44). The journal entries oscillate between the insurgent temerity of artistic resolve—"an intense fervor in the justness of my life"—and plaintive self-doubt: "Their whole lives are a continual escape," Partch remarks of his fellow travelers in "bumdom," but then admits (or "almost" admits), "I am myself almost exactly like them. I am escaping the meaningless, the stupid, the banal, in conventions, art, music, in a fantastic order of my own creation. But is mine so blind? I don't know" (38, 10, 37). On the whole, *Bitter Music* advances less a *romanticized* portrait of transient life than a radically *eroticized* one. A sensualist's regard for a world without "any tinge of morality," evident in the painterly descriptions of Western landscapes as much as in the frank treatments of same-sex desire, imbues a text playfully unconcerned with the notions of Christian salvation that it parodies throughout (42).[65] Take, for example, the scene in which Partch mollifies a regretful Pablo, who is nursing his wounded masculinity after "a few unchartered kisses" with another man (18-19). "Who cares who loves who?

It doesn't matter, anyway, so long as the music is good music." Charmed, Pablo wishes Harry a grateful "good-night":

> How warm it is—in this summer darkness—to have a "good-night" wished for me, so sincerely.
>
> It is late, and we tiptoe to our bunks opposite each other across the aisle.
>
> I hear him moving restlessly on hot blankets. But I do not think—I am carried into sleep by the sequential snorgles from forty human exhausts. (19)

The motivating concerns of the journal's daily entries—Partch's doubts, frustrations, and resolutions regarding music and those regarding his relationships with fellow transients like Pablo—unfold against a backdrop of richly observed social testimony. Thoroughly a product of the decade's vogue for documentary expression, *Bitter Music* informs us, for instance, of the rigidly enforced social hierarchies among transients, the mortal dangers and everyday drudgeries of life on the road, the bureaucratic operations of state and federal emergency relief agencies, the opinions and prejudices of the public regarding migrants and vagrants, and the keen humiliation of knocking on doors for meals after the Federal Transient Bureau shuttered its shelters in September 1935.

But if Partch's ambition to present an unflinching, unstinting account finally proves a rather familiar one, his commitment to speech-music distinguishes the book from all other cultural products of the Depression. While giving narrative shape to his personal experience, Partch was also training his ear on an entirely new social register: "I heard music in the voices all about me, and tried to notate it, and I tried to enhance the mood and drama of such little things as a quarrel in a potato patch. The nuance of inflection and thought of the lowest of our social order was a new experience in tone, and I found myself at its fountainhead—a fountainhead of pure musical Americana" (*BM* 5). Interspersed among the diary entries and ink sketches in *Bitter Music* are musical notations of speech—hurriedly recorded in a "rough way" *sans* instruments and then elaborated later at the piano—alongside quoted snippets of folk songs, hymns, and Wobbly anthems. *Bitter Music* is Partch's sneering answer to those "continental Europeans and upper-crust Yankees" given to opining that "America is not a musical nation"; he repeats this "untruism" with savaging irony throughout the text and takes to disproving it by sinking speech-music into the deepest structures of the work (10).

Combining this "art form as old as history" with literary narrative, the resulting hybrid enjoins its audience to a unique form of reading: "If possible the book is to be read at the piano, and the fragmentary music, on passages requiring emphasis and intensification of mood, occurs much as the incidental music might occur in a talking picture. The fragments are in no sense 'performers' music'; they are readers' music" (*BM* 6). Since Partch refined this "readers' music" at the piano and not in his Monophonic system (his instruments were stored safely with friends in Los Angeles), the project disarticulates speech-music from the principles of just intonation and microtonality, thus seriously compromising the theoretical positions he had been staking out for years. Despite this pragmatic recourse to the piano's "twelve black and white stiflers," Partch still accorded *Bitter Music* a decisive role in his career's trajectory (12). Not only did the journal's thematic concerns furnish a "faintly delineated canvas" for subsequent Monophonic compositions in the "Americana" vein, but modern speech-music itself—Partch's sustained exploration of lyric intermediality—apparently required for its further development new materials and new experiences: "my 'evolution' seemed to demand a sudden descent into hobo jungles" (*G* 323, *BM* 5).

The nature of this evolutionary descent can be clearly stated: *Bitter Music* is Partch's first attempt to set script-less oral expression. Up to this point, speech-music had meant—as it had for Yeats and Farr—the scoring of spoken recitations of prior texts like the Psalms or *King Oedipus*. In such cases, the transcriptions he would generate for compositional fodder effectively derived from the performance of other prior (alphabetic) transcriptions. In the shelters, camps, and railyards of the West, by contrast, Partch served American music as a kind of bard-ethnographer—one who listens, rather than sings, as he wanders, composing from the spontaneous utterances of his traveling companions. Significantly, as Partch pursues Yeats's bardic dreams of a "spiritual democracy" under the new-old auspices of his quasi-ethnographic practice, the cultural pessimism they share is momentarily eclipsed by an altogether different political valuation of speech-music, one staked on the sociality of speech.

Partch's critics are right to suggest that his stint at the Southwest Museum had prepared him to apprehend "the speech-music of his fellow hoboes," insofar as this work required transcribing from a script-less phonographic signal. But identifying melodic and rhythmic features from behind a machine's acousmatic blind proves a relatively simple task when compared to all that transpires in *Bitter Music*, where Partch, a lay ethnographer immersed in a "sea of chaotic humanity," appeals to speech-music as a technique for representing not just sonorous language but veritable *talk* (*BM* 5).[66]

Partch had long understood Monophony as a musical means of establishing singularly intimate relations between performers and audiences, but this intimacy is realized on a different order entirely when he takes speech-music "off book," as it were, to explore how sonorous address, in speech, establishes and sustains social relationships.[67] To borrow from Nicholas Harkness, this "readers' music" develops a critical aesthetic out of Partch's intuition that "voice is not merely a sonorous extension of an embodied individual or the natural expressive outlet or externalization of interior emotions"—though for Partch it is certainly those things, too—"but also, and centrally, a channel-emphasizing phatic mode of social contact."[68] As such, *Bitter Music* offers a study in the phaticity of speech: speech as a material channel facilitating communication between persons and positioning those persons relative to the social world.[69]

If before 1935 Partch had predicated Monophonic speech-music on the scientific observation and artistic exploitation of acoustic overtones, then *Bitter Music* finds the composer tuning in to what Mikhail Bakhtin styles the "contextual overtones" of speech, the way "each word tastes of context and contexts in which it has lived its socially charged life."[70] Linguistic anthropologists have long demonstrated how such "contextual overtones" instance the *social-indexical dimension* of language-in-use, those "elaborate meaning structures of speech behavior"—all the context-dependent social information—that strictly grammatical or semantic analyses would leave as mere "residue."[71] The concept of indexicality allows linguists to trace how relatively nonreferential aspects of language behavior, like the use of deferential pronouns or the sound of a voice, can disclose social relations and bring into pragmatic play aspects of identity or embedded position. Thinking like a linguistic anthropologist, or construing Partch as one, enables us to hear his notated voices as "embodied site[s] of both musical and linguistic expressivity, and of social distinction."[72] For everything else *Bitter Music* documents, it also records Partch's discovery of how one particular feature of language use—intonation or speech melody—can index an entire "micropolitics of emplaced, embodied, and voiced identity." What's more, the book demonstrates just how well suited his own speech-musical methods are for marshaling this social data into the space of an artwork.

At times the book's speech-music functions indeed as "incidental music," or aural mood lighting, not unlike the more conventional tunes of the hymns and pop songs also patched through the narrative; more often than not, though, the kind of "emphasis" these settings provide recalls us to that word's classical rhetorical sense ("to imply more than is said," "a meaning not inherent in the words used") or to its etymological link with the *phatic* function of language.[73] The stern command of a bureaucrat ordering Partch to strip for delousing, vulgar wisecracking, complaints and exasperations regarding the miseries of shelter life, conversations secreted in trembling tones of vulnerability, cries of drunken embarrassment, mocking impersonations of Partch's own speech patterns, even his own internal musings and the fancifully ventriloquized voices of the wind, stars, and apple trees—such melodies *situate* the diegetic speakers and their auditors, conjuring differentials of social class and identity, entailing complex terms of relation and association on which the text is semantically silent, and diffusing a wide panoply of affects and attitudes. Consider a very simple example. In the final pages of *Bitter Music*, Partch records his meeting with the man whose offer of a proofreading job affords him temporary respite from the road (fig. 1.5):

FEBRUARY 1, 1936—San Bernardino

"Do you drink?"
"No—that is, I don't make a habit of getting drunk."
"Well, I feel sure, after talking to you, that

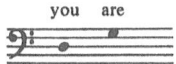

what you say you are, a practical newspaper proofreader."
"Huh— (I smile as the years crowd around) that I am."

Figure 1.5. Excerpt from Harry Partch's *Bitter Music*, 1991, p. 128. Used with permission of the University of Illinois Press on behalf of the Harry Partch Foundation

What the perfect fourth *emphasizes* in this moment of intersubjective confirmation (the kind of interpellation conferrable only by one with the power to relieve economic insecurity), and what Partch relishes enough to memorialize, is an inflection of mild astonishment swiftly, confidently acceded to: despite your circumstances, "you *are* what you say you are." And Partch—"Huh"—is surprised, too. Everyone in *Bitter Music*, and this includes the reader voicing this speech-music at the piano, would likewise be surprised at how precisely they are what they say they are—or *how* they say it.

Elsewhere in the text, speech-music arrives at points of intense pathos, suffusing the narrative with a melodrama that enlarges our feeling for the moment without flattening its emotional complexity. In the following scene, Partch says goodbye to the Hoodlum, a convict with whom the composer kindles a mutual infatuation when he meets the Hoodlum's road gang camped outside Big Sur. Here the Hoodlum, all mock bravado, has just promised to find him and "black both [his] eyes" should Partch forget to write him a letter (*BM* 95) (fig. 1.6):

> He laughs, but I know how desperate he feels, breaking the first bond that he has had with the outside world in years.
> "If I don't write, you can do worse than that."
> We stand against the earthy bank, saying nothing—wind in the chaparral, beating sea.
> Something comes to me out of the void.
> We shiver. He grips my shoulders. I feel the warm breath of his whisper,

Figure 1.6. Excerpt from Harry Partch's *Bitter Music*, 1991, p. 95. Used with permission of the University of Illinois Press on behalf of the Harry Partch Foundation

It's unclear whether the Hoodlum's whispered response refers to the "funny" meaning of Partch's jokey come-on—"you can do worse than that"—or the inflection of his speech, but of course *Bitter Music* points everywhere to the impracticality of such distinctions regarding language use, whether between sense and sound, content and form, or signal and channel. Speech-music often vitiates these distinctions in order to hail the complexities of an irreducibly inscrutable context. In the Hoodlum's reply, for example, we hear noise flickering through the channel as dynamics of social difference—those obtaining between the artist freely opting out of civil society and the convict forcibly removed therefrom—abrade desire and deep feeling to distend the word "funny" by an augmented fourth. Partch's "descent" into hobo jungles parallels another descent, then, at the level of aesthetic practice, from the bardic ideal of a universally intimate language to a conception of language shot through and sounded out with bitter contingencies.

Because *Bitter Music* is a book and because Partch forges its speech-music in symbolic notation, he can only engage the social-indexical dimension of language as far as the melodic contours of speech will take him. If Partch *had* used a phonograph, perhaps in a manner anticipating Steve Reich's own speech-musical experiments on magnetic tape in the 1960s or the tape-based "audiopoems" of Henri Chopin, then we would hear much more in the accentual features of the Hoodlum's voice—namely, its timbre, volume, rhythm, and all the social entailments thereof.[74] Once more, we find Partch plying his new-old art at the outside limit of its print medium, evolving an expressive practice by driving straight into the aporia of sound and writing instantiated by lyric's intermedial condition.

The Partchian project is marked, on the one hand, by compromising instability: "words *are* music," he insists, but the emphatic italics visibly devil the claim by reminding us of its mediation in print. Though the "readers' music" of his journal stresses the phaticity of speech as a sonorous mode of social contact, it remains a species of the silent page. On the other hand, insofar as speech-music is structured in tension between print and phonographic protocols, between scores and transcriptions, old and new media, writing and sound, it affords innovative possibilities for the literary "enregisterment" of sonorous speech. To suggest that individuals "enregister" speech is to notice, with Asif Agha, how interacting speakers "establish forms of footing and alignment with voices indexed by speech and thus with social types of persons, real or imagined, whose voices they take them to be."[75] By writing voices in musical notation, Partch shares the burden of just such acts of enregisterment, enjoining his readers to align themselves with respect to those textualized voices as socialized figures. When Partch privately imagines the homophobic mockery he expects from his fellow transients (fig. 1.7),

—with tongues sticking out obscenely on the *Lol*—

Figure 1.7. Excerpt from Harry Partch's *Bitter Music*, 1991, p. 64. Used with permission of the University of Illinois Press on behalf of the Harry Partch Foundation

readers playing along must themselves reproduce and audit the "lilting scorn" of a mimicked intonation that Partch himself reproduces from past experience, evoking in the process, if only briefly, a whole compacted order of masculine anxiety and iterable violence (*BM* 63). An entirely different register is at play—perhaps the tremulous footing of an eager acolyte—in Partch's memory of his final glimpse of Yeats (fig. 1.8):

> Although he is ill, and weak from long confinement, he insists on escorting me to the bus. He stands in the middle of the road with his hands upraised, his huge figure physically blocking its passage. And this is my final picture.[13]
>
>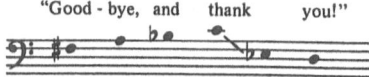
>
> That is all I can say.

Figure 1.8. Excerpt from Harry Partch's *Bitter Music*, 1991, p. 26. Used with permission of the University of Illinois Press on behalf of the Harry Partch Foundation

When these deployments of speech-music work hand in glove with Partch's diegetic structures, *Bitter Music* effects a conjunction of musical and literary practice that strains against the normative (normally silent) media channels of print narrative. This is a noisy, uneven road that prose writing has not since taken, and well worth our consideration for just that reason. But as I have suggested, the intermedial space visited here by narrative is where the printed poem, by contrast, permanently *lives*.

After *Bitter Music*, Partch would go on ostensibly to exhaust his speech-musical practice in a series of celebrated "Americana" compositions that submit the kind of socially saturated linguistic material found in the journal to the full theoretical and instrumental apparatus of his Monophonic system.[76] By the time Partch realizes his long-deferred ambitions to set Yeats's *King Oedipus* in 1951, the speech-music interdiction will have been essentially replaced by the "total theater problem": casting singers who dance, dancers who play instruments, and musicians who act (*G* vi). The Monophonic system, the construction of instruments, and the commitment to ancient and non-Western musics—these remain the animating elements of the composer's practice, but no longer does that practice hinge on preserving the "natural music" of speech. While Partch scholars have already underscored the significance of these earlier efforts for his later career, my own aim has been to position Partchian speech-music alongside its Yeatsian inheritance as an oblique case study in the media history of modern poetry. We are authorized to read Partch as a media theorist of the lyric because speech-music and printed poetry share the same ineluctably distressed media condition. Like speech-music, lyric writing can be understood as a compositional practice structured in tension between the taking of transcriptions and the making of scores. And like speech-music, this tension is driven to unprecedented extremes when poetry's historical claim as a technology for the representation of sonorous voices comes into contact with inscription machines that far outstrip its recording capacity. Modern poetry *does* try to make a finer, fuller record of itself—the "new art," Monophony, and experiments in speech-based free verse are all, in part, attempts to do just that—but the record is never as good as, never commensurate to, the score any poem also is. And so all poetry staked on the elaboration of language's sonic potential must in some manner confront this aporia of transcription and score, which is only the exacerbated crisis of lyric intermediality in the twentieth century.

When William Carlos Williams in 1950 reviewed the past half century in prosody and called it a "full dress return to the Greek," he was referring principally to modern poetry's struggle to approximate quantitative measures in English.[77] Meanwhile, the "dress return" orchestrated by Partch and Yeats involved returning ancient lyres to lyric, instruments that endeavored to solve, but inevitably pointed up, lyric's medial problematic and its sensitivity to changing technological environments. Misreading Williams only slightly, these efforts reveal modern poetry itself as a new-old medium, akin to the more marginal, though paradigmatic, new-old art of speech-music. And it remains so today. Wherever poems are still printed and read out from pages or from screens, the intermedial lyric has neither regressed nor advanced any further than Partch could take it. But poets are bound to try.

Objectivist Music

One Saturday in April 1952, Marianne Moore paid a visit to Louis and Celia Zukofsky in Brooklyn Heights, just up the road from her Fort Greene apartment. Moore had recently sent a small scholarship—an "Easter present"—to the Zukofskys' son Paul, a violin prodigy, to defray the cost of lessons at Juilliard.[78] In gratitude, the Zukofskys invited Moore to a private concert featuring Paul's performance of a "violin sonata + piano sonata + some Bach"—

though the evening was not all eight-year-old precocity.[79] With Louis singing and Celia on the piano, the pair capped the impromptu program with a contribution of their own, an excerpt from Celia's own setting of Shakespeare's *Pericles, Prince of Tyre* (1609).

Thus Moore was inducted into an elect group. Though the Zukofskys occasionally showcased "the *Pericles*" for friends, including William Carlos Williams, Mark Van Doren, and Edward Dahlberg, Celia's score would never be performed in full, and never for a public audience. Setting to music every syllable of Shakespeare's play, the *Pericles* is, by all accounts, a strange work of art. Its delicate vocal line, animated by a decorous undertow in the accompaniment, effects a "weird quietness" in Van Doren's phrase, or a "quiet intensity" in Gerard Malanga's, a music "vital and urgent" and faultlessly composed but curbed by a palpable quaintness, its "sentiments . . . distant, its formal peregrinations academic."[80] Listeners reach for paradox to describe the music's attested effects because Celia's subtle underscoring defers conspicuously to the text while working at the same time, by a kind of weak power, to concentrate, clarify, and enhance Shakespeare's rhythms.

The music's captivating diffidence provides a neat figure for its relationship to Louis Zukofsky's poetic oeuvre, with respect to which it radiates a similarly unassuming power. Though it was never performed in public, Celia did publish the two-hundred-page score in an augmented setting for chamber ensemble (oboe, horn, clarinets, lute, and violoncello) as the second volume of her husband's mammoth *Bottom: On Shakespeare*, prose centerpiece and summa of his poetics. When *Bottom* appeared in 1963, Zukofsky explicitly subordinated his own prodigious literary effort to his wife's musical one. The book designates a large swath of its highly wrought prose mere "notes for Her music to *Pericles*" and "the one excuse" for a substantial portion of the text.[81] Asked by Cid Corman whether he was serious about this latter remark, Zukofsky replied, "Yes, I mean that footnote . . . and will mean it more and more as the world understands less and less how much I mean it."[82] Taking Zukofsky at his word will mean recovering the intensely private meanings that he attributed to the *Pericles*—as well as those medialogics, instantiated in Celia's score, which plainly exceed the ideological frame of the Objectivist poetics that otherwise purports to claim her *Pericles* as its proof of concept.

Moore herself wasn't put off by the music's "weird" intimacy. Scratched edgewise in his report of the evening to Lorine Niedecker, an aside from Zukofsky beams with pride: "She loves the *Pericles*, of course."[83] Williams,

A Speech-Musical Modernism 73

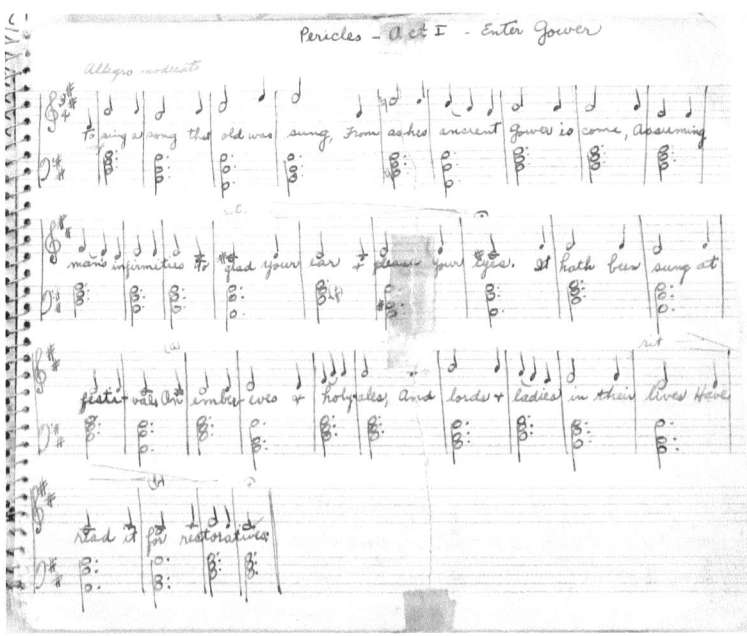

Figure 1.9. First bars of Celia Zukofsky's *Pericles* in manuscript setting for piano, 1943. Box 9, Folder 1-3, Louis Zukofsky Collection 1910-1985, Harry Ransom Center. Courtesy of Musical Observations

for his part, agreed. The experience of hearing *Pericles* led him to contract Celia Zukofsky for a setting of his own "Choral: The Pink Church" (1946) and to share his eloquent admiration in a letter to Louis: "Celia's music is something I could listen to for hours, the antithesis of all the shouting and spouting, distortion and clouding of words and phrases that is opera."[84] Inveighing against the operatic deformation of language, these lines remind us of Yeats and Partch's speech-musical interdiction, but the particular terms of Williams's praise lay a unique and counterintuitive emphasis on *visuality*. "The music holds the words in its amber," he writes; it "assists them to be seen." These observations touch on what is ultimately at stake in the *Pericles* project: namely, the co-implicating, enmeshed operations of sound and visuality—the "glad . . . ear" and the "please[d] . . . eye"—which for Louis composes the singular distinction of lyric language (fig. 1.9). The acuity of Williams's remark explains why Celia quotes this endorsement in the preface to her score, and the conceptual centrality of sight explains also why she and Zukofsky elected to publish the lengthy setting in a facsimile of her careful, visually appealing notebook hand. As we shall see, the music's enriching emphasis on the phonetic material of Shakespeare's language neither detracts nor distracts from the order of visuality, as in some zero-sum ratio of the senses. Sound in truth intensifies, substantializes, and lends phenomenal weight to the printed symbolic marks of the silent page. It "assists them to be seen." Or in Zukofsky's own formulation, as if to gesture at the melding of speech, song, and instrumentation at the center of speech-music's first principle, the *Pericles* score encourages us to "look at Shakespeare's *words* as if they were tuned objects that strike off tones."[85]

In what follows, I read Celia Zukofsky's *Pericles* as a critical intervention in the process by which the generative tension at the heart of lyric intermediality, which drove Partch to extraordinary lengths of musical invention, is peremptorily organized and resolved. In the final analysis, this speech-musical work is saddled with two contradictory meanings. We can view the score by the lights Louis Zukofsky himself preferred, that is, as the very apotheosis of his poetics, confirming his judgment that Celia's "music saves me a lot of words" in *Bottom*.[86] And yet, in the very act of substantiating this claim, we notice *Pericles* slip from this hypostasizing frame. Uncontained by the meanings Louis would apply to the score and performing something like the resonant material unconscious of his attempts to objectify lyric practice, Celia's music also insinuates an immanent critique of his poetic project. By restoring this neglected piece of music to its keystone position in Objectivist poetics, we can begin to grasp how the poet's career-long investment in the historical materiality of poetic language culminates in the ideological resolution of the tense accompaniment of sound and print. I conclude, then, by locating one absent referent for *Pericles*'s quiet weirdness in technologized sound. Though apparently never recorded, Celia's page-bound score indexes the cultural priority of sound recording as a latently remediating presence in Zukofsky's poetic thought.

Our first step is to credit Zukofsky as a singularly ingenious theorist of lyric intermediality. Writing in 1931 on the twin poetic values "sincerity and objectification," Zukofsky opined with respect to the latter, in his cagey, crabbed prose, "distinct from print which records . . . and incites the mind to further suggestion, there exists, tho it may not be harbored as solidity in the crook of an elbow, writing (audibility in two-dimensional print) which is an object or affects the mind as such."[87] This definition of the modernist poem as a "rested totality [that] may be called objectification" itself rests on the notion of an "audi*bility*" that inheres potentially "in" alphabetic print, writes Zukofsky. Readers may encounter this sonorous possibility as an "object" endowed with shape and autonomous substance such that, if it cannot be toted around in "the crook of an elbow" like a newspaper, it can nonetheless "affect the mind," rather than being simply effected by it. Elaborated across five decades of poetry and prose, Zukofsky's ambition on the one hand to "modernize/His lute" by foregrounding the phonic "melody" of printed verse and on the other to achieve the "objectively perfect" in poetry are two faces of the same extreme commitment to lyric materiality announced in the idea of a printed "audibility" specific to poetic practice.[88]

For decades poets and critics have worked to render explicit the meaning Zukofsky leaves undeveloped in his formula for the Objectivist poem.[89] What has passed largely unheralded is the way this early figuration of the poetic object presages the full-blown literary-media theory Zukofsky will pursue with uncommon if idiosyncratic resolve in the 1950s and 1960s. Eventually he will organize his thinking about lyric materials into an oddball historiography of language use (a "graph of culture") by fusing Ezra Pound's threefold schema of poetic modes—*phanopoeia*, *melopoeia*, and *logopoeia*, or image, music, and idea—with Henry Adams's concept of historical "phases," themselves keyed to respective elemental states.[90] "One is solid, another is liquid, and the other is gas." "It's the same with the materials of poetry, you make images—that's pretty solid—music, it's liquid; ultimately, if something vaporizes, that's the intellect."[91] Since these phases correspond to the embodied capacities of the eye, ear, and mind, they coexist and depend upon each other, but in each epoch any one element may prevail. Modernity is preeminently—and for Zukofsky, lamentably—gaseous, in a manner that echoes Partch's demand for a return to the premodern "Corporeal" (*G* 8). "I'd like to keep solid," Zukofsky admits, but "I was born in a gas age," suffering the heady abstractions of philosophical skepticism and modern science.[92]

Since the early modern period marks the gas age's fitful inception, when "late logic and science" begin to take root but memories of the solid and liquid ages, with their habits of mind and art grounded in the credulous eye and sensitive ear, remain culturally operative and vital, Shakespeare occupies a crucial transitional position in Zukofsky's "graph of culture."[93] "Shakespeare's *Works* as they conceive history regret a great loss of physical looking," Zukofsky writes. "The intellective propositions of their actions anticipate the present days' vanishing point, but unlike the present's propositions still sing an earthly underpinning."[94] Shakespeare wrote "the text of modern life," according to Emerson, and Zukofsky would have us respect the ways that text moves with a substantiality capable of ballasting the emergent "vanishing points" of abstractive representation and gaseous thought.[95] In other words, Shakespeare is an Objectivist avant la lettre, and his plays and poems are preeminent examples of the lyric object Zukofsky terms a rested totality.

I refer to Zukofsky's cultural-historical schema as a literary-*media* theory because the "gas age" is doubtless also the age of print. In positing a transformation from the sensorially adept cultures of the liquid eye and solid ear to the modern reign of the gaseous mind, Zukofsky echoes, in caricature, other more authoritative narratives of print culture's revolutionary emergence from a world of manuscripts and scribes.[96] Indeed *Bottom*'s obsession with the way print reconditions the aural and visual dimensions of language credits the book's genuine affinity with the efflorescent media-theoretical commotion of the early 1960s, when such titles as Marshall McLuhan's *Gutenberg Galaxy* (1962) and *Understanding Media* (1964), Eric Havelock's *Preface to Plato* (1963), and Albert Lord's *The Singer of Tales* (1960) were staking the nascent field of media studies around questions of writing, print, and sound. In *Flaubert, Joyce, and Beckett: The Stoic Comedians* (1962), an undercited contender for this list, McLuhan's student Hugh Kenner explains how vividly a "machine-novel" like *Ulysses* testifies to "a world fragmented into typographical elements."[97] Writing to Kenner in 1963, Zukofsky quotes from *The Stoic Comedians* to enroll himself into this same modernist project: "As for your theme it is not 'surprising' that 'some conatural awareness of the nature of a civilization structured by print' has *Bottom: On Shakespeare* looking (?) into the same area, site before sight with the voice still bouncing off it, until it diffuse voiceless in-sights into print."[98] With his deep attunement to the possibility that phonetic sound, bouncing off sighted words, can render "in-sights into print," Zukofsky intends in *Bottom* to restitute writing's full powers as a visual, sonic, and intellectual medium. This program smacks of Joyce's cognate ambition to offer a "verbivocovisual presentment" in literature, but it's worth parsing the difference between the "conatural awareness" of print that Kenner celebrates in Joyce and that which Zukofsky claims, via Shakespeare, for himself.[99] For Kenner, Joycean modernism—with its novels contrived of other books, chockablock with encyclopedic information, disparate lexicons, and self-reflexively permutating sentences frozen in "the stunned neutrality of print"—reflects the decadent zenith of print culture in the twentieth century.[100] Though impelled by this same "conatural awareness," *Bottom* nonetheless "record[s] with a backward look" a time when print was less an atmospheric condition deep-sunk into our modes of thought and perception, and more a material marvel, a febrile locus of sensuous experience.[101] Under the eyes of Zukofsky's Shakespeare, printed letters were solid typographical shapes dynamized by a phonetic music that for readers in succeeding centuries would melt into abstract air.

As artists of the alphabet, Zukofsky implies, most poets are "constantly seeking and ordering relative quantities and qualities of sight, sound, and intellection."[102] But *Bottom's* media-archaeological dig into the history of lyric intermediality finds Shakespeare's coagulating interplay of the solid, liquid, and gaseous dimensions of language ranging far beyond conventional conceptions of the signifier and semiosis more broadly. When Zukofsky speaks of "sound" in *Bottom*, he means something akin to Poundian *melopoeia*—the "music implicit in the movement and pitch of the words."[103] In referring to "sight," however, the poet favors a notable slippage between content and vehicle, "in-sight" and "site." The sighted object refers both to the *phanopoetic* "image," the mental picture evoked by the verbal structure of the poem, *and* the printed text-object itself. "The physical vision that Shakespeare suggests," Zukofsky writes, "often effuses like an old pictograph thru the syllabary or word it has become," cryptically implying that letters on the page can somehow transmit an image of the worldly referent by indexical or iconic means, overriding the arbitrary bar between signifier and signified.[104]

Thus the irreducible coactivity of eye and ear works to conflate mental and printed images—the objects a poem represents and the object a poem *is*. In the tradition Zukofsky is at pains to prize, the "interest of the arts" is to scramble any "hierarchy of the senses" such that words become "*tangible*": "This is after all a thoughtful word which has perhaps no closer definition than the casual sense of *substantial* or of *objective* intending *a solid object*. So when Dante 'thinks' a metric foot in *De Vulgari Eloquentia* a human foot stalks him like Cressid's. So the visible reference persists 'tangibly' as print, and the air of the voice in handwriting as notes."[105] The metrical foot is a substantial thing, "stalk"-like, Zukofsky telegraphs here, but also capable of stalking, and speaking, too. Like Cressida's feet in *Troilus and Cressida*, the matter of poetic meter has bodily voice:

> There's language in her eye, her cheek, her lip,
> Nay, her foot speaks; her wanton spirits look out
> At every joint and motive of her body.[106]

According to Zukofsky, the "visible reference"—Cressida's flirtatious feet, in Shakespeare's case, as they are bodied forth in audible metrical feet—endures "'tangibly' as print." Audaciously, here again Zukofsky suggests that a mental signified may become "tangible" in the signifier when the latter is sighted and sounded in the complex of the poem. Invested with a shapely solidity and musical liquidity that belies the age of "whistling gas" and the world of typographical fragmentation in which Zukofsky finds himself, the printed word can exceed its mandate of mere representation and become an object both sufficient to itself and materially responsible to its referent.[107]

I have suggested that Partch conceived lyric intermediality in largely antagonistic terms, a convoluted struggle between print and various kinds of sound. Zukofsky, by contrast, makes a bid to sublate this tension. For this particular lyric theorist, sound does not so much resist or unsettle the protocols of silent print as *enhance* them, conditioning the phenomenal experience of that poetic object Zukofsky first identified, in 1931, as the rested totality. Thirty years later, this aestheticized account of lyric intermediality stands at the very center of *Bottom: On Shakespeare*, where it's pressed into exceedingly peculiar service.

A trove of anti-epistemological polemic and wacky Shakespearian scholarship, this forbidding rhapsody—Zukofsky himself calls *Bottom* "a long poem"—in fact revolves around one single theme: the inability to love what we see with our eyes as a result of overactive reason.[108] In a baffling gesture of critical reduction, *Bottom* argues that the entirety of Shakespeare's works simply rehearse this tragic division of *sight* from *reason*. This division precludes the attainment of "Love's mind," that state in which "reason and love are an identity of sight," providing after the fashion of Spinoza's *Ethics* "a clear and distinct knowledge" of the loved object.[109] To redress programmatically the mind's tendency to overreach and undermine the verities of sight, "Shakespeare's text throughout favors the clear physical eye against the erring brain, and . . . this theme has historical implications."[110] These "implications" send Zukofsky into battle with the Western philosophical tradition; *Bottom* weaves readings of Shakespeare into a dense patchwork collage of Aristotle, Spinoza, and Wittgenstein, to name only the most frequently quoted interlocutors. As befits a book that bemoans gas age abstraction as well as the inability to trust one's own eyes, the poet's goal is to jettison the project of epistemology in toto, "tak[ing] exception to all philosophies from Shakespeare's point of view."[111]

Zukofsky scours the plays and poems for each mention of eyes and sight to expose the "definition of love" that "Shakespeare's writing embodies," though this definition is not manifest only at the level of plot and character.[112] "Love's mind" finds expression via Shakespeare's prosody, too. "That the definition is one with craft displayed by the action of the writing is also notable," Zukofsky understates. "The words show their task: a pursuit of elements and proportions necessary for invention that, like love as discerned object, is empowered to act on the intellect."[113] This sentence's interjected simile is a lynchpin of Zukofsky's late poetics. In a manner we have already previewed, Zukofsky links his Objectivist conception of the rested totality explicitly to his intermedial poetics of sensory "proportions." Now *Bottom*'s argument leads us to the further realization that the printed word, bodily invested with aural and sighted substance, is "like love as a discerned object." The lyric object that readers effect when they *see sound* in print is superimposed upon Shakespeare's longed-for union of reason and sight. To put the matter baldly, Zukofsky's definition of poetry is (Zukofsky's) Shakespeare's definition of love. "As I love:/ My poetics," the poet declares in his long poem *"A,"* and *Bottom* is occasionally as direct: "The basis for written characters, for words, must be the physiological fact of love, arising from sight, accruing to it and the other senses, and entering the intellect . . . for the art of the poet must be to inform and delight with Love's strength (and with Love's failings only because they are necessary)."[114] In the gaseous age of print, for Zukofsky love is intermedial.

Spelling out the equation linking lyric intermediality, the rested totality, and the theme of Love's mind sets in order the ideological backdrop against which the full significance of Celia Zukofsky's *Pericles* comes into view. Of disputed authorship and slipshod plotting, this particular play of Shakespeare's has enjoyed a wildly uneven reception among audiences and critics, but *Pericles*'s "mouldy tale" of incest, shipwrecks, prostitutes, and the serendipitous restitution of familial relations was a favorite of Zukofsky's.[115] Since the play centers on the titular hero's ability to *see* love—and to *know* he is seeing love—in the personal image of his lost wife and daughter, Zukofsky prizes this tale as an allegorical enactment of Love's mind.[116] But he insists that *Pericles*'s pride of place in his personal canon has equally to do with "the tragic insight of the poet's measure"—the way love's definition is bodied forth in Shakespeare's prosody. "In *Pericles* especially," he writes, "the rarefying as against the loveable constant eye of the definition of love of all the Plays acts thru the art of song, which with its 'waves' in space must always suggest that eyes can rarefy among sounds."[117]

It's hardly a simple matter to unravel in what sense love's definition "acts thru the art of song." Such an assignment would appear to require the temporalization—the extension across literary time—of the rested totality, of "love as a discerned object." Zukofsky sows a clue in *Bottom* when he explains how the intermedial dynamics of sound in print materially rehearse the tragic split between reason and love: as a consequence of its "audibility in two-dimensional print," lyric writing "suffers the passion that is explicit in the definition of love's qualities and impediments."[118] As if mimicking the excesses of a love that ignores the eyes in anxious deference to the intellect, at times the extraordinary phonetic music of Shakespeare's verbal texture slips from its moorings in the written word, and "the love of sound becomes excessively involved in an interplay of conceptual words. The tendency then is for the sound to persist as pun or tenuous intellectual echo, unless these words are spoken over and over again or, what amounts to the same process, unless the actual print preserves them for the eye to fathom but not to see." Like Pericles wandering the "waves" of the Mediterranean in search of his wife and daughter, the music of Shakespeare's language is overexcitedly adrift, seeking not so much an object as its very objecthood, its resting place as sighted, solid totality. The printed quarto page minimally "preserves" or objectifies this sound, allowing us to "fathom" its depth, but even in print "a good deal of the sound of this writing is thus gone as quickly as the processes of an imagination difficult to sound or to hear." Actually to "*see*" the sound of Shakespeare's words requires something else. Winking toward *Bottom*'s second volume, Zukofsky concludes, "It takes a competent musician to suspect their musical melody."

Celia Zukofsky, Music's Master

The two-hundred-page score by Celia Zukofsky, her husband's literary collaborator, "suspects" the "musical melody" of Shakespeare's verse, offering what Celia herself terms "a form of heightened speech, faintly trying to keep pace with Shakespeare's words, which in their rhythmic, tonal order are music in themselves."[119] In her aversion to vocal lines that distort the intrinsic "music" of poetry, Celia Zukofsky shares with Partch and Yeats a steadfast commitment to the interdiction against sung song, but as Williams intuited after hearing the music performed, the *Pericles* score is unique in its ambition to objectify these linguistic melodies—to "look at Shakespeare's *words* as if they were tuned objects that strike off tones"—in finely drawn

A Speech-Musical Modernism

notation where "the air of the voice" persists "in handwriting as notes."[120] The medialogic that licenses this collapse of word and tone, voice and note, involves a daring (if not exactly counterfactual) assertion of material continuity between alphabetic writing and musical notation, two conventional systems for representing sound. It's the same calculus that underwrites Zukofsky's formula in "A-12," which defines poetry by the integral "Lower limit speech/Upper limit music."[121] The chapter in *Bottom* devoted exclusively to music appears under the rubric "Musicks Letters," a reference to *Pericles*, and in Zukofsky's estimation, his wife's music shows the work, as it were, of that proposed conjunction.[122] The score does so by placing musical notation and alphabetic writing into such close proximity that they defamiliarize each in relation to the other. Alphabetic writing allows us to forget the possibility of sound because phonemes are bound to meaning; the notation of a vocal line, by contrast, peels off phonetic sound (or one representation thereof) from the linguistic stream running below the staff and objectifies, in the notes' black ink, what Ferdinand de Saussure termed the "sound-image" of language.[123] If notation visualizes phonetic sound, it reminds us as well of the aural phenomena potentialized in the visual shapes of alphabetic writing, which also scores speech for performance.

There is, of course, one all-important difference between "musick" and "letters"—namely, semiosis. Zukofsky is perversely agnostic on the subject. In *"A"* and elsewhere, this extraordinary investment in the opacity of the signifier seems a calculated embrace of nonidentity and noninstrumentality—it represents the kind of proto-language poetics for which Zukofsky has been celebrated by subsequent avant-gardes.[124] But in the context of *Bottom*, the meaning of his work's resistance to a narrow conception of semantic reference affects readers quite differently. There the poet's emphasis on scored language posits a sound/sight complex that makes possible a tangible, embodied experience of poetic language, but the goal is not to defamiliarize or deform linguistic codes so much as to establish the preconditions for a rested totality that is, to Zukofsky's mind, the very definition of poetry, and also of love.

As an object lesson in objectified writing ("audibility in two-dimensional print"), the *Pericles* score validates Zukofsky's concept of Love's mind, and not merely by "tag[ging] the insistent plays on *see* thruout the entire play" with conspicuous flourishes in the chamber accompaniment.[125] More significant than any cuing emphases in the musical content are the materially emphatic protocols of the score itself. A decade after hearing *Pericles* in the Zukofskys' living room, Moore's jacket blurb advises that *Bottom* "communicates its passion for music and its passion for Shakespeare," and that both "Louis and Celia Zukofsky are talented in the device of interruption to make text emphatic."[126] What smacks perhaps of faint praise for the book's mode of presentation also hints at its most insistent provocation. *Bottom* makes "text emphatic" by putting enormous discursive pressure on the phatic dimension of language, on the bare materials of the poetic channel, looking back to Shakespeare to recover an early modern regard for language's sensuous substantiality—as musical tones and as handwritten shapes—choked out by centuries of abstraction.

Thus far we have taken Zukofsky at his word by unpacking the aesthetic baggage piled atop the *Pericles* score—shining emblem of intellectual love and proof object of his Objectivist poetics. But now we must put some critical distance between ourselves and Zukofsky, for if *Bottom*'s first half reads its attendant volume as the apotheosis of the rested totality, its second volume shatters this interpretation in a manner almost too obvious to remark. In the score we encounter not any integral whole but three conspicuous parts: Shakespeare's language, presented in Celia's handwriting; a sequence of staved notes indicating the vocal line (whose rhythm and melody represent one interpretation of Shakespeare's verbal music); and another series of notes scoring the instrumental accompaniment. On the printed facsimile page these parts hold *in potentia* the music Moore enjoyed in 1952, which so few others, then or now, have been able to hear. Music's absence is too easily converted into its venerated ideality, and it's by means of just such a misprision that Zukofsky assembles these three moving parts into one rested totality.

Shakespeare himself, in the guise of his character Pericles, felicitously anticipates Zukofsky's mistake. In act 5 of the play, Marina, the daughter whom Pericles had presumed dead, appears—unrecognized and unrecognizing—on his ship anchored off the coast of Mytilene. Marina has been summoned by the king's friends to cheer the mute, grief-stricken Pericles with her prodigious talents in "music, letters"—"with her sweet harmony / And other chosen attractions" to "make a battery through his deafen'd parts, / Which now are midway stopp'd."[127] The "sacred physic" of her musical talents fails to stir her father ("Mark'd he your music? / No, nor look'd on us"), and it's only when Marina speaks, recounting her own calamitous encounters with "wayward fortune" that father and daughter blunder from the fog of sorrow

A Speech-Musical Modernism

into mutual recognition.¹²⁸ Then, just as Pericles embraces his daughter, another, less worldly music strikes up:

> PERICLES: I am wild in my beholding.
>   O heavens bless my girl! But, hark, what music?
>   . . . .
> HELICANUS: My lord, I hear none.
> PERICLES: None!
>   The music of the spheres! List, my Marina.
> LYSIMACHUS: It is not good to cross him; give him way.
> PERICLES: Rarest sounds! Do ye not hear?
> LYSIMACHUS: My lord, I hear.
> [*Music*]
> PERICLES: Most heavenly music!
>   It nips me unto listening, and thick slumber
>   Hangs upon my eyes: let me rest.
> [*Sleeps*]¹²⁹

Critics and directors debate the diegetic reality of this "music of the spheres," which contrasts so conspicuously with the real music that Marina actually—and ineffectually—sings.¹³⁰ Lysimachus at first hears nothing, and he urges Marina to indulge her daffy father. But it's hard to say, once the stage-directed "[*Music*]" begins to play, whether Lysimachus affirms "My Lord, I hear" in good faith or not. Does Lysimachus tell the truth when he attests these heavenly tones, or is it all in Pericles's head? If the latter, is this imaginary music some joy-addled delusion or a divine harbinger of the dream that will shortly reveal to him the location of his lost wife? And then, what of Marina, the only real musician present? What does *she* hear in Pericles's music? What does the audience?

In her setting of *Pericles*, Celia endorses the reality of this music. Would-be listeners can hear the English horn to which Pericles harkens, and the measure of solo horn that follows Lysimachus's declaration strongly implies that he hears it, too (fig. 1.10).¹³¹ Here we find one markedly straightforward example of Celia's influence on Louis's poetics, for her directional decision to make *actual sound* from this music of the spheres provides her husband, in the pages of *Bottom*, with a basis for identifying the rested totality as a music of the spheres. Writing of just this passage, Zukofsky acclaims both the phonic music of Shakespeare's language and the "heavenly music" that Lysimachus believes to exist entirely beyond the skeptical ken of epistemology:

Figure 1.10. Excerpt from Celia Zukofsky's *Pericles, Bottom: On Shakespeare*, vol. 2 (1963), pp. 216-17. Courtesy of Musical Observations

"The measure in *Pericles* is finally determined by a sense, following the tensions between eye and ear, that truth can never be confirmed enough when it consists of rarest sound and a question, Do you not hear? The heavenly music exists if a tentative *No* is so moved to love it is easier to say it hears as *tho* it sees. But that *is* music, It [*sic*] nips unto listening. Physical eyes may eventually close on it; in one's own head it goes on as notes in others' heads."[132] Coiled in these sentences is the poet's redescription of the fabled music of the spheres: heavenly music is the product of the clear and distinct sight entailed by Love's mind, which snatches ("nips") one to listening. A listener needs "physical eyes" to hear this "rarest sound," but once heard, it may go on "in one's own head." This passage explains, after a fashion, the connection Zukofsky strikes between actual phonetic sound and the ideal music composing the "upper limit" of his poetics, as well as the link between the intense visual experience of his wife's score and the absent music he attributes to it, as a powerful archetype of the rested totality. In both cases, sense certainty is achieved by means of an *objectification* of lyric materials. Zukofsky's reader will "hear as *tho* [they] see." Any actual music, meanwhile—be it Marina's song, the phonic textures of Shakespeare's language, or Celia's setting—gets silenced as mere precondition for a more genuine object of aesthetic concern.

To summarize: "the tensions between eye and ear" that this book terms lyric intermediality, Louis calls love, and then points to his partner's music to prove it. If not risible, necessarily, the poet's campaign to lard modern poetry's media condition with the full powers of the aesthetic forces a confrontation with the way Celia's volume is used and abused in the process. The poet's idea of music—and the claims for a rested totality he heaps atop it—rely upon her compositional labor. Zukofsky admits that the music of the spheres may go on "in one's own head" in volume 1 of *Bottom* only because it goes on "as notes in others' heads" in volume 2, and he is eager to bestow on Celia the title borne by Pericles, "music's master."[133] But once heard, the music of the spheres obscures the effort expended in the processing of that object's raw materiality—"physical eyes" in fact "close on it." We confront, in short, a familiarly gendered division of labor.[134] The discursive mill of Zukofsky's poetic thought works an ideological transformation on the *Pericles* score, the raw material of its handwritten notes. The result is a music of the spheres plainly contoured by these gendered dynamics. The poet's attempts in *Bottom* to give Celia mastery end up circumscribing the meaning of her music, silencing it on the page.

Figure 1.11. Cyril Satorsky's frontispiece from Celia Zukofsky's *Pericles, Bottom: On Shakespeare*, vol. 2 (1963). Courtesy of Dale Matthews

Yet the music does actually exist. And as we have seen, it deserves study for all the ways it points up the fissures in Louis Zukofsky's aesthetic ideology, resisting its own sublimation into a music of the spheres. Certainly the score is worthy of study and performance on its own terms, as well. The few to have heard its "quiet intensity" and profound reverence for the sonics of verse resounding in a midcentury Brooklyn Heights living room were, more surely than Lysimachus, convinced of that. The history of poetry is chockfull of idealized emblems of music and music making; like Florence Farr and Harry Partch, Celia Zukofsky is an indispensable guide to *actually existing* lyric music and new-old media. And as the frontispiece to *Bottom*'s second volume quaintly signals, the *Pericles* score is one more attempt in print, against media-historical odds, to restitute the lyre (fig. 1.11).

Why, then, we might ask, was it so hard for Zukofsky to see the *Pericles* for all it actually is? From what quarter issues the impulse to spirit it away into abstraction, to move from the raw materials of intermedial speech-music toward an ideal object. If Zukofsky looks at his wife's *Pericles* and sees

A Speech-Musical Modernism 87

not a musical score but a music of the spheres, it's because he looks past it toward something else. Working deductively, we can say this something else must be a model for the rested totality (an "objectification" of sonorous language), a tangible preservation of "audibility" or unheard music, and a concrete fusion of writing and score, speech and music. The object lurking just over the shoulder of Celia's *Pericles* is, I submit, the sound recording. The emergence of phonographic protocols collapsed sound and writing, driving lyric practices to their extreme symbolic limit. Whereas Partch responds by developing microtonal scales to extend lyric's lease, Louis responds to the latent competition of technologized sound with recourse to an idealizing aesthetics.

Ron Silliman suggests that *Bottom*'s notion of Love's mind is largely synonymous with Walter Benjamin's conception of the artwork's "aura."[135] Though Benjamin decreed that mechanical reproducibility effected "a decay of the aura," later historians of phonography have substantiated how certain auras—and in particular the auratic appeal of the authentic and embodied voice—are in truth *produced* by sound recording.[136] In the next chapter, we'll consider at length how this manufactured aura is registered and contested on the printed lyric page. In Zukofsky's case, we can trace sound recording's auratic effects displaced and channeled into the sphere of poetic production. The fact that Zukofsky's poems rarely mention phonography lends a particular frisson to the rare exceptions, such as "The Record," a birthday poem for Celia. The short poem commemorates a 1956 orchestral performance of Mozart's *Eine kleine Nachtmusik* in which their son Paul plays first violin in a program marking the composer's two hundredth birthday. But this public concert is lit by the glow of two private occasions: the performance takes place on Celia's own birthday, January 21, and the titular "record" refers to "a recording of the orchestral music of *Eine kleine Nachtmusik*," a gift from Celia to Louis on *his* birthday "some years before [they] were married."[137] Zukofsky replies decades later with a sweet, sprightly, Cold War–haunted valentine on steadfast familial love that belongs firmly in the orbit of *Pericles*. The poem draws a figure that seems to write small the larger direction of Zukofsky's Objectivist poetics, which is driven to abstraction by a fearsome regard for the sensuous and concrete. In another reversal of media history, the meaning of Paul's live performance, which he "bowed" on the "birthday" of "the girl who loved," assumes its full meaning, its glowing auratic significance, only in light of the recording that preceded Paul's own birth. The first recording is not superseded, exactly, by Paul's performance or the slight performance of Zukofsky's own "Record." It lives on, just out of sight in memory, lending the allure of its own permanence to a family unit, "Never to fall out," and to the delicate speech-music Zukofsky has woven to commemorate it, a record of love.

> Our valentine the heart proposed by three,
> Our vanity that talks and cannot see
> As it fell out—the girl who loved brought me
> The record we played for ourselves years way
> Before our son bowed it on your birthday,
>   Against fall-out
>   Never to fall out—
> Delight, Amadeus, to light
> The music of a little night.

2 Latent Remediation
Radio, Poetry, and Social Process

Poetry's Remedial Fantasy

At just the moment that Harry Partch was fashioning his Adapted Viola, others were dispensing altogether with the equipment of lyre and book for what seemed the next big lyric thing. By the late 1930s, as the country braced for war amid a national efflorescence of New Deal populism and antifascist fervency, as network radio reached peak levels of market saturation and cultural prestige, and as middlebrow literary culture was accommodating itself, in fits and starts, to an increasingly institutionalized version of aesthetic modernism, modern poetry saw its future—a dramatic, bombastic, star-spangled future—in broadcast radio. To many it was patently obvious that poetry was soon to leave the page for the airwaves. This chapter maps the conditions and charts the effects of the widespread techno-utopian surmise I term poetry's *remedial fantasy*. Though many of this fantasy's adherents would have agreed with Partch as to the infirmity of modern print, they needed only prick their ears to a radically transforming soundscape to realize that poetry's medicine lay not before print but beyond it, in the most modern technics. "In radio the poet finds an instrument much more amazing than the lyre ever was."[1]

The galvanizing event for US listeners was the April 11, 1937, broadcast on the *Columbia Workshop* of Archibald MacLeish's *The Fall of the City*, the first American verse-play for radio.[2] With hindsight, MacLeish's allegorical warning against fascist demagoguery propagandized at least as effectively for the possibilities of radiophonic poetry as it did for the virtues of deliberative democracy. Inspiring a cadre of innovative radio dramatists like Norman Corwin and Arch Oboler, and ushering established poets like Stephen Vincent Benét and Alfred Kreymborg into the ranks of major networks, the broadcast signaled the "real beginning of radio poetry in the United States."[3]

The Fall of the City may have proved "a kind of 'Armory Show' for radio," as Neil Verma observes, but the spectacle was less an avant-garde publicity coup than a calculated bid to level brows, the opening of a newly mediated chapter in modern poetry's anxious courtship of wider publics.[4] The play "marked an epoch," in the words of John Wheelwright, by urging poets to the microphone and "arous[ing] them to their duty towards their invisible audiences."[5]

As much as *The Fall of the City* drove poets to "storm[] the studios," prestigious network dramas by MacLeish and Corwin were themselves responding to a more extensive flourishing of expectations for poetry's imminent radio uptake in the period.[6] MacLeish's cultural capital as a poet and public intellectual made him only the most visible US advocate of poetry's historical appointment with broadcast. In his foreword to the published script of *The Fall of the City*, MacLeish articulates a number of widely professed warrants for this remedial fantasy. Foremost is "the great question of audience": "radio will reach an infinitely greater number of people" than either the stage or the page.[7] Those heralding their union are apt to figure poetry and radio as the two sundered halves of a cultural whole begging repair: "The situation of radio is the situation of poetry backwards," MacLeish observes elsewhere. "If poetry is an art without an audience, radio is an audience without an art."[8] Naturally, radio will vitalize and popularize poetry by means of the ear, since in radio, "a mechanism which carries to an audience sounds and nothing but sounds . . . There is only the spoken word—an implement which poets have always claimed to use with a special authority."[9] "Always" orients us toward a last salient feature of this remedial fantasy; anticipating the media-theoretical theme of a "secondary orality," poetry's listening audience will be a characteristically *restored* audience, returning the art to the oral/aural conditions of its origins.[10] No less a champion of print culture than Harriet Monroe, editor of *Poetry* magazine, put it succinctly in 1930: "For five hundred years poetry has been silent—appealing to the eyes instead of the ears. This is a wholly artificial condition, under which the art is hampered; and the radio is the poet's one best chance of escape from that condition."[11]

Thus poetry's remedial fantasy designates first of all an escape from the hieratic silence of print. Following Jay Bolter and Richard Grusin, I use the term "remediation" to describe the process by which new media forms constitutively subsume or incorporate their predecessors.[12] Poetry's remedial fantasy, by these lights, trumpets the succession of the printed page by radio. But proponents like MacLeish and Monroe teach us to hear in the word an additional meaning; they draw us back through the educational sense of "remedial" to its source in the Latin *remedium* (a means of treating illness or injury).[13] To indulge this particular fantasy for the future of verse is to prescribe a kind of listening cure for poetry's enervated, marginal cultural position at midcentury.[14]

This remedial fantasy comes into focus only if we peer beyond the myriad genres in which verse actually circulated on the airwaves (whether as radio drama, anthologized on popular middlebrow programs like Ted Malone's *Between the Bookends*, or otherwise disseminated by poets of widely disparate cultural affiliations looking to engage the medium's aesthetic and political affordances) to consider how radio operated as a site of explicit imaginative investment even for those poets who did not step up to the microphone themselves. Further still, we must clear conceptual space for the possibility that this radio fantasy and its cultural-material conditions shaped lyric practices in ways poets and their readers were not entirely conscious of. Though the first two decades of the twenty-first century have witnessed mounting interest in poetry's radio history, much of this pioneering work has concentrated on manifest contact points between literary and broadcast cultures, treating modern poets who wrote prodigiously for and about radio or who themselves appeared on air regularly.[15] Momentous as such episodes are, limiting our attention only to these most conspicuous confrontations with radio—to the matter of Ezra Pound's broadcasts, say, or to Edna St. Vincent Millay's—may fail to educe radio's significance at its fullest and subtlest. Writing the material history of poetry circa 1938 demands detailing some less apprehensible routes of media influence. It means, above all, tracing poetry and radio as they interact in and through social processes.

Lorine Niedecker's *New Goose* supplies a case in point. Though Niedecker's poems were not composed for broadcast, prevailing ideas about radio inform their making. Importantly, as discussed in the introduction, these ideas arise not directly from the radio apparatus itself but from its emergent practices—from the desires of rural listeners vis-à-vis the metropole, for instance, or from the anonymous orality of broadcast nursery rhymes. The present chapter foregrounds modern poetry's responsiveness to other dimensions of radio's social history, including the politics of shortwave and the racialization of commercial sound. Poetry's remedial fantasy is itself, of course, one discursive product of radio's broader mass-cultural construction in the 1930s. Insisting upon this link between literary history and the social construction of technology attunes us both to the contingently distributed influence of radio and the material effects of the ideological ferment it activates. Lyric's remedial fantasy did not result in print's usurpation by radio. Yet it would be a mistake to write off these techno-utopian hopes as mere evidence of a road not taken. Though only incompletely realized in the radiogenic verse-dramas of Corwin and MacLeish, these expectations for poetry's future reflect broader and subtler modes of media-ecological negotiation *already obtaining* in the cultural field. To go in search of "the real beginning of radio poetry," we must look beyond *The Fall of the City* to those subtending material circumstances that made poetry's date with radio seem, for a moment, so inevitable.[16] It was these historical conditions and not the machine itself that delivered commentators irresistibly to the conclusion that poetry was ripe and ready for wholesale remediation. If we can name these conditions, we can wrest into view the wider range of radio's cultural influence and historical meanings. And we can learn by what otherwise occluded processes printed poems come to bear the impress of their technological environments.

Poetry's radio future did not play out in the manner MacLeish and many of his contemporaries hoped and expected. And without sidelining or shortchanging eighty years of innovating excursions into new media as various as the LP, tape, film, CD-ROM, and social media, we might even venture to say that poetic practice as we know it remains predominantly oriented to the printed page and the page-like screen. This is the lesson transmitted by Partch, whose practical research into the modern lyre took lyric materials to their intermedial limit. But this hardly means that new technical-industrial conditions transforming the world beyond the printed page cease to shape and modify poetic practice in appreciable ways. To speak of poetic remediation, we must inquire how media themselves mediate *through* historical processes, latently spanning the distance between the phonic and the sonic, the page and the air, and for this some finer critical tools are in order.

What's needed is the capacity to read poems on the page, adduce the contexts of their production and reception, and meaningfully evaluate their shaping engagements with other media, like phonography or radio. This chapter gathers methodological resources for apprehending the modes of remediation operating *at a distance*, as it were, to produce the kind of *latent* remedial phenomena traceable in media-specific artworks like printed poems. A printed poem is *latently remediated*, I argue, if it reveals the constitutive imprint of sound technologies in the absence of any *actual* remediation on radio or a phonograph record. Since these cultural-industrial machines latently remediate poetry through the social processes responsible for their own contested institution, to unearth the lines of influence running from new media to poetry we need to study how these technologies are themselves socially mediated. At its furthest horizon, our goal is to fold the concept of remediation back into a robustly processual and material account of cultural mediation itself. When any single act of remediation (such as broadcasting a poem first written for the page) is grasped as one moment in a larger media ecology subsumable to more total social processes, we can begin to speak of the unpredictable ways that artworks bear material witness to the existence of *other* media, while also respecting the crucial specificities that distinguish media formats, one from the other.

I pursue the concept of latent remediation through case studies of two poets: John Brooks Wheelwright and Sterling Brown. In Wheelwright, our chief theorist of latent remediation, we discover one especially vigorous elaboration of radio's remedial fantasy. His heady engagements with shortwave broadcasting supply an object lesson in how struggles over radio's unsettled political meanings can materially shape a print poetics. I then spotlight a Norman Corwin program in which poems from Brown's 1932 collection *Southern Road* were abusively appropriated—in the name of poetry's radio future—by the racist protocols of golden age broadcast. His work is uniquely equipped to resist this minstrelizing treatment, I argue, because the poems were themselves produced through an earlier grappling with capitalized appropriations of Black culture on the phonograph record. Despite occupying rather far-flung positions in the cultural field, these figures are united in their shared receptivity to radio's influence. Because this influence is socially mediated, the labor of tracing the significant latency of technologized sound in their poetry becomes, as well, a means of exhuming otherwise hard-to-see features of US broadcasting in the 1930s. Wheelwright and Brown each invite us to read from the media-historical archive of their poetic forms some different chapter in radio's social history, the history of how radio *works* in the 1930s—as a political object, commercial entity, and artifact of daily life.

Verse + Radio = Poetry: John Brooks Wheelwright

We begin in 1938 at the University Club in Boston's Back Bay, where for several months that autumn the poet John Wheelwright ran a weekly poetry program on the World Wide Broadcasting Foundation's renowned noncommercial shortwave station, W1XAL (later WRUL). According to its primary funders at the Rockefeller Foundation, W1XAL had been since 1935 "the only station in the United States with national coverage . . . devoted exclusively to educational and cultural programs."[17] Transmitting 50,000 watts across four shortwave bands, W1XAL was by the late 1930s a leading "university of the air" disseminating "New England enlightenment" with the collaboration of faculty from Harvard and other regional universities to listeners across the country and, increasingly in languages other than English, to Europe and Central and South America.[18] To recite poetry into a W1XAL microphone was to hitch a ride on a powerful cultural apparatus, one that other entities like the Rockefeller Foundation, the Pan-American Union, and the federal government were simultaneously using, in Good Neighborly fashion, to secure US interests abroad.[19] It is high testimony indeed to W1XAL'S soft power that in May 1938, when congressional efforts to establish a government-owned shortwave station "to counter the heavy-handed propaganda of the Nazi and Fascist competition" were promptly squashed by the major networks, the *Daily Boston Globe* could point with beaming hometown pride to W1XAL, at that very moment pouring a patriotic "deluge of high-powered thought . . . into the ether for a world audience": "Already a channel to carry North American ideas and sentiments to South America is in active daily use. It is right here in Boston and the Government is sufficiently aware of its power and its international prestige to make frequent use of its world-circling short waves for messages to the neighboring Americas."[20]

Those listeners tuning in from Peru, Trinidad, or New Zealand were doubtless unfamiliar with W1XAL's poetry impresario, but Bostonians may have recognized Wheelwright, less perhaps from his poetry than from his minor cultural celebrity, the baffling figure he cut in a city so indelibly his milieu.[21] "What Dublin was to Joyce," recalled his friend Winfield Townley Scott, "Boston was to Wheelwright."[22] A direct descendent on his father's side of the antinomian rebel the Rev. John Wheelwright (1592-1679) and on his mother's side of the merchant Peter Chardon Brooks (1767-1849), once purportedly New England's wealthiest man, even as his family's finances deteriorated rapidly in the poet's lifetime, Wheelwright could boast of a near-perfect Brahmin pedigree.[23] But he also relished confounding this upper-crust exemplarity. As a young man grieving his father's suicide, Wheelwright rejected the Puritan faith of his ancestors for a High Church Anglicanism.[24] After his political radicalization in the early 1930s, he then set himself the task of squaring Anglican and apocryphal Christian traditions with a rigorous dialectical materialism. He describes his third and final collection of poems, 1940's *Political Self-Portrait*, which marshals St. Paul alongside Lenin and references the "apostolic orders" alongside exhortations to class war, as "a self-portrait of one who has found no way of turning, with Scientific Socialism, from a mechanical to an organic view of life than to draw from moral mythology as well as from revolutionary myth" (149).[25]

For Wheelwright, Marxism was not of intellectual interest only, however.[26] To a degree greater than many other committed artists of the period, Wheelwright threw himself into radical organizing, first with the Socialist Party and then as a founding member of the Trotskyist Socialist Workers Party, penning circulars, picketing, and stumping on the soapbox. This political education taught Wheelwright to inventory his own contradictory social position with incisive critical candor: "Scientific Socialism, Anglicanism, New England seaboard, Harvard College, the haut bourgeois, my family, have given me allegiances or rather prejudices which are revealed in my writing."[27] The poet's double-taking slip from "allegiances" to "prejudices" gropingly anticipates what Raymond Williams would later parse with his concept of "alignment," that manner in which a writer's "real social existence" dynamically determines their practice: "In any specific society, in a specific phase, writers can discover in their writing the realities of their social relations, and in this sense their alignment. If they determine to change these, the reality of the whole social process is at once in question."[28] For Williams, such a "conscious change of alignment" represents the only real meaning of political commitment, in the constantly renewing practice of which one realizes how thoroughly the barbed contingencies of one's historical situation condition their writing, and that "to write in different ways" would mean, first, necessarily and above all, "to live in different ways."[29] Wheelwright's ragged verse of the 1930s arises from an arduous dwelling in the contradictions which his own social alignments—in Williams's strong sense—threw up before him.

The poetry projects a "heterodox vision of a Marxist-Christian renewal" through a scrim of densely allusive and exuberantly rhetorical verbal action.[30] "The Word Is Deed" runs the title of *Self-Portrait*'s opening poem, and here Wheelwright means expressly to correct the infamous translation of the Gospel of John by Goethe's Faust, a "misread[ing]" endorsed by Friedrich Engels:[31]

> John begins like *Genesis*:
> *In the Beginning was the Word*;
> Engels misread: *Was the Deed*.

> But, before ever any Deed came
> the sound of the last of the Deed, coming
> came with the coming Word
> (which answers everything with dancing).
> . . . .
> Deeds make us. May, therefore, when our Last
> Judgment find our work be just
> all tools, from foot rules to flutes
> praise us; and our deeds' praise find
> the Second Coming of the Word.
> (Dance, each whose nature is to dance;
> dance all, for each would dare the tune.) (103-4)

Wheelwright was struck and killed by a drunk driver in September 1940, age forty-three, but in his last years he endeavored to choreograph this joyous dance of Word and Deed, of Christian thought and revolutionary politics, with uncommonly intense if idiosyncratic resourcefulness. To the extent that he managed to figure their reconciliation, he went further than almost any other modernist poet toward the elaboration of an innovatively *didactic* modernist poetics (Bertolt Brecht, W. H. Auden, and Ezra Pound are legitimate rivals in this respect), an achievement made possible by his engagement with shortwave radio. Like his correspondence course, The Form and Content of Rebel Poetry, his little magazines *Poems for Two Bits* and *Poems for a Dime*, and his engagements reading poetry before workmen's circles and labor schools, radio appealed to Wheelwright precisely as a medium for integrating his artistic and political efforts and for negotiating the contradictions attending his self-consciously politicized experience more generally. Wheelwright's voice, in the global sweep of its broadcast on W1XAL, strains for a redefinition of poetic action and public poetry under the sign of radio. My goal is to uncover the grounds of this intention—to show that by the time Wheelwright takes to the air, his poetics had already undergone a kind of remediation whose priority is pivotal. The mere *possibility* of putting poetry on international shortwave casts an enabling shadow across the page; in Wheelwright's estimation, it renders a politically efficacious poetry *impossible* in any other medium.

The incongruity of hearing Wheelwright, in a voice branded by "the struck-pie-tin, unresonant nasalities of Boston's upper class," hold forth over W1XAL on the subject of "super-national class struggle," doubtless put the city's literati in mind of other rumored spectacles of contradiction: the poet attending Socialist Party meetings in his "bowler hat . . . formal evening clothes . . . [and] luxurious raccoon coat," for instance, or the poet on the soapbox elucidating the capitalist system "by pointing to himself as an example of how certain members of society lived on unearned increment."[32] As Matthew Josephson recalls, "Though he looked like a dandy, he willingly appeared at the head of workers striking against wage cuts, or in confrontations with the tough Irish cops of Boston. Soon, in 1933, we heard that he had been arrested and jailed. Kenneth Burke wrote him then: 'I think you did right to be arrested, for I believe you better equipped to confuse the police in these matters than anybody I know.'"[33] These sartorial antics index the more serious, more intractable problem of the bourgeois intellectual's role in revolutionary struggle. Wheelwright elaborated the rather tortuous view that the vanguard intellectual's responsibility was not to *identify* with the working class—he savaged the Popular Front poses of "liberal persons [who] fall in love with ice men"—but to "address" them: "Our present function is to enrich Socialist ideas," he schooled a friend, and this duty demands that "one renounces allegiance [to] upper class politics and thought" (138).[34] But while "strong disciplined organization" is obviously required of a mobilized proletariat, Wheelwright, descendent of Puritan rebels, concludes that "it is the business of people like ourselves to be strong and disciplined individuals . . . Culturally, I'm all for rugged individualism." Here lies, in Gramscian terms, the one doggedly "traditional" aspect, the sticking point, of his otherwise "organic" intellectual position vis-à-vis the working class.[35] Its most vivid testament is not his bowler hat and racoon coat but Wheelwright's commitment to modernist poetry.

After all, a worked-out organizational role for the revolutionary intellectual does not solve the problem of what sort of poetry that intellectual might write. In reference to his own erstwhile adventures on the modernist scene, Wheelwright acknowledges, "I have had cultural allegiances. The athestic [sic] struggle within the upper class which was once vital to me is now dead. The victory of modern art is won, and I find myself perfectly indifferent to the fact. But if I live to see the cultural victory of Socialism I shall not be indifferent, and I am perfectly sure that whatever cultural sag may accompany the first stage of Socialism, I would like it better than the present Capitalist disintegration."[36] What kind of poetry does one write after the "victory of modern art" but before the "cultural victory of Socialism"? Friends referred to this quandary as the often-overweening Wheelwright's "one indecision": "*Political Self-Portrait* reveals reiterant doubt as to what *culture* a proletarian society would erect. It was something he could not know. He could go only so far as the belief that 'one can never be premature in arousing the masses to their cultural duties and to the joy of living.'"[37] Once "class restrictions are removed," will the "patents now withheld by technological taboo . . . command not only abundance but complete well being," Wheelwright asks, "or shall future proletarian patronage of a public standard of sophisticated living . . . be as half-hearted, almost, as present capitalist patronage?" (154). His anxiety about the "cultural sag" that might accompany a dictatorship of the proletariat betrays obvious condescension and unexamined class prejudice—lamenting the "philistinism of the working class," Wheelwright opines that "the longer capitalism continues the more of a saint an intellectual must be if he is to retain faith in the humanity of the working class"—but it also testifies to his conviction that given the bourgeois allegiances of modern art, the poetry of future societies would have to change utterly, and in utterly unpredictable ways.[38]

The answer, and the eventual aesthetic upshot of Wheelwright's political commitments, was a revolutionary didactic poetry on the radio. Wheelwright lit upon the didactic element first. In his 1933 collection *Rock and Shell*, he announces the program: "The uses of poetry are: to sound, to show, to teach" (57). Now that Imagism and free verse measures have submitted "the first two of these" to "complex clarification," "the didactic . . . must undergo proportionate handling" as poets press modernist techniques into the service of associative argument, moral thought, and political instruction. In a notorious review of his friend Muriel Rukeyser's *U.S. 1* (1938) and its long documentary poem "The Book of the Dead" on the industrial disaster at Hawks Nest Tunnel in Gauley Bridge, West Virginia—a poem Wheelwright censured for its focus on "the excrescences of capitalism, [and] not the system's inner nature"—he avers that a didactic mode is not necessarily a rationalistic one: "Poetry develops intuitive reason not by logical contrivances so much as by immediate sensual association. This is so delightful a process to follow that a poet easily wins readers over to his teaching, if only he be clear about what he has to say and whom he has to address . . . Where poems are stuttered and blurred, it is because their authors do not know what they are about."[39] It is just this quality of "knowing what you are about" to which Wheelwright appeals in drawing his (perhaps untenably) fine distinction between the "willful mystification" of "academic bourgeois decay" and revolutionary art that "recognizes mysteries and wrestles with them."[40] Elsewhere he describes this latter poetry as an "idealogical [sic] music," a mode of writing that may eschew logical argument in its sonorous, sensuous appeal—"disassociation of associated ideas and the association of the disassociated"—but that remains at all points answerable to some fundamental content, idea, or argument, such as one finds in the substantial prose glosses accompanying the poems in each of his three books.[41]

At some unspecified juncture in the late 1930s, Wheelwright determined that his "idealogical music" required the medium of radio, and W1XAL in particular, that "World Radio University" world famous for the unrivaled reach of its cultural exports. The poet began his broadcast career on the long waves, using a stint at independent station WORL in the summer of 1938 to hone the format of a weekly *Poetry Noon Hour*, with guest appearances by regional poets, actors, and academics, often his personal friends or acquaintances. By midsummer, Wheelwright had set his sights on the more prestigious W1XAL, pitching his program to executive Loring Andrews, who reported that Walter Lemmon, president of the World Wide Broadcasting Foundation, was "favorably inclined to your ideas" regarding a Boston-based poetry show.[42] The weekly broadcasts, which began on September 23, held to a straightforward format more akin in audio construction to middlebrow poetry shows like Ted Malone's *Between the Bookends* and Tony Wons's *Scrapbook* than the studio experiments of Corwin and MacLeish.[43] Wheelwright and his guests would read from their work or the work of other poets, making their "15 minute girdle round the earth," occasionally to musical accompaniment (as when a young Leonard Bernstein, then a Harvard undergraduate, played piano under a recitation of *The Waste Land*), and with brief critical commentary by Wheelwright himself.[44]

On both WORL and W1XAL, the guest list featured a decorous smattering of nascent New Critics and midcentury modernists (with fewer radical poets than one might have suspected), from Richard Eberhart, John Peale Bishop, and a twenty-one-year-old Robert Lowell, to Kenneth Patchen, Kenneth Porter, and May Sarton. Despite the well-credentialed inoffensiveness of the poets on offer, the favor of W1XAL's executives soured by early December, when Lemmon, "displease[d] with the conduct of the Poetry Corner," directed Loring to unceremoniously terminate Wheelwright's employ.[45] One assumes, on the basis of the few script sketches available, it was Wheelwright's political editorializing that turned an apparently respectable entrée to New England letters into a source of displeasure for Lemmon and his foundation.[46] Damon and Rosenfeld recall the airing of "anti-war poems, pieces of social satire, [and] political burlesques in verse," though it was also the case that Wheelwright framed otherwise innocuous poetic fare in ways that amplified its political meanings.[47] Contextualizing *The Waste Land* for WORL listeners, for instance, the noninterventionist Wheelwright suggests that Eliot's presumably "post war" poem "may however soon be called 'pre war'": "The undefined fear, the self-pity, the unrealizable nostalgia of our war-haunted era form Eliot's subject matter."[48] When May Sarton joins Wheelwright on W1XAL, the latter requests to hear a love sonnet from the title sequence of her widely noticed debut, *Encounter in April* (1937), but then shifts gears advisedly:

> Thanks, May Sarton, I thank you, on behalf of your audience . . . But I know that you, like so many other poets of our time, are not content with making your versions of the great themes of personal and individual affection, which are normal to poetry. For we are living in abnormal times, and you (quite undisturbed by the authorities who would deny a political voice to poetry) have written a poem of the public love for mankind. I think this poem makes the poets' appropriate contribution to the science of politics, and I wish you would read "From Men Who Died Deluded."[49]

Sarton's poem allegorizes the spectral voices of failed mountain climbers "who died deluded, far below the peak," but who resolve nevertheless "to send up one clear note from the edge of death" to embolden those who will come after to overtake their foiled progress.[50] The allegorical referent is poetry itself, of course, which contributes to "the science of politics" by orienting the desultory present toward a revolutionary future lying well beyond the pale of any individual lifetime:

> This is the time, this dark time, this bewildered
> To give our mortal lives that the great peaceful places
> May surely be attained by those who, when they falter,
> Must be confronted by the living vision on our dead faces.

"This is the time to set our lips upon great horns and blow"—Wheelwright considered W1XAL just such an instrument, which by trumpeting a didactic poetry could enliven dead letters and bestow future-tense meaning on present-day struggles.

Wheelwright was not alone in singling out W1XAL as an object of political utility. The station's institutional affiliations make Wheelwright's firing by Lemmon (an IBM executive shortly appointed to the National Defense

Communications Board, "presumably not for his purely educational work") a singularly unsurprising conclusion to the poet's short tenure on air.[51] Since W1XAL had been a key apparatus in the soft power campaigns of Roosevelt's Good Neighbor Policy, including those efforts to combat "Axis propagandistic inroads into the hemisphere," when the war began in September 1939, the station—redubbed WRUL—was the "first to go 'all out' in the active defense of democratic principles."[52] Perhaps most incredibly, beginning in 1940 the British Security Co-ordination (BSC), an undercover MI6 outfit operating in the United States, adopted WRUL, unbeknownst to its listeners and indeed apparently to its own staff, as one of its "chief venues for placing false news stories designed to outrage US public opinion against the Axis forces."[53] Not released until 1998, the official postwar BSC report reveals that "by the middle of 1941, station WRUL was virtually, though quite unconsciously, a subsidiary of BSC, sending out covert British propaganda all over the world."[54] As a result, well before the wartime State Department commandeered WRUL and all other shortwave stations in 1942, when the station began broadcasting for the Office of War Information and the Voice of America, a *Life* magazine profile could declare, with good cause, "the Nazis hate and fear Boston's WRUL."[55]

Though W1XAL's engagements with wartime propaganda postdate Wheelwright's poetry show, the station's enormous potential for political intervention must have been readily in evidence by the fall of 1938. Wheelwright's firsthand experience with this particular social application of radio technology—an institution authorized for cultural and educational objectives and operationalized for propaganda—strongly inflects his own version of modern poetry's remedial fantasy. The poet joins contemporaries like MacLeish in celebrating broadcast's promise of a larger and more democratic audience for poetry, and he repeated the ubiquitous claim that by obviating the mediation of print, radio would recall poetry to an orality which is the very soul and soil of the art. Radio "is the sternest and most refined test that poetry has ever undergone," Wheelwright notes, because it "compels poetry to sound" and thereby "return[s] [it] to speech from print," sweeping "the *calligramme's* unpronounceable arrangement of letters and syllables into that junk-heap which covers the hid side of the moon."[56] Wheelwright puts his own twist on this techno-optimism by arguing that radio's mode of acousmatic address uniquely suits the kind of associative argumentation at which good—that is, properly didactic—political poetry excels: "Exactly as it is hard to follow a logical argument over the air, so is an associational argument convincing. Here again is poetry so fitted for radio that broadcasting purifies it (almost automatically) by making its most characteristic elements the most effective."[57] Wheelwright's confidence in radio's amenability to poetic logic leads him to assert that on the air ostensibly difficult poets like "John Donne and Hart Crane are clear and simple."[58] Though far from persuasive, the claim is worth pausing over for how boldly it contravenes received ideologies of aesthetic modernism—namely, the anti-rhetorical autonomy of the artwork and its presumed opposition to mass media. For Wheelwright, poetry and radio meet precisely on the basis of their mutual serviceability as vehicles for the illogical but *interested* appeal increasingly associated, between the wars, with political propaganda.[59] In the spirit of swift caricature, we might call this Theodor Adorno's worst nightmare: the unholy marriage of all that makes radio perilous for deliberative democracy (its mystifying, sensuous, irrational address), on the one hand, with a loathsome didacticism that disengages what is authentically critical about lyric aesthetics, on the other.[60]

In the notes to *Political Self-Portrait*, Wheelwright defines verse as "an engineering of sound which when it rises to poetry, engineers meaning," and though this formulation baldly evades all the essential questions (How does sound rise to poetry? What does it mean to engineer sound or meaning?), the technical connotations of the word "engineer" significantly tip his hand (155-56). With the aim of securing for modernist poetry a plausible afterlife on the radio, Wheelwright sets up an overinvested concept of poetic sound as a tendentious means of papering over medial differences between print and broadcast. In other words, he pins his political hopes to an imaginative remediation of poetry, transforming by sheer force of rhetorical will the phonetic notations on the page into sounds that can be engineered on the shortwaves. Wheelwright believed that radio could very well save modernist poetry by enfolding poetic modernism itself into a revolutionary project, just as he believed that a committed poetry—the only poetry possible—was now insufficient without radio. As a result, even on the silent page poetry harbors radio's negative image as political *lack*. This lack is cut in the precise shape of a "socialist culture"—of a utopian, radio-backed resolution of the conflict between bourgeois art and Marxist politics splintering Wheelwright's own sense of social alignment. By 1938, the engineered passage from verse to poetry, from sound to socially salient meaning, proves impossible without radio's capacity to actualize poetry's didactic potential.

Wheelwright sketches this media theory of modern poetry in the unpublished essay "Verse + Radio = Poetry," a formula that neatly divulges the retroactive cultural logic I term latent remediation. By 1938, Wheelwright implies, radio is *already* constitutively involved with poetry, factored in as part of poetry's *equation* not only as a medium with which to imagine new formal possibilities for verse and new avenues for its circulation but, more specifically, as an object of tremendous political fantasy, a fantasy of such force that a poet like Wheelwright is led in its shadow to materially reconstrue the limits of his poetic practice. It follows, then, that before Wheelwright even begins broadcasting, his sonically persuasive and politically didactic poetics on the page is significantly enabled by the historical fact of radio.

Wheelwright's essay implies that we can also define radio in dialectically complementary terms: as an engineering of sound that, *devoid of something like poetry*, can only engineer meanings of potentially disastrous social consequence. For Wheelwright, it is exactly poetry's didactic, "idealogical music" that will make radio politically safe, by repurposing its powers for mass mobilization and persuasion. Like Walter Benjamin and Bertolt Brecht, Wheelwright understood that radio's revolutionary potential awaited social uses of the medium that would retool its one-directional, cultural-industrial, incipiently totalitarian protocols.[61] Poetry was for him such an application. And in its turn, radio would serve modernist poetry as a kind of halfway house on art's long march toward an as yet unimaginable socialist culture, altering in the meantime the very social relations of poetic production. "Radio gathers us . . . Sure as fate, radio will raise up multitudes of amateur poets of whom some will flower. The general level of proficiency in verse and of alertness to poetry will heighten. The level of genius will then rise."[62] Wheelwright doesn't give explicit account of the forces responsible for delivering this "poetic revolution"; radio offers an expanded audience, surely, but this "quantitative" change provides only the precondition for a vaguer, more radical "qualitative" dispensation "directed toward a music of idea and association"; namely, the manner in which a radio-sounded poetry will "weld into consistent style the whole of contemporary verse which is split highbrow-lowbrow."[63] Radio's acousmatic address, so well suited to the sensuous illogics of associative verse, will render Hart Crane easy to grasp, flattening class barriers with the microphone, but it will also encourage further prosodic experiment along recognizably modernist lines: "Poetry will invade the air by its seeming simplicity. It will rule the air by its inexhaustible complexity."

Recommended by Wheelwright's own experience with W1XAL and its massive apparatus for propaganda, radio becomes a deeply invested point of suture for his contradictory social alignments and a crucial last front in the development of his political commitments. We recall that for Williams, "the most interesting Marxist position, because of its emphasis on practice, is that which defines the pressing and limiting conditions within which, at any time, specific kinds of writing can be done, and which correspondingly emphasizes the necessary relations involved in writing of other kinds."[64] Nothing if not an interesting Marxist writer (for a long time he has been thought merely that), Wheelwright understood that under prevailing historical conditions he could no longer write the kind of modernist poetry incumbent on his cultural self-positioning, nor could he forecast writing of another kind in advance of the necessary social relations from which it would arise. A radical didactic poetry of the page is one response to this intolerable situation, but in the end only radio could catalyze this "idealogical music" into revolutionary art. When Wheelwright confirms his fantasy-laden investment in radio by declaring, in high Nietzschean style, that "radio works a transvaluation of values," he projects a messianic politics onto a technological medium—by no means an uncommon gesture in the twentieth century.[65] It may seem odd that Wheelwright makes *poetry* the vehicle of his radio-backed politics, but we shouldn't be surprised to find poetry surfacing in the currents of a sound technology's social history. To the extent that poetic practices thrive on a media problem, the generative lyric strain of sound in print, poems seek out such encounters with media technologies and the discursive struggles over their use and meaning.

Of course, radio did not work the transvaluation that Wheelwright expected for poetry. Nothing prevented W1XAL from its instrumental fate as a wartime propaganda machine, and radical poets did not seize the international airwaves. But "the social implications of broadcasting are as wide and subtle as the waves of the air," Wheelwright asserts, and by the late 1930s, no matter what happens on the actual air, the social fact of radio is baked in to the formula for poetry.[66] But where do we look on the printed page for evidence of these "wide and subtle" implications? We began this section listening for Wheelwright's voice on W1XAL. We are prepared now to avouch the material presence of that radio voice—sounding through the mediating discourses of propaganda and utopian thought—in the silent print of the poems themselves. Here is "Train Ride" from *Political Self-Portrait*, Wheelwright's "greatest poem" in the estimation of both Richard Howard and John Ashbery:[67]

>After rain, through afterglow, the unfolding fan
>of railway landscape sidled on the pivot
>of a larger arc into the green of evening;
>I remembered that noon I saw a gradual bud
>still white; though dead in its warm bloom;
>*always the enemy is the foe at home.*
>
>And I wondered what surgery could recover
>our lost, long stride of indolence and leisure
>which is labor in reverse; what physic recall the smile
>not of lips, but of eyes as of the sea bemused.
>
>We, when we disperse from common sleep to several
>tasks, we gather to despair; we, who assembled
>once for hopes from common toil to dreams
>or sickish and hurting or triumphal rapture;
>*always our enemy is our foe at home.*
>
>We, deafened with far scattered city rattles
>to the hubbub of forest birds (never having
>"had time" to grieve or to hear through vivid sleep
>the sea knock on its cracked and hollow stones)
>so that the stars, almost, and birds comply,
>and the garden-wet; the trees retire; We are
>a scared patrol, fearing the guns behind;
>*always the enemy is the foe at home.*
>
>What wonder that we fear our own eyes' look
>and fidget to be at home alone, and pitifully
>put off age by some change in brushing the hair
>and stumble to our ends like smothered runners at their tape;
>We follow our shreds of fame into an ambush.
>
>Then (as while the stars herd to the great trough
>the blind, in the always-only-outward of their dismantled
>archways, awake at the smell of warmed stone
>or to the sound of reeds, lifting from the dim
>into the segment of green dawn) *always*
>*our enemy is our foe at home,* more

> certainly than through spoken words or from grief-
> twisted writing on paper, unblotted by tears
> the thought came:
> > There is no physic
> for the world's ill, nor surgery; it must
> (hot smell of tar on wet salt air)
> burn in a fever forever, an incense pierced
> with arrows, whose name is Love and another name
> Rebellion (the twinge, the gulf, split seconds,
> the very raindrops, render, and instancy
> of Love).
> > All Poetry to this not-to-be-looked-upon sun
> of Passion is the moon's cupped light; all
> Politics to this moon, a moon's reflected
> cupped light, like the moon of Rome, after
> the deep well of Grecian light sank low;
> *always the enemy is the foe at home.*
> > But these three are friends whose arms twine
> without words; as, in still air,
> the great grove leans to wind, past and to come. (144–46)

The prose argument accompanying this poem informs us that "Train Ride" was composed "in palimpsest" upon an extraordinarily popular 1898 collection of poetry by the late Victorian Stephen Phillips, a poet once billed as "Tennyson's legitimate successor in his own line" (154).[68] It would seem that Wheelwright ranged across Phillips's *Poems* to select appealing conceits, diction, rhythmic patterns, and pastoral imagery, and then wove this repulped Victoriana into a meditation on "the decay of bourgeois democracy" and the "reassurance" one takes nonetheless in the fact of "unceasing social change." Nearly everything in "Train Ride" bears a relation to Phillips's moralizing blank verse, though unlike later avant-garde proceduralists or poets of erasure, Wheelwright refuses to limit himself to any single method for processing his source material. In certain instances he translates an image or phrase, word-by-word, into a fresh idiom. Phillips writes of those "who/ Panted toward their end, and fell on death/ Even as sobbing runners breast the rope," which Wheelwright delivers to the more fluid, "[w]e stumble to our ends like smothered runners at their tape," an altogether stranger proposition that swaps out panting and sobbing for the unlikelier "smothered," effecting a smoke-choked contrast with the pastoral motifs opening and closing the poem.[69] Elsewhere he lifts especially distinctive bits from Phillips's lexicon—"sidle," "surgery," "patrol," "garden-wet," "indolence," and so on—and sinks these down as posts around which to raise the poem's image structure.[70] Even the most apparently motivated figures were sparked by Phillips's text, from the complicated auroral simile compacted into the parentheses that turn the poem in its second stanza, to the final images of triply reflected moonlight and "the great grove" action-posed in the winds of history.[71]

"'Train Ride' resumes the pastoral in palimpsest" upon Phillips's text, Wheelwright tells us, and we ought to hear "resume" in its now-rare acceptation: to take back or reappropriate a possession.[72] The poem effects a kind of reverse enclosure of the commons, reclaiming the pastoral from its cultural reproduction as the alienated, fantasized other to urban centers of commerce and industry, those "far scattered city rattles." As a compositional metaphor, the palimpsest supplies a mode of textual "surgery" with which to "recover/our lost, long stride of indolence and leisure/which is labor in reverse," that is, reparatively to exhume a sense of the pastoral from the divisions of labor and the uneven development of city and country that gave rise to the pastoral itself as a space of bourgeois recreation and as a literary ideal. In one sense, the rather ghostly reverberations exerted by Phillips on Wheelwright's text can be taken to index the poet's own class alignments, haunting vestiges of those dead generations said to "weigh like a nightmare on the brain of the living."[73] I invoke the *Eighteenth Brumaire* because this famous passage in Marx's essay helps elucidate why a poet so committed to a "poetry of the future" should mine his textual material from an enervated past he also strives to repudiate. "The social revolution of the nineteenth century cannot draw its poetry from the past, but only from the future," writes Marx.[74] "It cannot begin with itself before it has stripped off all superstition about the past. Earlier revolutions required recollections of past world history in order to dull themselves to their own content. In order to arrive at its own content, the revolution of the nineteenth century must let the dead bury their dead. There the words went beyond the content; here the content goes beyond the words." Previous revolutions relied on dress rehearsals of the past, Marx observes. What better way to strip away these costumes than to show up putative "content" (*Inhalt*)—the fin-de-siècle Phillips—as something that can be scrubbed away or depreciated into mere "words" (*Phrase*) for future composition and new compositions of the future?[75] That Wheelwright's particular social alignments complicate and ironize this palimpsestic gesture—here is a person who wore tuxedos to party meetings; where indeed does the costume end and the skin you're in begin?—only intensifies the urgent search for a mode of literary "surgery." Ultimately the speaker of the poem realizes "[t]here is no physic for the world's ill, nor surgery," taking comfort instead in the relentless currents of historical change, but Wheelwright can attain to this solace only by submitting himself to Phillips-in-palimpsest. It is by means, for example, of an "odour full of arrows," the latter poet's polite description of how a sweet fragrance can seem to assault an overly sensitive soul, that Wheelwright arrives at "an incense pierced / with arrows," evoking not only St. Sebastian but also the inverted Mars symbol decorating the dust-jacket of *Political Self-Portrait*, and torquing Phillips's metaphor into a figure for class struggle as the crucible of history, that "fever" in which the world burns forever (fig. 2.1).[76]

This brings us to the poem's second source, the anti-militarist slogan for international class solidarity—*Always the/our enemy is the foe at home*—which Wheelwright attributes to the "elder Liebknecht," that is, Wilhelm Liebknecht, a founder of Germany's Social Democratic Party. The likelier derivation is Wilhelm's son, Karl Liebknecht (the only member of the Reichstag to vote against war in 1914 and a founder with Rosa Luxemburg of the German Communist Party), to whom Trotsky attributes the slogan in the 1934 manifesto "War and the Fourth International," where Wheelwright almost surely would have read it.[77] This mantra for the "civil war against war" serves the poet as a ringing refrain, while also serving us, readers listening for Wheelwright's radio voice, as a metonym for the propagandistic possibilities of broadcast speech (154). In effect, the unlikely pairing of the genteel versifier Phillips with the sloganeering Liebknecht neatly rehearses the formula Verse + Radio = Poetry, even as the contrast self-consciously textualizes the contradictory alignments determining Wheelwright's didactic poetics. "Train Ride" refuses any reconciling sublimation of verse and radio-friendly slogan, as we should expect from a poem whose intellectual drama is precisely the jettisoning of idealist hygienics—those "physics" and "surgeries" of conceptual thought that would cure the "fever" of history.

If not at all reconciled, this "fever" is acutely reflected in the intermedial dynamics of Wheelwright's prosody: the hortative music of the slogan—a triplicate rhythm that resolves rousingly into rising iambs—interrupts the flexible pentameter of the poem's main speaker, to whom Liebknecht's

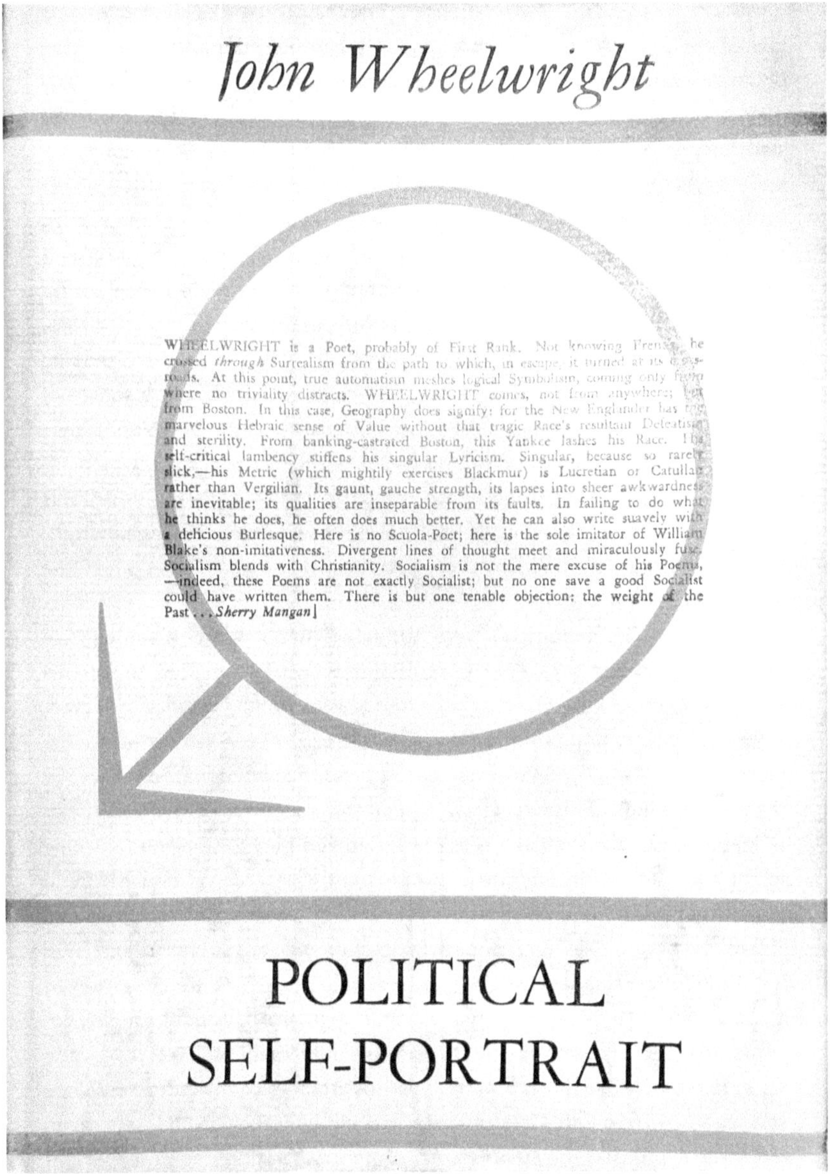

John Wheelwright

WHEELWRIGHT is a Poet, probably of First Rank. Not knowing French, he crossed *through* Surrealism from the path to which, in escape, it turned at its crossroads. At this point, true automatism meshes logical Symbolism, coming only from where no triviality distracts. WHEELWRIGHT comes, not from anywhere, but from Boston. In this case, Geography does signify: for the New Englander has the marvelous Hebraic sense of Value without that tragic Race's resultant Detestation and sterility. From banking-castrated Boston, this Yankee lashes his Race. His self-critical lambency stiffens his singular Lyricism. Singular, because so rarely slick,—his Metric (which mightily exercises Blackmur) is Lucretian or Catullan rather than Vergilian. Its gaunt, gauche strength, its lapses into sheer awkwardness are inevitable; its qualities are inseparable from its faults. In failing to do what he thinks he does, he often does much better. Yet he can also write suavely with a delicious Burlesque. Here is no Scuola-Poet; here is the sole imitator of William Blake's non-imitativeness. Divergent lines of thought meet and miraculously fuse. Socialism blends with Christianity. Socialism is not the mere excuse of his Poems, —indeed, these Poems are not exactly Socialist; but no one save a good Socialist could have written them. There is but one tenable objection: the weight of the Past... *Sherry Mangan*]

POLITICAL
SELF-PORTRAIT

Figure 2.1. Dust jacket for John Wheelwright's *Political Self-Portrait*, 1940, designed by the author. John Wheelwright papers, John Hay Library, Brown University

dictum never quite integrally belongs.[78] The result is a prosodic unevenness characteristic of Wheelwright's excessively baroque formal surfaces, often distinguished, as here, by protracted syntactic structures draped tortuously over irregular iambic meters. Free-floating across the poem, the slogan resists full prosodic assimilation. It's a choral refrain whose rhythmical difference marks the very limits of the speaker's consciousness. A fact the speaker both does and doesn't know, the slogan becomes referentially promiscuous, transforming in the various lights thrown off by the speaker's own transformations. Paranoiac and menacing in the first stanza wherein Wheelwright inventories an alienated bourgeois modernity, in the second stanza the slogan shifts its burden into a decidedly more affirmative and battle-readying key: the enemy's proximity, not to say its immanence, becomes a source of power not unlike the comfort Wheelwright takes in the protean unmasterability of social processes. What he had taken for an idyllic "railway landscape" is already shot through with the strife and struggle the pastoral is propagated to exclude; all walls have anticipatively fallen for Wheelwright the Anglican Marxist, including those between what we call "Love" and what we call "Rebellion."

In the poem's penultimate image, Wheelwright configures his three Ps—Passion (Love/Rebellion), Poetry, and Politics—with an analogy to solar irradiance: as moonlight glancing off a mirror is but sunlight twice reflected, so too are poetry and politics indivisible projects, of the same stuff fundamentally, and both fundamentally determined, like wind-warped trees, by the greater winds of history. I have argued that radio was central to Wheelwright's poetic achievement because it allowed him to imagine how one might go about politicizing the fact—apparently clearer than ever before in 1938—that poetry and politics, those booming "double guns of Word and Deed," are potentially contiguous practices of mediated sound (138). "Train Ride" allegorizes this process via intermedial sound. Their difference marked by Wheelwright's meter, slogan and poetic speech are brought roughly together in a negative synthesis whose positive fulfillment radio seemed, for a time, to promise. No doubt a certain pathos attaches to the fact that we are still reading "Train Ride" in alphabetic writing, whether on the page or on the screen, and that the "whole body of poetry" was not, as Wheelwright expected it would be soon, "broadcast and recorded as a matter of course."[79] But whether we regard this late 1930s remedial fantasy as a foiled but plausible revolutionary project or as the spirited delusion of poets bewitched by a new cultural technology, we should not be so quick to understand it merely as a road not taken by "modern poetry" writ large, as a curious but negligible blip in a literary history whose medium is the enduring printed page. It would be better to understand these investments in poetry's radio future as the chief signs of changes already in process and as an invitation to refine our media-theoretical tools for writing literary history. Perhaps then we can read the history of modern poetry the way Wheelwright reads his leaning trees, which, only insofar as they seem to resist the wind, can tell us what is really past and what is to come.

Broadcasting Dialect: Sterling Brown

At 5:00 p.m. on Sunday, January 15, 1939, radio listeners tuning in to their local CBS affiliates enjoyed the seventh installment of *Norman Corwin's Words without Music*, a new sustaining program that professed "to vitalize poetry for a larger audience."[80] "This experimental program is based on the theory that words, when arranged in the right way, are music in themselves," ran the show's lead-in, "and to support this theory, Norman Corwin has taken a number of poems and applied them to the special uses of radio through the combined techniques of orchestration and augmentation."[81] Prior to joining CBS as a producer-director the previous spring, Corwin had distinguished himself as an innovating poetry jockey on New York's WQXR, but the major network, home to the *Columbia Workshop* and hot off the 1938 premier of Orson Welles's *The War of the Worlds*, offered this self-described "frank propagandist for poetry" the institutional resources necessary to flex and expand his auditory imagination on the proving ground of a soon-to-be unrivaled career in radio dramaturgy; in just three years' time, Carl Van Doren would crown Corwin the Christopher Marlowe of American radio, a wartime radio-wright capable of commanding, in one instance, an audience of sixty million listeners.[82]

While *Words without Music* furnished the platform for Corwin's first smash hits on the air, *The Plot to Overthrow Christmas* (December 25, 1938) and *They Fly Through the Air with the Greatest of Ease* (February 19, 1939), these original verse dramas were exceptions to the programming rule. The majority of episodes adapted familiar poems of the printed page, a blend of canonical classroom fare (Jonathan Swift, Henry Wadsworth Longfellow, Matthew Arnold) and popular light verse (Edward Lear, Lewis Carroll, W. S. Gilbert) with the demotic strains of the "new" American poetry (Carl Sandburg, Edgar Lee Masters, Vachel Lindsay). Corwin's procedures of "orchestration and augmentation" involved freely editing and elaborating the text, installing sound effects, and distributing the poem among a cast of voice actors whose expressive coordination of "timbre, rhythm, tempo, dynamics, and general pitch range" effected a kind of dialogized or "choral speech," bursting open monological utterances to synchronized multitudes in a style consonant with the populist ethics of the decade.[83] These medium-specific adaptations distinguished Corwin's show from "the sing-song programs of straight poetry readings" in ready supply elsewhere on the air by appealing both to highbrow "poetic bugs" *and* "the ordinarily literate with no particular lyric sense," while also drawing no less deeply on poetry's cultural capital.[84] As Andrea Selch has argued, this aesthetic tightrope walk allowed a competitive and prestige-seeking CBS to pitch Corwin's program, and poetry itself, as a democratic good.[85]

But when Corwin elected to dedicate his January 15 show to "Negro poetry" specifically, *Words without Music* offered its listeners something even more improbable than poems "vitalized" for liberal democracy. In the act of applying his radiophonic procedure to poems in Black dialect—and to the poems of Sterling Brown in particular—Corwin orchestrated, quite unwittingly, nothing less than a crash course in African American literary history and media studies. Corwin's remediations from the page to the airwaves amounted in certain cases to a racist deformation of these poems, but the media-ecological conditions of Brown's achievements in Black dialect in the late 1920s rendered his poems potentially resistant to their exploitative appropriation by Corwin a decade later. The poetry's immanent resistance to network broadcast can be grasped as a function and measure of its originally intermedial conditions—that is, its historical provenance in a print medium unsettled by other technologies for the reproduction of speech, including folkloric transcriptions and commercial phonography. Like Wheelwright's didactic verse, Brown's *actual* confrontation with radio invites us to consider the ongoing, latent work of medial influence informing poetry's dimensions on the page. No recording exists of these poems as they were heard by CBS listeners in 1939, but we have good reason to suspect that Brown's dialect verse could refuse the exclusionary terms of golden age radio, precisely insofar as these works represent constitutive expressions of the literary-media effect I'm calling latent remediation.

Before turning directly to the January 15 broadcast, we ought to briefly consider, by way of introduction to Corwin's remedial method and its racializing dynamics, the opening poem on the first episode of *Words without Music*, Vachel Lindsay's "The Daniel Jazz."[86] One of Lindsay's "singing poems"

Latent Remediation

or performative "poem games," in all its print versions this ragged, comic retelling of Daniel in the lions' den is framed by glosses indicating imaginary, racializing accompaniment (*"a strain of 'Dixie,'" "a touch of 'Alexander's ragtime band'"*) and sing-along directions for the audience (*"The audience sings [Go chain the lions down] with the leader, to the old negro tune"*).[87] Though certainly we ought to class "The Daniel Jazz" among Lindsay's "Negro sermons" ("To be read in your own variety of negro dialect"), the poem itself does not resort to orthographic dialect in its depiction of Daniel and Darius the Mede.[88] As his ultra-famous "The Congo" illustrates most egregiously, Lindsay's aspirations toward a "higher vaudeville" often relied on strategies of "multilayered racialized mimicry," but in "The Daniel Jazz" this mimicry is effected largely *paratextually*, as a potential, fantasized performance to be indulged, if only imaginatively, by white audiences.[89] Insofar as this fantasy begins by turning its audience into media machines for sound effects and playback—*"Let the leader train the audience to roar like lions, and to join in the refrain 'Go chain the lions down,' before he begins to lead them in this jazz"*—the poem exercises in stark terms what Mark Goble has described as that tendency of modernist texts to imagine "the experience of recorded sound as a vicarious experience of race."[90] For white readers, the antic pleasures of "The Daniel Jazz" hinge on the possibility of becoming a performative *medium* for Black music, and if the poem opens by invoking the very musical emblems of white appropriation in Tin Pan Alley ragtime and minstrel "Dixie," by poem's end Lindsay has projected this jazz back, thanks to the phonographing audience, to its putatively authentic roots in an "old negro tune."[91]

On *Words without Music*, Corwin's arrangement swaps these silent paratextual "strains" of cross-racial identification for the outright "aural blackface" of exaggerated dialect, as if the means of translating the racializing significance of merely figurative music is to make actually audible a dialect never present on the page.[92] Lindsay largely eschews dialect, for instance, in the following exchange between Darius, Daniel's "sweetheart," and his mother:

One Thursday he met them at the door:—
Paid them as usual, but acted sore.

He said:—"Your Daniel is a dead little pigeon.
He's a good hard worker, but he talks religion."[93]

On the radio, by contrast, Darius is made to say a bit more and more offensively, as Corwin fleshes out the poem's dramatic action with stereotyped dialogue that his performers deliver in minstrel style: "Now here's your wages as of de weekend and Thursday, which is today . . . And by the way! Your Daniel is a dead little pigeon. He's a good hard worker, but he talks religion, he talks religion all the live-long day! Now y'all come wit me—just step this way!"[94] As for Daniel himself, whereas Lindsay had cast the biblical hero as "the chief hired man of the land," Corwin parlays this suggestion into plantation stereotype; to God's command, "Go chain the lions down," Daniel responds in the unmistakable key of the Old South: "Yes, Lord, right away, Lord."[95]

To remediate Lindsay's poem for radio, then, is apparently to retrace, in burnt cork, what Jennifer Lynn Stoever has theorized as "the sonic color line." Examining the historical construction of racialized sound and racializing listening practices from antebellum slavery through Jim Crow to postwar ideologies of "color blindness," Stoever details how the sonic color line has long disciplined "American listening . . . to 'match' certain sounds, voices, and environments to visual markers of race" and thus to essentializing hierarchies, positing "racialized subject positions like 'white,' 'black,' and 'brown' as historical accretions of sonic phenomena and aural stereotypes that can function without their correlating visual signifiers and often stand in for them."[96] Because radio absents visual regimes of racial identification, network broadcasting operated "as a technology of the sonic color line, propagating racialized aural representations, mediating racial discourse, and practicing racial exclusion while depicting itself," in avowal of its acousmatic color blindness, "as incapable of racialization."[97] The suffocating racial "predicament" of so-called golden age broadcasting was such that, as Barbara D. Savage has also documented, "national network radio remained virtually inaccessible to African American influence and control," even while "fascination with African Americans and African American culture permeated radio's early programming and spurred the medium's popularity."[98] For evidence of radio's own specific encounter with the constitutive "black mirror" of American popular culture, we need look no further than network radio's very first serial, *Amos 'n' Andy*, that neo-minstrel show "as essential to American domesticity as a car in every garage."[99]

Corwin did strike out against his industry's investments in exclusion and stereotype: he cast Black actors and performers more frequently than did his peers, argued publicly for a politics of racial inclusion, and featured in his radio dramas of the 1940s what one writer for the *Negro Digest* recognized as "unmistakable, militant, poetic references to discrimination and equal rights."¹⁰⁰ But the fact that Corwin could enjoin Black listeners aggrieved by racist stereotypes to "phone the studio, write letters, publish editorials!" and "go after everything that is caricature, bad taste, incorrect even at the risk of being excessive in criticism," while also giving the *Amos 'n' Andy* treatment to "The Daniel Jazz" underlines all the more emphatically what Savage and Stoever identify as the constitutive racial exclusions of commercial radio and the racializing aural protocols according to which the medium of broadcasting almost inevitably enforces and polices a sonic color line.¹⁰¹ Successfully adapting poetry to "the special uses of radio" circa 1939 involves acceding not only to the technological and dramaturgical demands of the medium but to its social entailments as well. In the case of a poem like "The Daniel Jazz," structured as it is by white fantasies of a musically mediated Blackness, "making poetry talk" on *Words without Music* involves making it talk in the timeworn voices of grotesque stereotype, voices which seemed to issue involuntarily from the very "makeup" of radio itself.¹⁰²

A few weeks later, Corwin would unwittingly articulate the medium of commercial radio to another technology of the sonic color line: written Black dialect. The January 15 episode of *Words without Music* featured poems by three writers variously engaged with dialect: Irwin Russell ("De Fust Banjo"), James Weldon Johnson ("Go Down, Death" and "The Creation"), and Sterling Brown ("Southern Road" and "Memphis Blues").¹⁰³ A gospel group, the Deep River Boys, participated in the show alongside a cast composed mostly of Black actors but that also apparently included Corwin himself, aurally blacked up.¹⁰⁴ As for the poets, the selection would have struck any listener familiar with the cultural politics of African American literature as an ironic précis of the contentious history of dialect poetry across three generations. Far from the "Negro writer" some listeners understood him to be, the white Mississippian Russell was in fact an important forerunner of white dialect writers like Joel Chandler Harris and Thomas Nelson Page, those local colorists most responsible for the "plantation tradition" of postbellum Southern writing that nostalgically idealized slavery by trading in minstrel stereotypes, visually enforcing the sonic color line in print.¹⁰⁵ The original script for Corwin's free adaptation of Russell's poem *does* indicate the poet's race, marking some social-contextual distance between "De Fust Banjo" and the poems of Johnson and Brown, but this framing context was cut for time, as indicated by the large red X drawn through the introduction (fig. 2.2).¹⁰⁶ In this way, Russell's poem—a rendition of the folk tale in which Noah's son Ham strings the first banjo by skinning a possum's tale, a poem that appeared on *actual* minstrel stages in the twentieth century—is implicitly subsumed without a word under the rubric of what many listeners heard as an "all-Negro poet show," erasing along with Russell's own whiteness the white origins of Black dialect poetry.[107]

Joel Chandler Harris, of Uncle Remus fame, admiringly accounted Russell "among the first—if not the very first—of Southern writers to appreciate the literary possibilities of the negro character," although he admitted that owing to careless writing, Russell's "dialect is not always the best."[108] Significantly, in his groundbreaking study *Negro Poetry and Drama* (1937), Sterling Brown would similarly register Russell's primacy by reversing the terms of Harris's verdict. Russell is "considered the first American poet to portray the Negro in dialect with any degree of success," notes Brown; he was "a good humorist who knew Negro speech, and something about Negro life in the delta. But his poems treat only those aspects of Negro life and character agreeable to a white Mississippian in reconstruction days."[109] Brown was often punishingly critical of that "limited gamut of characteristics allowed to Negroes" in the work of Russell and other plantation traditionalists, past and present, but he was also quick to acknowledge the technical accomplishment of these early dialect writers, whose "study of Negro speech . . . was fruitful and needed."[110] In Brown's view, there was "nothing 'degraded' about dialect"—it was not the orthographic medium itself but rather its radically limited articulation to racist representational conventions that posed problems for Black readers.[111]

Scholars have long judged Brown, with Zora Neale Hurston and Langston Hughes, centrally responsible for vindicating the written representation of Black vernaculars as a politically relevant and aesthetically "supple mode of expression."[112] "In a remarkably tangible sense," writes Henry Louis Gates Jr., Brown's *Southern Road* "marks the end of the New Negro Renaissance as well as the resolution, for black poetic diction, of a long debate over its mimetic principles."[113] In his poetry and literary criticism, Brown not only re-sources written dialect from its originary associations with minstrelsy to the authenticating provenance of a Black folk tradition; as Gates indicates, he also answers dialect's emphatic rejection by a Black bourgeois establishment for whom it was "thought to be a literary trap."[114] In Corwin's own incidental survey of African American cultural politics, this latter position is epitomized by the towering figure of James Weldon Johnson. An erstwhile dialect poet himself, in the first preface to his *Book of American Negro Poetry* (1922), Johnson famously announces that Black dialect is a medium too

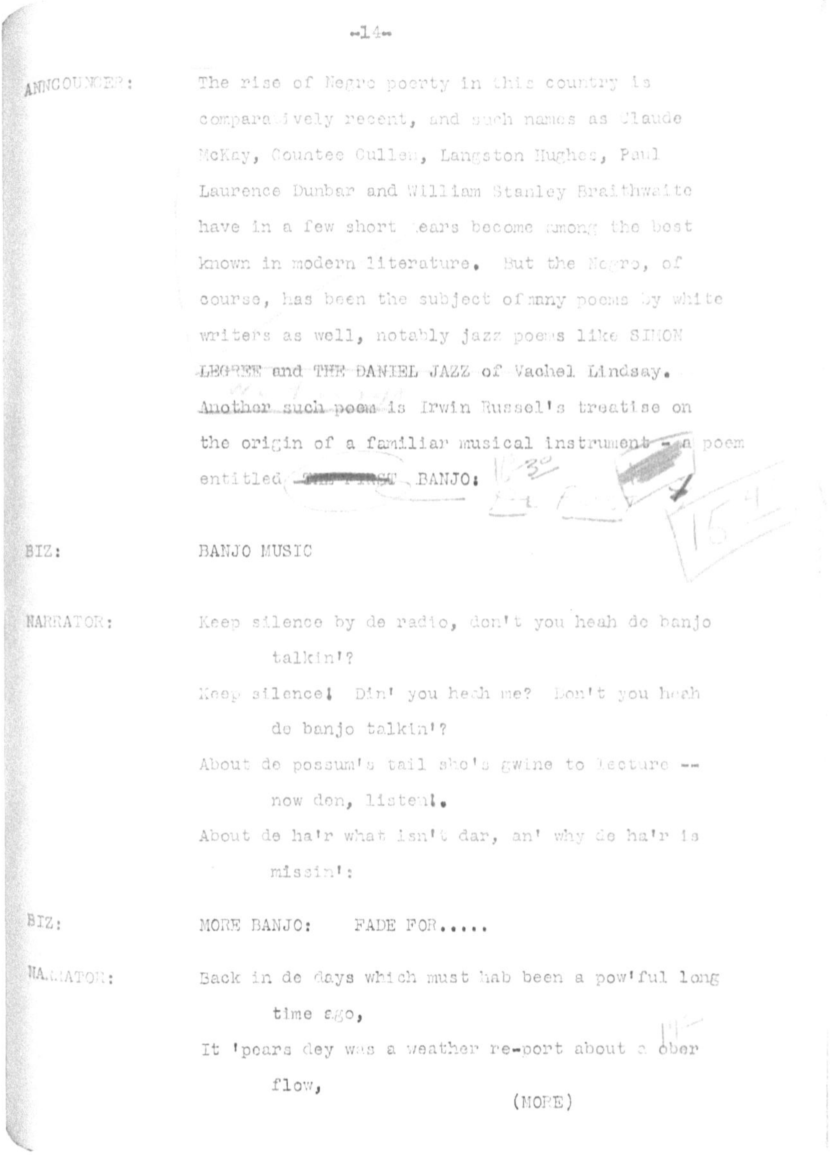

-14-

ANNOUNCER: The rise of Negro poetry in this country is comparatively recent, and such names as Claude McKay, Countee Cullen, Langston Hughes, Paul Laurence Dunbar and William Stanley Braithwaite have in a few short years become among the best known in modern literature. But the Negro, of course, has been the subject of many poems by white writers as well, notably jazz poems like SIMON LEGREE and THE DANIEL JAZZ of Vachel Lindsay. Another such poem is Irwin Russel's treatise on the origin of a familiar musical instrument — a poem entitled THE FIRST BANJO:

BIZ: BANJO MUSIC

NARRATOR: Keep silence by de radio, don't you heah de banjo talkin'?
Keep silence! Din' you heah me? Don't you heah de banjo talkin'?
About de possum's tail she's gwine to lecture --
now den, listen!
About de ha'r what isn't dar, an' why de ha'r is missin':

BIZ: MORE BANJO: FADE FOR.....

NARRATOR: Back in de days which must hab been a pow'ful long time ago,
It 'pears dey was a weather re-port about a ober flow,
(MORE)

Figure 2.2. Excerpt from the script of Norman Corwin's *Words without Music*, January 15, 1939, p. 14. Norman Corwin Collection, University of California Santa Barbara Library. Courtesy of the Estate of Norman Corwin

crudely programmed by white fantasies of racial difference, "an instrument with but two full stops, humor and pathos," and thus wholly unsuited to the representational needs of a modern Black poetry.[115] When he updates his preface in 1931, and in his celebratory introduction to *Southern Road* one year later, the recent innovations of Brown and Hughes lead him not to recant, as one might expect, but to declare his claim "as sound today as when it was written ten years ago": "The passing of traditional dialect as a medium for Negro poets is complete."[116] Johnson can double down on his position, while simultaneously commending the younger poets for their orthographic mobilization of "the common, racy, living, authentic speech of the Negro," by drawing a newfound distinction between the latter and the "traditional" or "conventionalized" dialect of the plantation tradition.[117]

Johnson and Brown agreed that dialect writing was mired in supremacist fetish, but Brown evinced a superadded faith in the flexibility of the medium itself. As we shall see, to the extent that Brown succeeded in positing new aesthetic relays between orthographic representation, aural reference, and his "folk" sources, he did not have to wait for the waning of racist conventions to renovate dialect writing. Far from being "capable of only two stops," Brown declares in direct response to Johnson, "dialect, or the speech of the people, is capable of expressing whatever the people are. And the folk Negro is a great deal more than a buffoon or a plaintive minstrel. Poets more intent upon learning the ways of the folk, their speech, and their character, that is to say better poets, could have smashed the mold."[118]

On *Words without Music* Corwin orchestrated two pieces from *God's Trombones: Seven Negro Sermons in Verse* (1927), Johnson's celebrated attempt "sincerely to fix something" of "the old-time Negro preacher [who] is rapidly passing," and whose "fusion of Negro idioms with Bible English" the poet approximates in a syncopated free verse designed to script the reader's intonation.[119] *God's Trombones* should have been the last word on dialect, a poetic exemplum of its utter dispensability for adequately channeling Black folk expression in a written medium. And yet it was Johnson's sermons, as Mark A. Sanders notes, which ultimately provided "the closest, most instructive model for Brown's experimental sense of Afro-modernism"—his own attempts "to smash the mold."[120] In the latter poet's approving estimation, Johnson had realized the programmatic designs expounded in his prefaces: "Convinced that dialect smacks too much of the minstrel stage, [he] attempts to give truth to folk idiom rather than mere misspellings."[121] It's of course just this "folk idiom"—here the strategically essentialized other to white-authored dialect—that will mobilize Brown's reinvigoration of dialect writing. Surefooted scholars have already underscored the mediated complexity of Brown's attitude toward vernacular "folk" materials.[122] And yet, the relation between Brown's orthographic practice and this "folk idiom," between printed dialect and the speech sounds to which it somehow refers and that it somehow produces, remains a critical puzzle, largely because we have yet to attend carefully enough to the particular *media* material to these mediations.

To this end, I suggest we follow the illustrious Howard professor briefly to school—that is, to Brown's *Outline for the Study of the Poetry of American Negroes* (1931), a slender booklet of annotations, study questions, and appendixes designed to accompany the second edition of Johnson's *Book of American Negro Poetry*.[123] Sonya Posmentier has framed *Outline* as an instance of Brown's lifelong commitment to stimulating "affective, interpretive, and contextual ways of reading" modern Black poetry.[124] At least once, in the discussion questions devoted to Johnson's own sermon on "the creation," this implicit lyric theory unfolds as a kind of introductory course in medial analysis. After directing students to the sermon's formal, figurative, and thematic aspects, Brown advises them "if possible [to] take down the words of an 'old-time' Negro sermon, either from actual experience or from one of the many phonograph records"—at just this moment an explosively popular subgenre of "race record"—and to compare this transcription with Johnson's poem.[125] Then students are advised to repeat the comparative analysis, this time with two recent white-authored renderings of rural Black Christianity, Roark Bradford's *Ol' Man Adam an' His Chillun* (1928) and Marc Connelly's *The Green Pastures* (1930). Connelly's Pulitzer Prize-winning play, a treatment of the "living religion . . . of thousands of Negroes in the deep South," adapts Bradford's collection of short stories, which themselves adapt episodes from the Old Testament into minstrelized folk fables.[126] Elsewhere Brown cites Bradford as an example of how the "Negro folk preacher has long been burlesqued" and in fact expressly links Russell's "De Fust Banjo" with "Bradford's biblical recreations," which are lamentably "like it in approach."[127] In contrast, Brown regarded Connelly's play, which featured Broadway's first all-Black cast and the Hall Johnson Choir, much more positively, praising its "tenderness and reverence," though not without noting the departures from "accurate truth" observed by many Black audiences.[128]

By this point, Brown has invited students to orient their own reception of "The Creation" toward a diverse media-ecological array of representations of the "old-time Negro preacher"—in relation to a sermon from the pulpit, or a "real" performance thereof on a commercial phonographic record, to folk preaching's gross debasement in white-authored prose, and to the ambivalently successful rehabilitation of that prose for Black voices on the Broadway stage. Only now can Brown raise for his students the crux of his own poetic experiments, the question of written dialect's relationship to the "folk idiom": "Read carefully James Weldon Johnson's introduction to *God's Trombones*," and "notice that the poem is not in dialect. Does this lessen the folk quality of the poem? What does the poet mean when he speaks of the 'folk idiom'?"[129] Clearly Brown is here encouraging his students to dissociate dialect writing from any essential relationship to a "folk idiom." But even more consequentially, the study guide likewise implies that whatever the "folk idiom" finally *is* can only be decided by considering the various representational media in and against which it is both instantiated and circulated.

To meet Brown's challenge, and to specify the nature of this "folk idiom," we might link up the poet's own media analysis with recent critical activity around what Katherine Biers has termed the "black phonographic voice."[130] Much of this scholarship has sought to register the deconstructive challenges posed by various modes of Black aesthetic practice to Western representational regimes, though it has done so often by attending specifically to phonic and sonic materialities, correcting for the "paradoxically phonocentric deafness" that sometimes hobbles poststructuralist reading procedures.[131] As Biers usefully synthesizes, such interventions demonstrate how "black writers, musicians, and singers . . . have historically explored the expressive potential of the zone in between articulate speech or melodic song, on the one hand, and the materiality of rhythm and 'pure' sound, on the other," to "bridge and articulate the gap between the phone and the graph, speech and writing."[132] Most germane to Brown's case are those theorists who emphasize how Black aesthetics are conditioned by modern sound technologies—how, in Alexander G. Weheliye's analysis, "the phonographic technologization of black music and black speech" presents "a condition of (im)possibility for modern (black) cultural production."[133] But whereas Weheliye links the radical fecundity of Black aesthetic practices chiefly to the paradoxical "ephemoromateriality of sound" offered by the phonograph's splitting of sound from source, Bryan Wagner takes up the example of early twentieth-century folklorists to show how the phonograph's "principle of live fidelity"—the projected coincidence, in the moment of recording, of signal and source, voice and body, song and consciousness—effectively *produces* a notion of authentic expression: "Alienating the voice from the body, in this instance, creates rather than disrupts speech's capacity to stand for subjectivity, producing a new opportunity for face-to-face immediacy between collector and informant."[134] According to Wagner, "the aura is made, not destroyed, by the phonograph."[135]

Evidence suggests that Brown understands the "folk idiom," à la Wagner, as a pure product of the cultural and technological mediations in every instance preceding it. In a superb discussion of Brown's tenure as head of the Federal Writers' Project's Office of Negro Affairs, Todd Carmody has illustrated how Brown's editorial direction, which often involved "foregrounding the artifice" of dialect, "actually worked to denaturalize the African American voice" and interrogate its instrumentalizing construction as authentic Black expression under the aegis of those "mechanisms of social recognition by which the performance of blackness takes on meaning in the larger public sphere."[136] Wagner, for his part, reveals how closely such mechanisms are bound up with the technological protocols and cultural logics of sound reproduction. These strategies of recognition include not only the rapacious appropriations of the culture industry, nor only the liberal agendas of the New Deal state (Carmody's interest), but also the practices of folklorists like Howard Odum and John Lomax who, in Wagner's words, "gleaned their new ideas about cultural authenticity from the phonograph, whether they knew it or not."[137] It would appear that Brown, an avid reader of folklorists like Odum and John and Alan Lomax, knew it; in other words, he apprehended the "folk idiom" toward which his poetics strove *just as* a species of the phonographic voice. "Believing one's ears, especially where folk-music is concerned, is probably better than believing the conventional notation of that music," Brown opines, but "believing phonograph records, as recent scholars are doing, is even better."[138] As an ethnographic document, the record trumps the real thing when the real thing is a product of the record, and going directly to the source means attending first and foremost to the signal.

Remarking the phonographic dimension of the "folk idiom" affords a crucial foothold for comprehending Brown's dialect verse on the page. Let it be said that white-authored dialect of the local color variety constitutes the diametric obverse of the Black phonographic voice, understood both as that which sounds the gap between *graph* and *phone*, foiling any assumption of their identity, and that which indexes the technological construction of Black authenticity. Indeed, Russell's poetry denotes a conception of dialect that takes "dialect" at its word, insofar as that word itself aspires to collapse speech ("dialect" as spoken idiom) and writing ("dialect" as orthographic technique), stabilizing the radical difference or untranslatability that is any idiom's special prerogative. Paradoxically, as the deformed other of standard English, the printed textualization of the sonic color line that we find in a poem like "De Fust Banjo" offers a spectacle of absolute difference at the same time as it pretends absolutely immediate access to, and control over, Black speech. Such are the print media conditions subtending the plantation tradition that Johnson and others so vehemently reject. Brown's ability to "smash the mold," as it were, stems from his willingness to open up the practice of dialect writing to a horizon *beyond* print, to articulate dialect to something like the Black phonographic voice. Decades ago, Gates announced that "the use of dialect in poetry . . . must be seen in the context of the music from which it springs—black speech and black music (especially the spirituals), which is the final referent."[139] We are in a position now to recognize what Brown himself intuited: this "final referent" is itself a mediated construct ultimately referable to a dense historical context involving the technological mediation of interracial fantasies at the hands of ethnographers, artists, and cultural industrialists alike.

Questioning the temptation to apply standards of "authenticity and realism" to written dialects, Gavin Jones reminds us that such "dialects are, at best, gestures toward a spoken reality that the reader can bring to the text from his or her recollection of heard experience."[140] For Brown and for other renovators of dialect verse, this "spoken reality" signifies not so much speech itself as the reality spoken into being by socially determined media increasingly technologizing the "heard experience" of twentieth-century life. Brown's consummate skill as a technician of dialect verse stems principally from his self-aware, self-reflexive use of folk materials. In "Southern Road," for instance, one of the two poems adapted by Corwin, we can observe how Brown's formal innovations index the cultural processes furnishing the poem's sources. The poem's stanzaic structure hybridizes two cultural forms: the communal work song, with the strong-stress swing and grunt of its time-keeping chant, and the traditional blues with its three-line development.

> Swing dat hammer—hunh—
> Steady, bo';
> Swing dat hammer—hunh—

Steady, bo';
Ain't no rush, bebby,
Long ways to go.[141]

"Southern Road" gestures back toward the shared origins of these two expressive practices. But the poem's generic blend also points forward to Brown's own contemporary moment and the practices of folkloric collection, transcription, recording, and interpretive categorization that have produced these genres as recognizably distinct stanzaic forms, as genres that exist to be artificially blended. Howard Odum and Guy Johnson's *The Negro and His Songs* (1925) distinguishes "work songs" from the blues songs and ballads collected in its "social songs" chapters, and *American Negro Folk-Songs*, a prominent 1928 anthology compiled by Newman Ivey White, also features separate chapters headed "Work Songs" and "Blues and Miscellaneous Songs."[142] But for many of the working-class musicians themselves, whose blues practices were just then furnishing a new field of capitalization for an emerging culture industry, "music was not a way to pace work," as Karl Hagstrom Miller observes.[143] "Music was a way to stop working—or at least to work at a task over which they had much more control." This dialect poem's generic synthesis of the work song and the blues reverberates with that painful discord at the heart of the phonographic voice: namely, the ambivalent yoking of exploitation and free expression. Riding conspicuously astride the sonic color line, this poetic speaker of dialect knows a great deal about the cultural machinery enforcing that line.

The fact that Brown could conceive of the phonographic voice of his folk idiom as the final referent of his poetic structures—the fact that a poem like the hybrid "Southern Road" can reach beyond the borders of the printed page to reflexively entail a whole cultural apparatus of folkloric collection and technologically conditioned transmission—means that the resulting poetry is a product of Brown's negotiation with capitalized appropriations of Black culture, and with ethnographic and commercial uses of phonography. The poem's stanzaic structure and its coordinated use of dialect are impossible to imagine outside of a milieu in which songwriters are busy fitting the blues to "race records" and folklorists are eagerly entextualizing an oral archive of work songs. Before it ever hits the airwaves on *Words without Music*, then, "Southern Road" is latently remediated on the page. A print artifact, it nevertheless bears the constitutive traces of nonprint media, and this has important consequences for how we measure the poem's critical resistance to Corwin's adaptation. Even in its relatively straightforward "vitalization" for the air (Corwin cuts one stanza and adds a "real sledgehammer for effect"), we have reason to suspect that, as in "The Daniel Jazz," Corwin applied a degree of minstrel makeup to "Southern Road" when he prepared the poem for a broadcast medium strictly operationalized along the sonic color line.[144] Nevertheless, one might also wonder whether Brown's poetry, though written ten years earlier, was homeopathically prepared to parry these remedial deformations, precisely because the poems themselves had arisen from a prior grappling with that sonic color line and with the phonographic voice.

Consider the second Brown poem featured in the *Words without Music* broadcast, "Memphis Blues," arranged by Corwin for a speaking chorus of nine voices. Like "Southern Road," this poem can be read as a dynamic formal hybrid, except its interlacing of blues balladry with the conventions and concerns of sacred song and preaching presents such a fine weave that teasing out or exactly taxonomizing its folk sources is beside the point.[145] By aesthetically mobilizing historical linkages between the sacred and the secular, Brown offers something more on the order of a *genealogy* of contemporary blues practice; in Stephen E. Henderson's words, the poem stands as "a brilliant exploration of the song-sermon form in which the blues are historically and formally grounded."[146] Or, to borrow Robin D. G. Kelley's formulation, "Memphis Blues" points up Black "folk" culture as so many variegated modes of "*bricolage*, a cutting, pasting, and incorporating of various cultural forms that then become categorized in a racially or ethnically coded aesthetic hierarchy."[147] And yet, the poem is also interested in internally processing the very means by which those mediating categories and hierarchies get installed. Brown signals this ironic self-reflexivity in the poem's title. Though W. C. Handy self-published "Memphis Blues," the first commercially successful blues composition, in 1912, it exploded across vaudeville stages two years later with lyrics by the white songwriter George A. Norton. These lyrics anxiously perform their own linguistic superfluity with respect to Handy's music by attempting to establish the music's provenance—"Folks I've just been down, down to Memphis town"—with a cartoonishly racializing *instrumentalization* of the spirituals:

> when the big Bassoon
> Seconds to the Trombones croon,
> It moans just like a sinner on Revival Day.

That melancholy strain, that ever haunting refrain
Is like a darkie's sorrow song.[148]

Brown may have chosen "Memphis Blues" for his title to supply an alternative, reparative lyric for this landmark tune, but as a consequence, he willfully commences the poem under the vexed auspices of white appropriation. As throughout *Southern Road*, here Brown assumes the pose of the "utopic ironist," nimbly querying the indeterminate relations between Black style and white stereotype, inexorable pessimism and vital affirmation.[149] To the extent that Brown honed this strategy by critically engaging media representations of Black life, its aesthetic products are built to parry the racializing operations of Corwin's program.

The title also provides Brown with the poem's motivating trope. As we learn in the swung dimeter quatrains of its opening section, "Memphis Blues" preaches the contemporaneity of biblical catastrophe:

Nineveh, Tyre,
Babylon,
Not much lef'
Of either one.
All dese cities
Ashes and rust,
De win' sing sperrichals
Through deir dus' . . .
Was another Memphis
Mongst de olden days,
Done been destroyed
In many ways . . .
Dis here Memphis
It may go;
Floods may drown it;
Tornado blow;
Mississippi wash it
Down to sea—
Like de other Memphis in
History.[150]

If the mood here is blues Ecclesiastes, the method is exegetical. Brown adapts the traditional double voicedness of the spirituals by locating allegories for contemporary experience in the Old Testament and then openly explicates that tropic strategy by *literally*—as a function of the lettered word "Memphis"—linking the ancient Egyptian capital to the Tennessee city lately devastated by just the sort of natural catastrophe he invokes. As the first poem in *Southern Road*'s second section, "On Restless River," "Memphis Blues" serves as a prologue of sorts to Brown's other poems on the Great Mississippi Flood of 1927. Posmentier has brilliantly described how these works formally mediate "the sometimes circular historical and environmental experience of catastrophe," and clearly in this instance, too, Brown is interested in fathoming "History" as the ruinously recursive record always threatening to subsume—or shall we say, subsumptively *historicize*—the present.[151] But what distinguishes "Memphis Blues" from other blues-based flood poems like "Ma Rainey" and "Cabaret" is the spectral presence of those "sperrichals," which force readers to reckon with the possibility of transcending this cycle of catastrophe.

In the poem's middle section, stanzas of anaphoric call-and-response deliver a parodic re-sourcing of the blues in their deflating treatment of a familiar theme from the spirituals: "What you gonna do" on Judgment Day?[152]

> Watcha gonna do when Memphis on fire,
>   Memphis on fire, Mistah Preachin' Man?
> Gonna pray to Jesus and nebber tire,
>   Gonna pray to Jesus, loud as I can,
>   Gonna pray to my Jesus, oh, my Lawd!
>
> Watcha gonna do when de tall flames roar,
>   Tall flames roar, Mistah Lovin' Man?
> Gonna love my brownskin better'n before—
>   Gonna love my baby lak a do right man,
>   Gonna love my brown baby, oh, my Lawd!
>
> Watcha gonna do when Memphis falls down,
>   Memphis falls down, Mistah Music Man?
> Gonna plunk on dat box as long as it soun',
>   Gonna plunk dat box fo' to beat de ban',
>   Gonna tickle dem ivories, oh, my Lawd!
>
> Watcha gonna do in de hurricane,
>   In de hurricane, Mistah Workin' Man?
> Gonna put dem buildings up again,

> Gonna pum em up dis time to stan',
> > Gonna push a wicked wheelbarrow, oh, my Lawd!
>
> Watcha gonna do when Memphis near gone,
> > Memphis near gone, Mistah Drinkin' Man?
> Gonna grab a pint bottle of Mountain Corn,
> > Gonna keep de stopper in my han',
> > > Gonna get a mean jag on, oh, my Lawd!
>
> Watcha gonna do when de flood roll fas',
> > Flood roll fas', Mistah Gamblin' Man?
> Gonna pick up my dice fo' one las' pass—
> > Gonna fade my way to de lucky lan',
> > > Gonna throw my las' seven—oh, my Lawd!

Heeding Ralph Ellison's oft-cited definition of the blues as "an autobiographical chronicle of personal catastrophe expressed lyrically," we can read these stanzas as performing a kind of blues heuristic, the apostrophizing call of the spirituals prompting "personal," "lyrical" responses from the various addressees.[153] Though the section opens with the pious "Mistah Preachin' Man," the exchanges tend increasingly toward the secular end of experience; eventually, the stoic constancy on playful display modulates into a no less certain, no less constant commitment to dissolution and to chance—to misters Drinkin' and Gamblin' Man. Compare Brown's use of this particular topos from the spirituals with Johnson's own exhortations in his "Judgment Day" sermon:

> Oh-o-oh, sinner,
> Where will you stand,
> In that great day when God's a-going to rain down fire?
> Oh, you gambling man—where will you stand?
> You whore-mongering man—where will you stand?
> Liars and backsliders—where will you stand,
> In that great day when God's a-going to rain down fire?[154]

Johnson's speaker intends by these inquests to work a spiritual change in the "gambling" and the "whore-mongering man," who are expected to shun the allegations of their epithets. Conversely, as Sanders observes, in the ironic economy of "Memphis Blues," these preacherly interpellations work all too well for such evangelical results.[155] Who is called the "Drinkin' Man" will continue to drink, just as surely as the "Preachin' Man" will continue to preach, end times be damned.

Henderson submits blues *style* as the distinguishing resource of these stoic responses: "The language of the poem in these six blues vignettes reflects Black life's emphasis upon style," which represents "an assertion of meaningfulness, not of frivolity or illusion": "These are not poor, benighted, superstitious foundlings in the hands of Fate . . . but ordinary blues people heroic in their own way."[156] Such a reading emphasizes the oppositional fortitude of these characters, who exercise both grit and wit in the face of devastation. One can agree with Henderson while also observing how Brown allows some ironic daylight to flicker between the affirmational avowals of these heroic figures and the poem's wider authorial frame. After all, readers familiar with Brown's critical writings of the 1930s would be hard pressed to deny a slightly discomfiting resemblance between these six stylized mock-ups of Black life and Brown's own pathbreaking analysis of literary stereotype. In "Negro Characters as Seen by White Authors," published a year after *Southern Road*, Brown embeds his own seven-pronged taxonomy of racist caricatures—the Contented Slave, the Tragic Mulatto, the Exotic Primitive, and so on—in an account of the ideological forces sustaining these stereotypes in American letters. In addition to anticipating cultural studies critiques of media representation, the essay also presciently forestalls certain essentializing tendencies in the poet's own academic reception, since the burden of Brown's argument is not only to condemn narrow, white-authored depictions of Black experience but also "to show how obviously dangerous it is to rely upon literary artists when they advance themselves as sociologists and ethnographers," those writers "too anxious to have it said, 'Here is *the* Negro,' rather than here are a few Negroes whom I have seen."[157] By the lights of Brown's critical realism, "the sincere, sensitive artist . . . will be wary of confining a race's entire character to a half-dozen narrow grooves," and this fact should caution our approach to the poem's six ostensibly heroic blues respondents. Characteristically, we locate Brown once again probing that social margin between what counts as oppositional style and what amounts to oppressive stereotype, between what stoically resists catastrophe and what is itself the catastrophic travesty of a mass-mediated cultural imaginary.

After this raucous bluesification of the folk sermon in section 2, the poem returns to the plaintive stoicism of its opening, reinvoking the "sperrichals":

> Memphis go
> By Flood or Flame;
> Nigger won't worry
> All de same—
> Memphis go
> Memphis come back,
> Ain' no skin
> Off de nigger's back.
> All dese cities
> Ashes, rust . . .
> De win' sing sperrichals
> Through deir dus'.

This duple chant reasserts the poem's an-apocalyptic vision of history—that is, history as an endless cycle of end times, without transcendence or revelation—though now this vision is tinged with the ironic, slightly bitterer note picked up in the previous blues parody. Sanders locates an inference of indomitable "black continuity" here: "The final stanza makes explicit the dichotomy between mutability and permanence . . . signaling an African American presence beyond God's judgment and one independent of (perhaps transcending) hegemonic forces."[158] But the striking manner in which Brown figures this "independence" ends up raising quite conspicuously the whole question of emancipation. The charged deployment of the idiomatic phrase *no skin off my back*—of only recent currency in the early 1930s—cannot but index the brutalities of slavery and its ongoing trauma, just as it recalls the obvious fact that, like the ancient Egyptian capital, King Cotton's Memphis was built on the backs of slaves.[159] The idiom prepares readers to countenance the full perplexity of the poem's concluding refrain, "De win' sing [wincing?] sperrichals / Through deir dus'," an image that complicates any straightforward relationship between those "hegemonic forces" wreaking ruin and an "African American presence" that survives nonetheless. The lines echo and retool W. E. B. Du Bois's description of the sorrow songs as "the siftings of centuries," orally composed in transmission from generation to generation by the "black and unknown bards" famously celebrated by Johnson.[160] As if to emphasize these "seemingly miraculous" origins, Brown's figure attributes the songs' anonymous authorship not even to any collective source but to the wind itself, though as a consequence the spirituals lose all substantiality as cultural artifacts, no longer the objects of centuries-long "siftings" but the very process of sifting itself, and so less particularized, less *there*, than even that sifted "dust" they sing through.[161]

Stranded in this wasteland where only the spirituals live on beyond all ruin, one might be tempted to read "Memphis Blues" as a rejoinder to T. S. Eliot's monumental modernism, in which "[t]he wind / Crosses the brown land, unheard."[162] In conclusion, I would like instead to propose another literary referent for Brown's final lines, one that permits a return to the matter of Corwin's radio. Brown's image recalls nothing so much as the Romantic trope of the aeolian harp, that agentless instrument played by inspiring winds. In "Memphis Blues," it's the dust of civilization that figures this lyre, suggesting that only "through," or by means of, these sounding ruins can music be made at all. Herein lies the problem with claiming for Black cultural expression an "independence" from historical time. The spirituals do not transcend their past or their present—they are not impervious to the passage of history; rather, they testify, with Saidiya Hartman, to how "the distinction between the past and the present founders on the interminable grief engendered by slavery and its aftermath."[163] All models of history, even those patterned nonprogressively—neither typologically nor cyclically— prove insufficient to account for the understanding of history on which Brown's poetics appears so restively to rest. In this final image, then, "Memphis Blues" leaves its readers with a vision of a kind of cultural product that improbably transpires out of its own ceaseless processing of its relation to historical trauma.

The appearance of "Memphis Blues" on *Words without Music* discloses the *media*-historical dimensions of this trauma. We cannot hear the orchestrated remediation that Corwin produced when he scripted the poem for his nine-person chorus, any transcription recording having long gone the way of all Memphises. On the page, the dialected "sperrichals" hints at the foundational deformations conditioning Brown's poetry, those historical ruins he is howling through, though we cannot know exactly how it was sounded out for CBS listeners in 1939. Most everything we *do* know about golden age broadcasting would lead us to fear that the subtle cultural ironies of "Memphis Blues" were flattened in the performance of Corwin's version, that when heard alongside Russell's "De Fust Banjo" and Johnson's sermons, the poem was pressed into the service of an ultimately limited, racist caricature of Black religious life, scarcely different from the preprogrammed response of "Mistah Preachin' Man." But perhaps not. My goal has been to show that a certain dynamic "phonographic" adroitness characterizes Brown's cultural poetics, and that this critical fitness for making art that directly engaged the media ecology laying down the contradictory terms of his cultural moment ought to be accounted a veritable instance of latent remediation.

Brown published his poems in print, but even page bound, this verse is a reservoir of medial experience. As indexed by their formal structures, their self-conscious use of vernacular materials, and above all by their orientation toward a sonorous "folk idiom" that is itself the acknowledged product of a varied media landscape, these print artifacts were forged in the crucible of encounters with *other* technologies for representing, reproducing, circulating, and capitalizing upon Black cultural expression. And if, as a poem latently remediated by technologized sound, "Memphis Blues" had already well-nigh anticipated its use and abuse on commercial radio, then who knows what listeners heard. Maybe just as the "sperrichals" survive on the wind, Brown's poem survives on the air.

3 Langston Hughes's Songwork

Lyric Burdens

Langston Hughes understood lyric in just the terms this book seeks to describe—that is, as a media condition. What's more, his deep practical knowledge of lyric media ordains his career an indispensable vantage point from which to survey the material history of modern lyric intermediality. But the first hurdle in substantiating this point is, admittedly, a humbling one. We must parse the spirit from the letter of Hughes's own comments on the subject, since for him, "lyric" functioned as a byword for conventional modes of poetic expression. It was the generic category under which he gathered the short, imagistic cries of interiority that marked departures from his more widely acclaimed poetry of social and racial import. Hughes memorably portrays this distinction between poems of the "roses and moonlight school" and "social poems . . . about people's problems" in his 1947 essay "My Adventures as a Social Poet": "I have never known the police of any country to show an interest in lyric poetry as such. But when poems stop talking about the moon and begin to mention poverty, trade unions, color lines, and colonies, somebody tells the police."[1] As it happens, the year 1947 also saw the publication of Hughes's first "more or less completely lyric book," *Fields of Wonder*.[2] Critics have followed Hughes himself by lamenting this "book of lyric poems" as a book of disengagements, a field of retrenchments. "In spite of its lyric ambitions," writes Arnold Rampersad, "*Fields of Wonder* negatively endorses the poetic power of Hughes's racial and political sense, which endowed him with almost his entire distinction as a poet."[3]

And yet when we frame lyric as a media condition, *Fields of Wonder* itself admits the centrality of lyric intermediality to Hughes's "distinction as a poet," "social" and otherwise. These "gloomy, brackish" poems of bruised yearning insistently figure their complaints via medial difference. Their speakers mourn the unavailability of song, probing like a wound the sundering of song's promised union of word and sound. In "Fragments," for instance, a proliferating music upsets the proper proportion of sound to articulate expression:

> Whispers
> Of springtime.
>
> Death in the night.
>
> A song
> With too many
> Tunes.[4]

The similarly epigrammatic "Burden" laments the disparity between sound and word as a matter not of "too many/Tunes" but of none at all:

> It is not weariness
> That bows me down,
> But sudden nearness
> To song without sound.[5]

The "weariness" the speaker begins by dismissing recalls us to that signature blues poem of 1925, "The Weary Blues," which by directly incorporating blues verses inaugurates the encounter between poetry and song that would occupy Hughes, in various forms, for the rest of his career.[6] But this "sudden nearness" is one the poem itself also performs. The title's pun ("burden" is another word for a refrain or chorus) points up the quatrain's own "nearness," as a ballad stanza, to song. This particular meta-lyric implies that poems like "Burden" are *nearly* songs, or else near *to* song—that song will sidle suddenly up at the scene of reception. In any case, as a matter to be grieved, the poem figures this "sudden nearness" to silent song as the sufferable burden, and fundamental condition, of lyric reading. If Hughes elsewhere assumed the normative characterization of "lyric" as depoliticized meditation on nature and affective experience, "roses and moonlight," his poems themselves in this most "lyric" collection self-represent the genre instead as a matter of an unbridgeable, historically fateful distance between sound and word, song and poem.

Hughes's conception of the "burden" borne by printed poems in the shadow of veritable song is yet another thematization of lyric's constitutive intermediality. Whereas Harry Partch's speech-music explores intermediality at the level of material inscription, Hughes's encounter with popular song transposes the lyric tension between sound and writing to the level of *genre*, where modern poetry and popular music exfoliate a problematic inherited from the ballad tradition and exacerbated by a culture industry structured by racial capitalism. Indeed, to raise the issue of genre is at the same time to illuminate modern poetry's negotiations with the ruinous stumbling block of the commodity form. Hughes helps us to write the media history of twentieth-century lyric because his capacious practice forces a confrontation between poetry and that catchy commodity we call the popular song.

From the fragmented welter of Hughes's archive, this chapter develops a media-historical account of the poet's own intuition about the determining "burden" of twentieth-century poetry—its "nearness" to song. Rather than proceeding by the familiar route of Hughes's printed poems, I address poetry's vexed proximity to pop music by redressing one specific literary-critical symptom of this larger problematic: the relative neglect in Hughes scholarship of the poet's remarkably prodigious career as a practicing lyricist and collaborating songwriter. The first section of this chapter surveys several critical approaches to poetry and popular song by asking whether their salient differences are best apprehended as functions of media or genre. In the second section, I elaborate the category of *songwork*, inclusive of Hughes's song lyrics and lyric poems both, as a ruined or riven concept capable of reflecting back the social contradictions fracturing an impossible lyric whole to which, repurposing a phrase from Theodor Adorno, the torn halves of pop songs and poems simply do not add up.[7] The songwork concept will require us to recognize how the antinomies of writing and sound are mediated by the pop song's commodity form. My primary exhibits, taken up in the chapter's third section, are drawn from the voluminous song drafts preserved by Hughes for the Beinecke Library's James Weldon Johnson Memorial Collection. Tracing Hughes's sometimes inscrutable shifts in political affiliation during the 1940s, I focus especially on the song lyrics produced during the poet's wartime quest to write a patriotic hit. Daniel Tiffany's renovated notion of poetic *kitsch* provides the theoretical scaffolding for a reading of Hughes's songwriting as a historically revealing encounter between lyric and commodity forms. Out of the fragmentary wreck of this collision, Hughes crafts a super-flexible practice for navigating a racializing culture industry bent on capitalizing lyric's burdens.

Lyric/Lyrics, Media/Genre

Hughes is far from alone in his concern with the lyric "burden" of popular song. Since at least the Romantic uptake of the ballad, and even more certainly with the advent of the modern music industry, printed poetry's "nearness" to pop has provoked the cultural-political interest of poets, songwriters, literary critics, musicologists, and pop fans most of all. The question *are song lyrics poems?*—we might call it the lyric/lyrics question—proves as old as the theorization of mass culture itself. Already it was available for easy rhetorical exploitation when Clement Greenberg opened his influential 1939 indictment of popular culture, "Avant-Garde and Kitsch," by gravely announcing the scandal at hand: "One and the same civilization produces simultaneously two such different things as a poem by T. S. Eliot and a Tin Pan Alley song."[8] For Greenberg, these two artifacts are sociological ciphers, antithetical emblems of genuine art and its corrupted and corrupting mass copy. If historicist critics and cultural theorists have been working since at least the 1980s to repair the "Great Divide" that Greenberg presupposes, it has fallen more recently to scholars associated with the new modernist studies to account in fastidious detail for the demonstrable links between, say, *The Waste Land* and its pop song intertexts.[9]

To these historicist approaches we must also add those countless others undertaken by scholars of a more formalist cast who, impelled by rock, hip-hop, and the leveling protocols of postmodernity more broadly, have brought literary methods of close reading to the task of puzzling out poetry's relation to pop.[10] Diverse as they may be, these formalist approaches tend uniformly to pass over in silence the very same meta-critical question, one that reliably bedevils attempts to say much that is definitive on the subject; namely, is the relationship between modern poetry and popular song better framed as a question of *media* or of *genre*? Are songs simply remediated poems and vice versa, as a media analysis might suggest? Or is it the processes of generic recognition, our protocols for sorting song lyrics from lyric poems, that deserve chief emphasis?

Asking what ought to take critical priority—media or genre?—throws into high relief the central occlusions motivating the lyric/lyrics problem. The question directly engages a conceptual knot of assumptions about materiality and reception, the unsnarling of which devolves upon a decision about the role that generic analysis, or a heightened attention to scenes of reception, ought to play in literary-media theory. Not only is the lyric/lyrics problem irresolvable by recourse to media or genre alone, but we'll see that keeping both in view at the same time proves a curiously difficult charge. And yet the very persistence of the question—genre or media?—is what specifies our object, demanding in answer new heuristics for a lyric archive intent on including lyrics and poems both. The following pages canvass a few critical approaches to song and poetry with an eye to tracking how genre and media are handled therein. This retreading of some familiar positions will force us to postpone our address to Hughes, but a measure of meta-critical preambling is necessary to establish the concept that I call songwork, the transmedial, transgeneric category against which Hughes's achievement comes into focus.

Among formalists one finds surprisingly little disagreement when it comes to the difference between poetry and song lyrics. According to Charlotte Pence, editor of *The Poetics of American Song Lyrics*, the most reliable distinction between these "two genres . . . may be the one we intuit: a song needs music to complete it."[11] If we accept a rather minimal definition of "music" as any sonorous expression beyond speech, then Pence can be heard to echo the sensible dictum laid down by rock critic Robert Christgau in 1968: "Poems are read or said. Songs are sung." That's why even the choicest "tidbits from [Bob] Dylan's corpus . . . don't look so tasty on a paper plate."[12] For Christgau, the printed page tends to *embarrass* a song lyric, and musicologist Simon Frith explains why: "Good song lyrics are not good poems because they don't need to be: poems 'score' the performance or reading of the verse in the words themselves, words which are chosen in part because of the way they lead us on, metrically and rhythmically, by their arrangement on the page . . . Lyrics, by contrast, are 'scored' by the music itself. For a [song] lyric to include its own musical (or performing) instructions is, as Ned Rorem observes, to overdetermine its performance, to render it infantile."[13] Here Frith unpacks for us Paul Valéry's famous quip that putting a good poem to music is akin to lighting a painting with a stained-glass window.[14] Whereas song lyrics require musical accompaniment, the argument goes, poetic language achieves an autotelic plenitude by accompanying itself, prosodically. Confronted by printed song lyrics without enough prosodic interest to compensate for their absent music, Pence, Christgau, and Frith are each responding to a *medial lack* conspicuously aroused when an artwork is felt to be out of its element or ill-suited to its medium. Though this perceived lack can be described and accounted for in any number of ways (the suggestion of a missing "music" is only the most common), and while it may appear readily as a formal or generic difference (the slack or underdetermined prosody of song lyrics, for instance), it refers ultimately to the medial gap between print and sound.

It would seem, then, that these intuitive conclusions about the difference between poems and song lyrics are informed by a largely unacknowledged notion of medial difference, one that is itself indebted to midcentury theories of orality.[15] The critic Mark Booth, for his part, comes close to acknowledging this medial lack as such when he draws a cybernetically inflected contrast between the "low density of information" distinguishing song lyrics and the semantic richness afforded poetry by its print medium: "The existence of songs in sound, in time, is the simplest distinction between them and written verse"; whereas writing allows one to linger over certain words, go back, drop in, and speed up, "words only heard are not forgiving in this way."[16] Strangely forgetting the existence of recorded music, which endows listeners with just this capacity for time axis manipulation, Booth contends in any case that transcribed song lyrics reflect in their simplicity and redundancy both "the burden of oral communication" and their reliance on an absent music. Poetic writing, by contrast, exploits the permanence of the visual page to encode an informatic richness.

But medial difference ramifies in both directions. Alongside the whiff of lack suffusing transcribed song lyrics, we can cite another often-remarked deficit—namely, the poverty endured by printed poems in the light of song. If song lyrics seem insufficient next to poems, as Jahan Ramazani observes, "the comparison to sung verse also makes literary verse seem deficient in melody and harmony, the physicality of the embodied voice, and the thick social and performative contexts" of musical practices.[17] An instance of medial envy, this overdetermined desire for the condition of music is what Hughes bemoans as lyric's signature "burden." Song's sensuous charm for poets and literary critics alike is one local species of the larger and "profoundly conventional" appeal of "song" as a poetic trope.[18] I mean of course the wholly familiar manner in which *Songs of Innocence and of Experience*, "The Love Song of J. Alfred Prufrock," and *Song of the Andoumboulou* all figurally aspire to music by way of what Daniel Albright calls "pseudomorphosis": the attempt by an artwork in one medium to do the work of another medium.[19] A media theorist of the *longue durée* might identify poetry's envy of song as an effect of medial-systemic transformation: printed lyric poems are shrouded in the perduring aura—what Albright terms the "*eidolon*"—of an anterior orality in song.[20] In this account, formal and thematic gestures toward "song," including the brash anachronism of the lyre, supply poetic traditions with tropes that compensate for the alienating silence of modern print.

This wide-angle media history grows more complicated at the end of the nineteenth century, when, in Friedrich Kittler's account, the revolutionary capacities of the phonograph fully usurp poetry's ancient function as a technological supplement to human memory. Because phonographic inscription bypasses the spiritualizing sieve of the written alphabet and its "imaginary voices" with the physically indexical capture of actual voices or "the unimaginable real," Edison's invention effectively splits the historical trajectory of lyric poetry into "lowbrow and highbrow culture[s]." On the one hand, "the effects that poetry had on its audience migrate to the new lyrics of hit parades and charts," where once again "anonymous" bards cater to the "illiterate" masses. On the other hand, modern poets—Mallarmé among the first—"exorcize the imaginary voices from between the lines and inaugurate a cult of and for letter fetishists." This latter formation is twentieth-century poetry as we have come to know it, a "professional" poetry that disavows its envy of song in favor of "a new typewritten materiality" that "can do without all [the] supplementary sensualities" with which Romanticism had imbued it.[21] In Kittler's view, then, the "informatic density" of the printed poem ultimately expresses large-scale, irreversible changes in the media-technological situation, but this semantic richness is paid for by the hieratic asceticism we have come to call poetic modernism. These historical processes mean that a song lyric served up on a "paper plate" today finds itself doubly stripped—of the "supplementary sensualities" accorded nineteenth-century poetry by readers willing to entertain "imaginary voices," and also of the "sensualities" it acquired by virtue of culture-industrial accompaniment.[22]

But even when their medial dimensions are acknowledged, as in Booth, or deterministically foregrounded, as in Kittler, all the approaches surveyed above eventually run up against a rather predictable critical limit, and here Christgau's gustatory trope gets directly to the point: efforts to locate the medial lack animating the lyric/lyrics question on the page—as a matter of relative formal complexity or semantic density—run the permanent risk of merely reflecting a critic's ideologically refined "taste" for a certain kind of poetic fare. The measure of a category like "density of information" is, after all, constructed finally in *reception*, and one would be hard pressed to argue that the same semantic economy characterizing the relation between, say, *The*

Waste Land and "That Shakespearean Rag" obtains for a poem by Allen Ginsberg or Amiri Baraka or Sharon Olds when held up against a set of hip-hop verses; the informatic density of the latter may very likely trump the former, but that doesn't mean the former is any less a poem and the latter any less a song lyric.

Kittler's literary history also fails to account for developments in lyric intermediality such as those discussed in the previous chapter, in which poems were seen to reflect, index, and critique the social construction of technological forms. Media-focused frameworks are instructive, then, insofar as they draw our attention to the medial lack circulating between poems and songs, and because they lend a finer material grain to a historicizing position that might otherwise throw up its hands at the sublime variety of historical practices around poetry and popular song in the twentieth century. But as soon as one emphasizes medial difference, whether in the psycho-aesthetic forms of lack or envy or as the symptoms of large-scale media-systemic change, the particularizing matter of *reception* emerges to scuttle any conclusion that ignores the significance and even the priority of a specific interpreting audience. We must deal again, in other words, with determinations of *genre*. By heeding its relevant return, we bear witness to the second movement of a meta-critical dialectic intrinsic to the lyric/lyrics problem: the difference between a poem and a song lyric may first present itself as a question of genre, but generic descriptions are quickly seen to mask dynamics of medial difference, which are then discovered to be structured in reception in such a way as to renew the issue of generic recognition. The distinctions between poems and song texts that win collective assent are specific and socially contingent, though that contingency includes—even if it is not determined by—the absence of music or the presence of print.

Hughes's songwork will require us to dwell in the dizzying vicissitudes of this dialectic, ping-ponging between matters of reception and medial makeup. As critical lenses, genre and media interpenetrate, of course, and sufficiently robust theorizations of either will likely entail the other by supplying an account of its necessity. We readily accept that material media play a constitutive role in the life of genres when the latter are grasped not as formal or artifactual strictures but as forms of "social action" that render cultural practices legible and repeatable.[23] This is to understand genres, with Lisa Gitelman, as "modes of recognition," or "socially realized sites and segments of coherence within the discursive field."[24] In a similar fashion, and to borrow again from Gitelman, when we conceive the media concept to involve not only technological forms but also, and just as centrally, a medium's associated "social protocols," we also encompass the *genres* of social practice that construct and pragmatically sustain media as "socially realized structures of communication."[25] It follows from this conceptual rapport that to examine an artwork by toggling between generic and media-specific analyses ought to be as simple as shifting one's footing back and forth, from a view that discloses the social mediations of reception to one that emphasizes the medium's materiality. And yet those who consider the subject of poetry's relationship to song from a formalist perspective often end up insisting on either media or genre at the expense of the other: song lyrics are likened to (or distinguished from) lyric poems on the basis of how they do (or do not) conform to our ideas about popular song and poetry, or else their relation is explained, at bottom, as an expression of medial differences between writing and various forms of sound.

Why is it structurally so difficult to keep both medial and generic analyses in view simultaneously? For one, as soon as the social-pragmatic conception of genre introduced above gives way to a more familiar notion of a genre system, poetry and popular song retreat by long cultural habit to fully asymmetrical positions. The "Great Divide" rears up, in other words, and song lyrics become "paraliterary" again. In Samuel R. Delaney's formulation, the paraliterary designates "those texts which the most uncritical literary reader would describe as 'just not 'literature,'" but which nevertheless form the constitutive outside of the literary itself, from which they are excluded by the "material practice[s] of social division" endemic to classrooms, the book market, prize committees, and the like.[26] Delaney traces the "literary/paraliterary" distinction, an "abyssal split," to print-technological developments in the late nineteenth century, though Kittler reminds us that in poetry's case specifically the contemporaneous emergence of sound recording is at least as significant.[27] A signal effect of the literary/paraliterary split and its uneven distribution of cultural capital is that poets and readers come to invest medial difference with just those psychological valences—*envy* and *lack*—sighted above.

Meredith L. McGill has suggested that prior to the twentieth century it was the *ballad* that formed the paraliterary outside to literary poetry, its "counterpart or shadow image."[28] After the advent of commercial sound recording and most certainly on the other side of rock music, hip-hop, and cultural studies, these tensions between the paraliterary ballad (thickly mediated and socially embedded) and the literary lyric (transcendent and formally expressive) are subsumed inside a more capacious, tendentiously transmedial concept of lyric itself. Recent scholarship has productively framed lyric as a problematically "super-sized" modern poetic genre, more the abstract product of literary criticism than its object of study.[29] It remains to be observed, nevertheless, that the poetry we have called lyric in the twentieth and twenty-first centuries also shoulders the "burden" of an even larger, and more demonstrably fantasized, lyric genre—one inclusive of short poems and popular songs both. The habit of referring to popular song texts *as* lyrics announces both the plausibility and appeal of this overarching, transmedial genre, even though the unity implied by the shared term "lyric" is only an afterimage of fundamentally divergent cultural practices. However specious and ideologically laden it may be, poets like Langston Hughes work quite consciously in the "ruins," as it were, of this conception of lyric. Songwork will be our name for the way such poets negotiate this generic ambition, passing its "burden" back and forth from poem to song across media, and across more local generic boundaries. This chapter posits an object of study that doesn't really exist but one to whose possible figure, emerging with gestalt-like integrity, lyric readers and poets like Hughes nonetheless cleave. This transmedial lyric genre persists as the negative sum of the discrete social and aesthetic practices that seem inevitably to constitute it—collaborations between poets and composers, the writing of poems that masquerade as song lyrics, the writing of songs that double as printed poems—what I term the traffic of songwork.

The Traffic of Songwork

Langston Hughes may be the twentieth-century poet most keenly sensitive to the dynamic relations between song lyrics and lyric writing. Throughout his career, from his earliest blues poems of the 1920s to his midcentury jazz masterworks, *Montage of a Dream Deferred* (1951) and *Ask Your Mama: 12 Moods for Jazz* (1961), Hughes wrote his poetry by worrying the "sudden nearness" of song to poetry and by paying heed to the burdensome shadows cast upon his printed verse by adjacent lyric media like scores, phonograph records, radio, and tape. As a result, Hughes's practice poses an irreducible challenge to the critical approaches surveyed above, along just those lines we have sighted: his lyric career testifies to the riddlingly complex interpenetration of genre and media entailed by the question of poetry's relationship to song. While Hughes is far from the only twentieth-century poet to write both songs and poems, and far from the only poet to refuse any essential distinctions between the two, no other poet maintained such constant traffic between these categories of lyric production. In confronting Hughes, we need a new vocabulary with which to describe a lyric practice that remains, even today, unassimilable to the disciplinary grooves of literary and cultural history.

For a quick entrée into what I will shortly term Hughes's songwork, consider "Hard Luck," one of the seventeen poems "written after the manner of the Negro folk-songs known as *Blues*" that scandalized the Black bourgeois establishment when they appeared in Hughes's second collection, *Fine Clothes to the Jew* (1927):

> When hard luck overtakes you
> Nothin' for you to do.
> When hard luck overtakes you
> Nothin' for you to do.
> Gather up yo' fine clothes
> An' sell 'em to de Jew.
>
> Jew takes yo' fine clothes,
> Gives you a dollar an' a half.
> Jew takes yo' fine clothes,
> Gives you a dollar an' a half.
> Go to de bootleg's,
> Git some gin to make you laugh.
>
> If I was a mule I'd
> Git me a wagon to haul.
> If I was a mule I'd
> Git a wagon to haul.
> I'm so low-down I
> Ain't even got a stall.[30]

Like all Hughes's blues poems, and indeed like the poems by Sterling Brown taken up in the previous chapter, "Hard Luck" is the synthetic product of a historically specific process of cultural mediation. The poem formally reflects contemporary struggles over the blues as an aesthetic practice in the throes of commercialization. Specifically, *writing* the blues in this instance involves not merely the (re)mediation into print of an oral form but also a self-conscious negotiation between the compositional practices of two genres vying for dominance in the 1920s: an improvisational or piecemeal "folk" mode in which the song stanza is the primary unit of composition, and a nascent cultural-industrial practice associated with the "classic" or commercial blues, which privileges coherence at the level of the popular, recorded song.[31] The clearest index of the former, the rather abrupt shift in point of view and imagery between the second and third stanzas, also encodes the geographical distance between these two logics of construction, as the urban pawn shop gives way to the rural mule and wagon. Setting the northern, more commercial iterations of the blues in subtle tension with its southern "folk" provenance, the poem reverses the path of the Great Migration while also rehearsing—at the level of form—the itineracy and homelessness with which the poem concludes.

Attention to these competing subgenres of blues music clarifies the radical achievement of Hughes's blues poems. The three- or six-line blues stanza did not exist as a printed poetic form prior to 1927. And yet to claim, as most critics will, that Hughes *invents* the blues poem as a poetic structure is to move too swiftly past a significant question: What does poetic invention mean vis-à-vis other media for writing, playing, hearing, collecting, and studying the blues in the mid-1920s—most germanely, perhaps, Tin Pan Alley or "classic blues" compositions, phonograph records proliferating in the explosive "race record" boom, and folklorists' transcriptions?[32] What does it mean, in other words, to *write* contemporary *song*—to entextualize a fragment of this wider cultural ferment by recording, in printed verse, a song lyric shorn of music?

The audacity of Hughes's blues poems may be somewhat diluted by readers' long familiarity with the intermedial operations of literary ballads. Like the ballads so zealously imitated, expropriated, copied, and forged during the ballad revivals of the eighteenth and nineteenth century, blues poetry is after all a "distressed genre," Susan Stewart's term for the self-conscious, historicizing artifactuality that stamps literary imitations of folk forms.[33] In her own work on balladry and the "romance of orality," Maureen N. McLane has paid exhaustive attention to the complex "mediality" of the distressed ballad and its significance for British poetry.[34] McLane documents how the ballad was constructed at the end of the eighteenth century as a multimedial object of study by antiquarians and collectors called to the collective discursive effort of "balladeering" by a range of scholarly, commercial, and cultural-nationalist interests. From McLane's ballad collectors of the Scottish highlands like Sir Walter Scott and Robert Jamieson, who pioneered the conventions of fieldwork and citation that established oral informants as privileged sites of cultural authentication, we can trace an analogy (if not also a disciplinary lineage) to the ethnological efforts of those responsible for establishing the "folk" blues self-evidently as a cultural referent, from New Negro intellectuals like James Weldon Johnson and Zora Neale Hurston to white ethno-musicologists and folklorists like John and Alan Lomax.[35] Hughes's poetic adaptation of the blues in the 1920s was made possible by (and perhaps also contributed to) the construction of the blues itself via so many variegated processes of collection, performance, and capitalization.

But the analogy between ballads and the blues highlights a telling contrast. What ought to strike us in poems like "Hard Luck" is the pointed absence of evidence of the cultural work McLane calls "balladeering." In the hands of Romantic poets like Wordsworth, ballads and tropes of balladeering become staging grounds for "complex encounters between oral and literary *poiesis*" in which the "romance of orality" comes in for conspicuous representation and even theorization.[36] Lyrical ballads like "Simon Lee" and "The Solitary Reaper" foreground the supercharged *distance* between a poet-informer and their rustic, ostensibly illiterate subjects.[37] In the former poem, the work of cultural mediation intercedes in the guise of the "gentle reader" to whom the poet-informer addresses the moral burden of his encounter with the ailing huntsman. In the latter, the legacy of Romantic balladeering, its cultural politics, and the investments in a notional orality that those politics underwrite are all signed and certified by stratifications of class and language, and by the speaker's aestheticizing incomprehension of the "solitary Highland Lass" singing in the fields—"Will no one tell me what she sings?"

We can identify a similar structure of vexed identification and performative appropriation in Hughes's first blues poem so called, "The Weary Blues," where the captivating spectacle of a Harlem blues performance intimates a sympathetic merger of identities between a racially ambiguous speaker and the bluesman he has come uptown to hear. The poem includes quoted verses from a "sad raggy tune"—the very first blues song, purportedly, that Hughes heard as a child in Kansas—though the speaker's distance from "those Weary Blues" remains enforced by a pentameter frame.[38] By contrast, less than two years later, Hughes takes the crucial step and begins—in David E. Chinitz's phrase—actually "*writing* blues instead of writing *about* blues."[39] In *Fine Clothes to the Jew*, the Romantic "I" or poet-as-informer drops away, such that disappointed, suicidal lovers and hard-luck bad men sing in dialect for themselves, in printed stanzas mimicking a lyrics sheet or blues transcription. Save for the short prefatory "A Note on Blues" that goes only so far as to inform readers that "the first eight and last nine poems in this book are written after the manner of the Negro folk-songs know as *Blues*," which prescribe a "strict poetic pattern" and a paradoxical "mood" ("almost always despondency, but when they are sung people laugh"), *Fine Clothes* lacks the obvious traces of what we might call, after McLane, blues "balladeering": the groundwork of performing an investment in the blues as cultural referent and expressive practice, the necessary preparation, we might assume, for asserting the blues as a poetic genre.[40]

The absence of this groundwork—either in the form of frame narrators or a paratextual apparatus—means that Hughes's critics continue to argue over just what *kind* of blues, "classic" or "folk," Hughes meant to imitate. And regardless of the controversy attaching to the blues itself in the 1920s, surely one reason *Fine Clothes* provoked such scandal upon its release was that readers were relatively unprepared for its claim to a new genre. In writing his *Lyrical Ballads* (1798), Wordsworth had recourse not only to the live ballad culture of the working class and the contemporary literary vogue for peasant poetry but also to a concept of the literary ballad forged by the culturally accrediting work of proto-ethnographers, historians, collectors, imitators, and forgers of traditional English ballads.[41] With Hughes, by contrast, we witness a strikingly sped-up version of the "distressing" process by which ballads were artifactualized: the poet wrests a written form from an oral tradition only recently subjected to collection, documentation, and study.

What accounts for this accelerated process? What authorizes Hughes to pick a song from the air and turn it, ex nihilo, into a poetic genre? Of the many differences between Wordsworth circa 1800 and Hughes circa 1925, the latter's experience of a rapidly expanding commercial music industry seems particularly decisive. Record companies first awoke to an untapped market of Black audiences in 1920 when Mamie Smith recorded "Crazy Blues" for OKeh Records; by 1927, the year of *Fine Clothes'* release, Jeff Titon estimates that "10 million race records were purchased by African Americans . . . one for every black person living in the United States at that time."[42] In short, Hughes could eschew the paratextual trappings and frames of blues balladeering because *pop music* had arrived to render his poems legible to his readers—if not as poems, exactly, then as the words to songs. Whereas establishing the literary ballad involved supercharging the cultural distance between poetry and oral balladry, borrowing the forms and language of traditional ballads, and thematizing encounters with figures of Romantic orality, Hughes created his new poetic genre by a much simpler, though somewhat contradictory, cultural operation: he simply wrote song lyrics.

Needless to say, Kittler's sweeping account of media differentiation shines scant light on this convergence between poem and song. "Hard Luck" precisely does not beat a retreat to the typographical materiality of the page—unless it does so negatively, the sensuous poverty of the printed page thrown up by the poem's aspirations toward blues song. Neither do the formalist approaches mentioned above explain much about Hughes's blues poems. To be sure, generations of critics have resourcefully substantiated these poems' sufficiency and formal interest on the page, but these readings proceed against the grain of the fact that poems like "Hard Luck" aim directly to travesty the distinction between lyric poem and song lyric. To specify all the ways in which, as Alain Locke cautioned long ago, "these poems are not transcriptions"—to notice, that is, how literary entextualization and Hughes's "self-concealing poetic craft" have fashioned blues material into hermeneutically assailable artworks—is not fundamentally to alter the fact that "Hard Luck" *could* be a song lyric.[43]

As indeed it was. Composer Herbert Kingsley set the poem to music in 1938 for a suite of dance works by choreographer Felicia Sorel (fig. 3.1).[44] In June of that year, Kingsley and Sorel appeared with Hughes himself and W. C. Handy for a "Blues Symposium" on WQXR's *Music and Ballet* series. There, for a small in-studio audience, the white Kingsley performed his own revised version of "Hard Luck" to Sorel's accompanying dance. In the epitomizing description of the program's host, Sorel's invented character, "Mr. L," is by song's end staggering and hiccupping in grotesque minstrel fashion. As an absent signifier of race, which radio's acousmatic medium must anxiously materialize in sound, the invisible Mr. L supplies an object lesson on the racializing force of the sonic color line.[45] Kingsley's performance also offers an example—an especially ugly though not necessarily atypical one—of the painful compromises involved in the career-long traffic between Hughes's poetry and his musical collaborations.[46]

I'll have more to say about Kingsley in a moment. For now, I would like to emphasize that his setting represents just one specific, concrete realization of a categorical *possibility* attaching to the blues poem as such. In his subtle charting of the blues poetics at work in Hughes and Sterling Brown, Brent Hayes Edwards has discussed poems like "Hard Luck" in terms of a "poetics of transcription," which operates by "formal mimicry" of blues

Figure 3.1. Langston Hughes's "Hard Luck Blues," n.d., with adapted words and musical setting by Herbert Kingsley, undated. Box 394, Folder 7496, Langston Hughes Papers, James Weldon Johnson Memorial Collection, Beinecke Rare Book and Manuscript Library. Courtesy of the Langston Hughes Estate

music to instantiate an "unresolved tension" between the notion of poem as "score"—the poem awaiting its "music" in performance—and the more unexpected notion of the poem as "transcription"—calling out in "formal apostrophe" to the "absent music" forever in its wake.[47] One merit of Edwards's theorization is to show how a blues poem manages to reincorporate, as a *formal effect*, its own medial lack, or its own "distressed" character with respect to the song it desires to be. Both a transcription and a score, a printed blues poem is poised between imagined musics of the past and future. We can plausibly extrapolate this formal dynamic to sketch the relations that any blues poem maintains, if only proleptically, with the historical media ecologies of which it's surely a part: just as "Hard Luck" is latently mediated by the cultural marketplaces of which it's a fragment—a blues ecology dominated by phonograph records and folklorist transcriptions—so too is the poem anticipatively mediated by its potential settings and remediations, its future accompaniments. Even on the page, a poem like "Hard Luck," a "song without sound," is also a "song/With too many/Tunes," including Kingsley's.

Tracing this verbal matter across the borders of song and poem underlines the knotty meta-critical dialectic of genre and media described above: the generic recognition of a poem or song lyric is conditioned by its medium, but the difference that medium makes is itself the product of a socially instituted generic regard. "Hard Luck" began its life as a printed poem pretending to be an entextualized song lyric. To observe as much is already to exercise a few assumptions about song lyrics and poetry—namely, that *Fine Clothes to the Jew* is a book of poems, and song lyrics require musical accompaniment; to call the blues poem an *intermedial* genre is to publish those assumptions under the sign of medial difference. When "Hard Luck" becomes an *actual* blues lyric consequent to Kingsley's remediation for musical score and radio, a veritable change of medium underwrites a generic shift—print into sound and poem into song. But remediation alone cannot account for the transformations in meaning underway when Kingsley sounds Hughes's text for radio. Generic recognition—that is, interpretation at the scene of reception—clearly shapes a listener's regard for "Hard Luck." On WQXR, the *Music and Ballet* host expressly distinguishes the songs by W. C. Handy—"authentic" blues—from Hughes's poems, presumably because the latter have been set to white-imagined music and dance.[48] When Hughes's lyrics are set to music on the radio, meaningful differences in media format and generic uptake conspire. Then and now, listeners to Kingsley's "Hard

Luck" hear something different, but they also hear that something differently too.

As a genre formulated ex nihilo, staked on and riven by its transmedial character, the blues poem rather clearly explicates the tensions structuring what I term Hughes's songwork. But the songwork concept does not extend only to the song-like poetry of this "most musical poet of the twentieth century."[49] This speculative category aims to gather the busy traffic *between* this poetry and the various song forms in which Hughes worked throughout his career. In addition to the myriad poems set to music by a broad stable of composers working with and without the poet's express collaboration, Hughes also penned song verse for an astonishing number and variety of theatrical productions—including revues, musicals, cantatas, operas, and of course the sui generis gospel plays of the 1950s—as well as for hundreds of individual popular songs.[50] "I write my lyrics like I do my poetry," he claimed, and meant it, as we can see from the numerous lyrics that served double duty as poems and songs, remediated back and forth from books and magazines to songsheets and records.[51] As easily as "Hard Luck" could acquire musical accompaniment, or a poem like "Ballad of the Girl Whose Name Is Mud" could be parlayed into a pop song for Burl Ives, in the course of his songwriting Hughes would sometimes spin off a "poem version" of a lyric he otherwise intended for music.[52] This transformative, instrumental use of lyric material is the substance of songwork.

By any standard, Hughes's songwriting represents a major dimension of his prodigious output as a poet, novelist and short-story writer, dramatist, children's author, journalist, critic, anthologist, and translator. If it has largely escaped the attention of scholars, we may attribute that neglect to the unmistakeably mercenary character of this work: addressing Hughes's song verse demands a reckoning with the most "professional," most overtly commercial aspects of his career.[53] The transgeneric, transmedial operations of Hughes's songwork cannot be frankly appraised until we admit and specify the significance of the cultural-industrial marketplace to Hughes's lyric career. In 1950, the poet's friend Arna Bontemps inaugurated the tradition of crediting Hughes as the first "Negro poet since [Paul Laurence] Dunbar who has succeeded in making a living from poetry."[54] Tellingly, Bontemps feels immediately impelled to qualify what he means by "poetry," since "a poem must be used in many ways to yield enough sustenance to keep a hearty individual like Mr. Hughes in the kind of food he likes. Therefore it is not surprising to find his poems being danced by Pearl Primus on the stage while they are sung by Juanita Hall in night clubs and on radio and television and by Muriel Rahn in Town Hall concerts and while Paul Robeson is reciting 'Freedom Train' in the United States, the West Indies and Central America."[55] To make his poetry pay, Hughes trafficked his lyrics across generic and medial borders, and the cultural distance these lyrics traveled—transforming all the while in the light of various paying publics—marks out the wide territory of songwork, one coextensive with the international reach of Hughes's career more broadly.

It was the American Society of Composers, Authors and Publishers (ASCAP), as the institution responsible for monetizing this traffic, that ordained songwork a viable financial endeavor. "The Lord has blessed me with ASCAP," Hughes exclaimed in 1941.[56] After applying for membership in the mid-1930s, ASCAP royalties were an increasingly significant income source as his rating with the organization—and his payment rates—rose steadily throughout the 1940s. Instructively, assignations of copyright are aesthetically blind, respecting no difference between song lyrics and lyric poems. In Hughes's lengthy ASCAP repertory, art song settings of the classic anthology piece, "The Negro Speaks of Rivers," for example, are listed alphabetically right alongside Hughes's 1942 pop tune "A New Wind a-Blowin." If we understand Hughes's songwork to designate the fragmentary, heterogenous material archive corresponding to an impossible lyric whole—to a notion of twentieth-century lyric that would somehow comprehend both popular song and poem—it's tempting to regard ASCAP, after a fashion, as perhaps the only perfect reader of this discrepant archive. And if we grant that what's at stake in the traffic of songwork, in the various "uses" to which a lyric can be put, are the possibilities of commodification proper to lyric, then one might go further still and claim that ASCAP's indifference to the lyric/lyrics question supplies an image of how the pop song's commodity form mediates the generic and medial differences between poem and song.

What ASCAP does *not* reveal, however, is the way Hughes's songwork contends with a culture industry structured in racist domination. We must address his blues poetry in particular in the context of one intensely contested cultural form, that is, the "classic" or "commercial" blues of women artists like Bessie Smith and Ma Rainey.[57] The "race records" boom in the 1920s saw white-owned record companies exploiting Black performers and capitalizing on discourses of racialized authenticity and the objectification of Black female sexuality.[58] At the same time, critics like Hazel V. Carby and Angela Y. Davis have taught us in what sense the classic blues was also a space of "representational freedom" and cultural-political struggle, one where women's bodies were "reclaimed . . . as the sexual and sensual objects of women's song."[59] The commercial blues sounds out from a double bind of exploitation and expression.

Affixed to the printed page, Hughes's blues poetry formally refracts this struggle. We have seen how "Hard Luck" yokes together two divergent strategies of composition, disrupting the sculpted coherence of the industrial pop song with a nod to the patchwork assembly of an oral folk tradition. Overlaying this formal structure, raveling pain with stolen joy, a host of fine ironies speaks to the uneven processes by which blues music is commercialized. "When hard luck overtakes you," your goods are bought up cheap in racialized exchange ("the Jew," for pawnbroker, was a bit of Harlem slang Hughes would come to regret adopting for his title).[60] The only relief available comes by way of wry self-arraignment. In "Hard Luck," the speaker knows they're getting a raw deal from the pawnbroker, but they also know they're making it worse by using the $1.50 not to pay bills but to buy "some gin to make [them] laugh." In the face of foreclosed agency—"Nothin' for you to do"—the only course of action is the defiant, jubilant embrace of misfortune. Hence the ludicrous resignation of the last stanza, where further burdens are not so much shouldered as self-punishingly, sardonically sought out: "If I was a mule I'd / Git me a wagon to haul." In drawing from "Hard Luck" the title for his collection of blues poems, Hughes flaunts the allegory: if blues songs can redeem the experience of hard luck, the hard luck of an exploitative cultural marketplace prevents the songs themselves from easily transcending those conditions in circulation; one must sell those fine blues cheap and then "laugh to keep from crying."

Whether or not we accept the allegorical reading, Kingsley's setting for radio does its best to scuttle Hughes's craft. The latter's well-modulated ironies depend on the gin and laughter arriving in the second stanza, where they prepare the entrance of the winking first person and their subjunctive self-identification as a mule. Kingsley, by contrast, rearranges the stanzas, beginning with the line "If I was a mule," which now seems more a dehumanizing joke than a bit of mordant self-mockery. Meanwhile a new fourth stanza strains to lend the poem a pat coherence but lands it, instead, in clichéd nonsense:

I'm stubborn as a mule
when I'm full of gin,

> Stubborn as a mule
> when I'm full of gin.
> But I sold my clothes
> so look what a hole I'm in[61]

Just a few years later, in "Note on Commercial Theatre," Hughes would lament how white appropriators had "taken my blues and gone— . . . you mixed 'em up with symphonies / And you fixed 'em / So they don't sound like me."[62] Kingsley's "fix" is itself a fixture of the racially capitalized cultural landscape, one primed to mar Hughes's lyrics when they leave the page and accrue the motility of song. This particular dimension of songwork's burden has as much to do with racializing uptake as it does with poetry's nearness to sound, but as we'll see, Hughes's extremely fungible lyric practice—his way of *doing* songwork for pay across lyric media—is designed to unfix just these deforming strictures.

Wartime Kitsch and the Dummy Lyric

Because Hughes's 1940s do not lend themselves to easy biographical or cultural-political shorthand—unlike, say, the Harlem Renaissance of the 1920s, his left radicalism in the 1930s, or the civil rights struggles and intensifying international-diasporic consciousness of the 1950s and 1960s—it can escape notice just how extraordinarily productive these years were for the poet. Not only did he publish as much poetry in the 1940s as in any other decade of his career, but Hughes toured and lectured at a relentless pace, committed himself to the multiform patterns of "literary sharecropping" that would characterize the rest of his professional life (including the beginning of his weekly *Chicago Defender* columns, numerous musicals and other dramatic productions), and with major translation and anthology projects continued to lend his hand to the forging of literary networks across the Black diaspora.[63] To the extent that his 1940s *do* signify in the biographical abstract, many readers have followed Rampersad's influential account of Hughes's "public repudiation of his alignment with radical socialism" in 1941, after which year Hughes began his long retreat from the left, embracing what has been called a "poetics of indirection" or "ethics of compromise."[64] A wider-angle view of Hughes's literary activities in the 1940s does not necessarily contravene these ascriptions of an emergent political quiescence, but it does complicate the picture. Recognizing the sheer breadth of Hughes's cultural production enjoins us to attend in a reparative spirit to the ways a certain politics may work itself out by means other than protest, and in unlikely places.

One such place is Hughes's songwork. Throughout the decade he invested considerable time and effort in writing lyrics with and for composers across the musical spectrum, including Margaret Bonds, Elie Siegmeister, Duke Ellington, William Grant Still, Elliot Carpenter, Kurt Weill, Emerson and Toy Harper, and W. C. Handy. In the early 1940s, Hughes's songwriting efforts were directed to a goal that may seem rather unlikely given his past political affiliations and his erstwhile anti-war position: he aimed to pen a patriotic hit, a morale-building tune to galvanize the stateside war effort and inspire GIs abroad, as George M. Cohan's anthem "Over There" had managed during the First World War. A year prior to Pearl Harbor, and just six months after he had felt inclined to hold the US Communist Party's line by endorsing an anti-war statement by the League of American Writers, Hughes anticipated his own later campaign with "America's Young Black Joe," a "kind of Negro GOD BLESS AMERICA," written with Elliot Carpenter.[65] But wholehearted participation in Tin Pan Alley's "quest for the Great American War Song of World War II" would have to wait until 1942.[66] That summer, at the newly integrated Yaddo, the exclusive artist's community in Saratoga Springs where Hughes was the first Black writer to take up residence, he devoted himself to churning out fighting songs like "Go-and-Get-the-Enemy Blues," flag-wavers like "Rights of Democracy," and war-themed novelty numbers like "Honolulu Yaka-Hula Dixie," in which a "Singing Sam/From Alabam," a GI stationed in Hawaii, "[m]eets Yaka Lula Acky-Oues" and the two harmonize their respective musics with the goal of letting "drop those 'Honolulu Yaka-Hula Dixie Combination Blues.'/I've got to fight and this war has got me all enthused" (fig. 3.2).[67]

This patriotic songwork represents only one facet of Hughes's more various contributions to the war effort: he also served on the Writers' War Board, lent his pen to Treasury Bond drives and the Office of Civilian Defense, contributed work to mass rallies and pageants like 1943's "For This We Fight" at Madison Square Garden, composed radio propaganda like *The Man Who Went to War*, a 1944 ballad opera for the BBC, and kept up a demanding schedule of speaking engagements across the country. Hughes undertook much of this work under the polemical auspices of the Double V campaign—victory against fascism abroad and Jim Crow at home—promulgated in the mainstream Black press.[68] He left his starkest articulation of this position in the 1943 poetry pamphlet *Jim Crow's Last Stand*, as well as

```
                                                    1st draft,
                                                    Yaddo,
        When Singing Sam                            August 6, 1942.

        From Alabam

        Meets Ya-Ka-Lu-La-Ack-Ay-Oues,
miss
        MMM Ya-Ka-Lu-La

        Starts to hum a hula,
                 croons
        MMMM Sam/those Dixie blues.
   and          at Waikiki
  their catch
        When those two tunes blend,

        The melody is guaranteed to send:

        Oh, those Honolulu Ya-ka-hu-la Combination Dixie Blues.

        A southern pal, Hawaian gal, a moon---and what you got to loose?
                                                         low.
        Guitars playing, grass skirts swaying, dusky voice singing blues
                                             but, honey babe,
        Hate to leave you, know 'twill grieve you, but I got to go.
Repeat
        Uncle Sammy, and my Mammy, and the folks in Alabamy
                      out
        Sent me here to get the fmax enemy---

        So Miss Yaka Lula, when I come back, Miss Yaka Lula,

        When x I x came x back x I think x any x your x hula
        I'll have time to learn your hula,
        And you can learn the blues from me.
```

now I've got to fight,
Take wrong and set it right,
This war has got me all enthused.

```
        When there's no more stormy weather
              our two songs
        We'll put/them two together

        In a Honolulu Yaka-Hula Dixie Combination Dixie Blues.



             blues         news
             coos          pews
             dues          rues
             dews          strews           Langston Hughes
             enthuse       two's
             fuse          too's
             goos          views
             loose         whose
             news          who's
                           youse
```

Figure 3.2. Worksheet for Langston Hughes's "Honolulu Yaka-Hula Dixie," August 6, 1942. Box 394, Folder 7527, Langston Hughes Papers, James Weldon Johnson Memorial Collection, Beinecke Rare Book and Manuscript Library. Courtesy of the Langston Hughes Estate

in wartime essays like "Democracy, Negroes, and Writers" (1941): "To us democracy is a paradox, full of contradictions . . . Negroes, like all other Americans, are being asked at the moment to prepare to defend democracy. But Negroes would very much like to have a little more democracy to defend."[69] Quick to impugn the hypocrisy of fighting Hitler with segregated armed forces, Hughes also had words for those who mistakenly supposed that white supremacists like Georgia governor Eugene Talmadge and Mississippi representative John E. Rankin "are more devilish because they are closer at hand and holler so loud," or that simply because they have "slapped a few white faces in Hong Kong . . . the Japanese are friends of ours": "The Germans and the Japanese are *really* devils, and very dangerous ones, at that." "The truth of the matter is, of course, that they are *all* devils—Hitler, Mussolini, and Hirohito abroad, plus their Klan-minded followers at home"; "it is the duty of Negro writers to reveal"—to Black and white alike—"the international aspects of our problems at home, to show how these problems are merely a part of the great problem of world freedom everywhere."[70]

Such was the position Hughes staked out on the lecture circuit and in the press as a leading Black intellectual. His songwork, however, covered a wider ideological ambit. Some tunes straightforwardly embraced their purpose as nationalistic stirrers of home-front morale ("Bonds for All," "Song of the Defense Workers," "Rationing Blues") or as baubles for the entertainment of GIs themselves, "the lads who're fighting for me and you" ("Broadway's songs make a singing light / For the boys of U.S.A. / As you go to meet the foe—A salute from old Broadway").[71] Of the several numbers reflecting a racial politics, some, like "America's Young Black Joe" ("inspired by Joe Louis") and "Matt Henson" (about the polar explorer), showcased Black achievement, while others gave voice to the Double V campaign by folding anti-Jim Crow rhetoric into anti-fascist agitprop. Attempting to enumerate all these evils occasionally risked bursting the seams of Hughes's ballad verses:

> They call you the Fuerher, Il Duce,
> Hiroito [*sic*]—Heaven's Son—
> But Japanese, Italian, or German,
> You're the same piker rolled into one.
>
> They call you Gene Talmadge or Dixon,
> Beloved of the Klan and the Bund—
> In Georgia, Alabama, or Berlin,
> You're the self-same son-of-a-gun.[72]

Still other wartime songs adopt stances more expressly critical of the federal government, soft-pedaling patriotic fervor in the interest of denouncing segregation more explicitly, as in "Dixie Negro to Uncle Sam" ("How can you/Shake a fist at tyranny/Everywhere else/But here?") or "Message to the President," an "American Negro National Defense Song" ("Mr. President, let me hear you say:/No more segregation in the U. S. A.").[73] Hughes also circulated a poem version of the lyric to "Message to the President," and many of these protest songs are close in texture and theme to the poems contemporaneously gathered in *Jim Crow's Last Stand*, which brings a biting, radical edge to the Double V cause. In the poem "Red Cross," for instance, Hughes levels a compact salvo at the American Red Cross's policy of donor segregation: "The Angel of Mercy's/Got her wings in the mud,/And all because of/Negro blood."[74] "Good Morning, Stalingrad," meanwhile, took the then increasingly risky position of affirming the Soviet Union not merely as an expedient ally of the United States but also as an inspiration to African Americans battling Jim Crow, an opinion likewise hinted at in "Salute to Soviet Armies," a lyric set by Elie Siegmeister in 1944 and published, as a poem, in *New Masses*.[75]

Unsurprisingly, it was not in songs like "Salute to the Soviet Armies" or "Message to the President" that Hughes placed his hopes for a home-front hit. No matter how scholars account for his complex, potentially contradictory ambition to balance a retreat from left radicalism with an unwavering commitment to anti-racist struggle, as Rampersad details, Hughes's wartime songwork was guided from the first by plainly commercial motivations. He spent much of his stays at Yaddo in 1942 and 1943 writing songs "in a deliberate gamble that one would bring him wealth."[76] But if the nation's demand for "THE war song" presented Hughes with an especially lucrative target for his songwork, he would nevertheless face the same daunting challenges awaiting lyricists in any popular genre.[77] These are vividly synthesized in "The Song Writing Game," a *Defender* column of 1945 in which Hughes, by then a wised-up veteran of the culture industry, adopts a deflating tone to chasten aspiring songwriters: while "songs are comparatively easy to write," songwriting itself "is just about the most difficult field of all forms of writing for an amateur and an outsider to break into." The thrust of the article is to lay a disillusioned emphasis on "the plugging end of the profession," that is, the indispensable but difficult work of "getting [one's] song played by a band, or sung by a good entertainer, or into a show or a picture ... To achieve that is worth your life, and may take years—unless you are very good or very lucky."⁷⁸ Eventually Hughes himself relied on the professional "plugging" services of a publicity representative, Leavia Friedberg, but the advice he dispensed to *Defender* readers was hard earned indeed; Hughes worked tirelessly to get his song lyrics to move like commodities, and one need only tally the number and variety of musicians, publishers, and performers with whom these lyrics came into contact, all the various media formats in which they addressed their multifarious publics, to grasp how songwork comprehends a good deal more than setting words to music.⁷⁹ Hughes's words were sung, for instance, by students in high school and college auditoriums around the country, at the Stage Door Canteen by Rosetta LeNoire, at Radio City Music Hall and the legendary Café Society Uptown, on Asch Records by Josh White, by Paul Robeson during a special "CIO-Labor for Victory" program on New York's WEAF, and on network broadcasts like the *Treasury Star Parade* and *The March of Time*. Throughout the war years, the materials of Hughes's songwork were plugged in and routed through the extensive cultural channels of a mobilizing home front.⁸⁰

To foreground the work expected of a pop lyric—and it's expected, above all, to move—is to light up the intractably riven nature of songwork. The insistence in Hughes's article on the necessity of *plugging* songs points to a certain production-side gap in remuneratively successful songwriting. Adorno's notorious critique of popular music, elaborated at just the moment Hughes's songwriting reaches fever pitch, usefully lays out the terms of this disjuncture.⁸¹ In the course of his invective against pop's ubiquitous standardization, Adorno attributes the "stimulatory" semblance of originality in the newest hit songs to processes of "pseudo-individualization," whereby a pulverizing standardization disguises itself behind "the illusion and, to a certain extent, even the reality of individual achievement."⁸² It's significant that Adorno acknowledges the "reality" of the situation. Effects of pseudo-individualization are "grounded in material reality itself," since, as a function of the "'backwardness' of musical production," popular music circa 1940 remains "on a handicraft level and not literally an industrial one." In the age of mass media, songwriting is as yet a craft. Only something on the order of algorithmic automation will seal pop music's full industrialization; so long as song lyrics are written like poems, by individuals laboring at self-expression or the simulation thereof, pop music remains a residual practice with respect to the culture industry.⁸³ It's just this developmental lag that song hawkers must surmount. The plugging of lyrics expresses, at the level of the market, the necessity of plugging the gap between modes of production, of inserting handicraft lyrics into the cultural-industrial stream and overcoming the distance—a distance measurable by media as well as by genre—between lyric writing and remunerative song. As a central activity of songwork, song plugging testifies to the basic *unevenness* of the concept, as well as to its historical specificity. Adorno reminds us that when pop music can no longer be considered a handicraft practice because its production is wholly automated, this tension will vanish.

The practice of plugging lyric commodities also bears directly on the language of Hughes's lyrics. I've already mentioned these songs' resemblance to poems Hughes was writing concurrently—those collected in *Jim Crow's Last Stand*, for instance. Expressly associating these poems with Hughes's songs and his early *Defender* articles, Rampersad observes that these verses in "deliberately unpoetic language" aspire to evoke "the verbal nonchalance and semi-literacy of a putatively 'typical' member of the black masses."[84] He means of course to distinguish such efforts from Hughes's more obviously "literary" achievements, which include both his conventional "lyric" works and his blues poems, insofar as "the latter are a product of heightened moments of pleasure or pain, and thus attain levels of art that transcend the question of literacy." To adduce "this semi-literate, topic type" of poem, Rampersad quotes a version of "Total War":

> The reason Dixie
> Is so mean today
> Is because it wasn't licked
> In the proper way.
>
> I'm in favor of beating
> Hitler to his knees—
> Then beating him some more
> Until he hollers, *Please!*
>
> If we let our enemies
> Breathe again—
> They liable to live
> To be another pain.

Understandably, Rampersad supposes that many readers will wonder "how a poet of Hughes's ability, or a poet of any ability, could write such bad verse," but we might ask whether judgments like these would be at all mitigated if readers knew that Hughes at some point considered this poem a song lyric.⁸⁵ More pointedly, we might ask whether Rampersad's imputation of "semi-literacy" can be refunctioned so as to refer not merely to the poem's demotic aspirations but also, and more accurately, to its lyric intermediality, its material status as *semi-writing*. After all, the crudities on display in "Total War" are clearly linked to the lyric's capacity for songwork beyond the printed page. Set to music, its irregular meter, bathetic colloquialisms and simplistic wit may prove quite appropriate, not to say effective.⁸⁶

But to determine by what measure this "bad verse" qualifies as a pure product of songwork, we need first to render audible an aesthetic category glimpsed at the beginning of this chapter: the category of *kitsch*.⁸⁷ If one feels emboldened to raise the question of kitsch in connection with a poet like Hughes, it's largely thanks to Daniel Tiffany, whose study *My Silver Planet: A Secret History of Poetry and Kitsch* persuasively unearths this surprising secret history by revising the historical origins of kitsch away from its alleged bases in either Romanticism or mass industrial culture toward the ballad scandals and "pseudo-vernacular" poetic impostures of the English eighteenth century. "In the most precise historical terms," writes Tiffany, "the correlation of poetry and fraudulence identified by modernist theories of kitsch can be traced to the eighteenth century's affinity for—and contamination by—poems that are quite literally *counterfeit*."⁸⁸ To make this history legible, Tiffany advocates a return to the now seldom invoked category of poetic *diction*. Kitsch has its roots in a particular register of a cliché-ridden, "arrest[ed]" language that emerges at a specific moment—the ballad matrix of the eighteenth century—when, so Tiffany argues, "the reproducibility of beauty becomes its most salient feature."⁸⁹ The concept of diction lights up this "heightened and restricted vocabulary associated specifically with poetry," itself understood as "a recursive genre designed to elicit certain generalized poetic *effects*."⁹⁰ But because diction is a collective possession, it also invites us to apprehend the "verbal substance" of kitsch as a possible "progressive model of social totality": "The authority of poetic kitsch lies not in its powers of representation, which are in fact extremely weak, but in its capacity to *express* through its artificial phraseology an impersonal, social 'substance' concealed by ideology."⁹¹ What Tiffany identifies as kitsch's revolutionary potential amounts to a politicization of the subterranean link between poetic kitsch, as synthetic, replicative, collective, and impersonal poetic diction, and mass culture.⁹²

Because Tiffany is principally interested in outlining a *poetics* of kitsch, the genealogy reckoned by *My Silver Planet* follows kitsch's "ruinously dialectical" movement straight into the heart of modernism (with a reading of Ezra Pound's attempt to write "les Paradis artificiels" in his *Cantos*), and then to the postwar avant-gardism of the New York School poets.[93] The exclusive focus on poetry itself is suspended only for brief discussions of such things as lullabies and advertising jingles, though even these serve the argument mainly to epitomize certain aspects of poetic kitsch's ballad matrix, most especially the *refrain*, a device for repetition that further "arrests the language of the poem," dislocates sound from sense, and "gives expression to the sentiments and commentary of its anonymous listeners or readers."[94] Notwithstanding the brilliance of Tiffany's case studies, his emphasis on the refrain implies a license to depart from his own analysis and discover the purest twentieth-century terminus for poetic kitsch nowhere in poetry itself, but rather in the lyrics of the midcentury hit parade. Indeed, no linguistic artifact so extremely exemplifies the glitteringly formulaic, the cheaply reproducible, and the sentimental aspects of kitsch as the pop song.[95]

As a purveyor of "predigest[ed]" or "prefabricat[ed]" aesthetic effects, Tin Pan Alley in its interwar "golden age" was a natural target for modernist invectives against kitsch, and Tiffany's description of poetic kitsch as "arrested," compulsively repetitive diction surely embraces Adorno's diagnoses of popular music's "frozen" standards circa 1940.[96] Crucial to Tiffany's argument is the observation that the eighteenth-century vogue for antique or "distressed" ballads was an "inaugural artifact" of mass culture as subsequent centuries have known it, and the project of "expos[ing] the veiled affinities between 'serious' poetry and popular culture" through poetic kitsch may supply the literary-historical condition of possibility for a future poetry's radically "reciprocative and newly equalized relation to an emerging—and perhaps unrecognizable—conception of popular culture."[97] Hughes's own no-less spurious ballads—and his own "distressed" poems—constitute an unacknowledged chapter in this prehistory of poetry's kitsch future.

Somewhat paradoxically, the voluminous song drafts collected in the Hughes Papers attest to the *arrested* diction of poetic kitsch and the *frozen* conditions of the culture industry by way of nothing so much as Hughes's astonishing *flexibility* and the *fungibility* of his lyric material. To a degree, this adaptability reflects the dynamics of Hughes's collaborative practice. When his collaborations proceeded through the mail, as they often did, Hughes would draft lyrics in total ignorance of the music (a departure from common practice on Tin Pan Alley).[98] Thereafter, as if to compensate composers for the imposition of having to begin with, or adapt their music to, the lyric, Hughes would give them a free hand "to cut, fit, add to, and change as you choose" when drafting the lead sheet.[99] In the case of "My Heart Is a Lone Ranger" (a number inspired by Hughes's Yaddo pal Carson McCullers), for instance, Hughes could take it in stride when his composer Irving Landau wrote to say that, while he liked "the sentiments of [Hughes's] second chorus very much," he couldn't use it since he did not "want to change the music of the original chorus," enclosing a lead sheet with circles around all the words that "do not fit the music at all" and double underlining those "which fit but which are wrongly accented."[100] Hughes also invited revisions dictated by extramusical concerns, as when the industry savvy W. C. Handy (to whom Hughes had given carte blanche: "feel free to cut, combine, or tighten it up as you choose") argued that "Go and Get the Enemy Blues" required a "love or sex interest" to counterbalance its patriotic urgency. Provided "a happy combination of the two values" could be struck—"love for the sake of selling it commercially and realization of the Nazi menace for the sake of morale"—Hughes was willing to accede, just as he was willing to adopt Handy's mischievous suggestion of amending the lyric's "Hitler" to "Hitler's carcass"—because "you know what they are going to put in the place of carcass."[101]

But Hughes's flexible accommodations to the demands of the cultural-industrial song form are in full evidence even prior to the give-and-take of collaboration. Reading through the draft materials, above all one is struck by the countless different *versions* of individual songs. Either in the interest of experimentally finding his way to finally sufficient forms or alternately, as a tactic for securing a single song's optimal pluggability on the market, Hughes left behind a genetic record which reveals his songs as irreducibly plural and polyvalent artifacts. In addition to composing alternate endings and substitutable verses for his lyrics, Hughes would produce "commercial" versions and "long" versions; "race" versions and versions "suitable for whites, or mixed groups to sing"; "male" and "female" versions; "Mother's" versions and "religious" versions; wartime versions for otherwise apolitical numbers; Spanish-language versions; and of course, as we have seen, "poem" versions.

A lyric like "Freedom Road," for instance, is putty in Hughes's hands. Written with Emerson Harper in 1942, the song was the poet's closest brush with a wartime hit.[102] In the version made popular by folk singer Josh White, the verses offer a tempered rehearsal of the Double V position, ostensibly for an integrated audience:

> That's why I'm marching, yes, I'm marching,
> Marching down freedom's road.
> Ain't nobody gonna stop me, nobody gonna keep me,
> From marching down freedom's road.
>
> Hand me my gun, let the bugle blow loud,
> I'm on my way with my head a-proud,
> One objective I've got in view,
> Is to keep a hold of freedom for me and you.
>
> Ought to be plain as the nose on your face,
> There's room in this plan for every race,
> Some folk think that freedom just ain't right,
> Those are the very people I want to fight.
>
> Ain't nobody gonna stop me, nobody gonna keep me,
> From marching down freedom's road.
> Now, Hitler may rant, Hirohito may rave,
> I'm going after freedom if it leads me to my grave.
>
> United we stand, divided we fall,
> Let's make this land safe for one and all.
> I've got a message, and you know it's right,
> Black and white together unite and fight.

Hughes also drafted other versions of the lyric for the ears of more specific publics. Whereas White's rendition adopts the soldier's perspective, a "female version" focalizes the lyric from the home front: "Mothers all know / To keep this nation free, / Ev'ry son must / Fight for liberty." Likewise, a "race version" for Black audiences intensifies its anti-racist message with additional verses on Jim Crow and segregation in the armed forces ("Jim Crow regiments / Jim Crow Navy is not right. / Who believes in Jim Crow / Ain't my allies in this fight"), and a "church" version for distribution to religious organizations furnishes an Old Testament allegory ("Moses said, Go down, / Yes, way down in Egypt's land, / And tell old wicked Pharaoh, / Turn a loose of Israel's band"). These proliferating versions increase Hughes's returns of investment by making "Freedom Road" just that much more pluggable. They also afford him a measure of proleptic control over the cultural-industrial career of his lyrics. By anticipating their reception across media forms and in certain cases even forestalling the kind of appropriative deformation performed by Kingsley, Hughes installs a supersensitivity to generic uptake at the center of his songworking practice.

One could make sense of Hughes's handicraft virtuosity, his ability to ring the changes on a single theme to appeal to an array of publics, as an expression of the inflexibility of the midcentury music industry—all these are so many versions of the popular same, as it were, pseudo-individuated for success in a rigidly standardized, if socially stratified, marketplace. Just as plausibly, one might grasp these proliferating iterations as a reflection of the popular song form itself, since musical works are conventionally understood to exhibit a more distributed ontological status than poetry, given the latter's pragmatic attachment to the printed page.[103] Yet Hughes's practice of lyric versioning, as a staple feature of his songwork, also has something important to tell us about *verse*, with which it shares an etymological connection. It's possible to read these flexible made-to-order versions as enacting a poetics of contingency in the historical shadow—perhaps as the negative image, or even the constitutive unconscious—of contemporaneously ascendant ideologies of lyric form as necessary, autotelic verbal icons. Versioning on Hughes's scale reverses that New Critical motto, which W. K. Wimsatt considered "worth quoting in every essay on poetry": "A poem should not mean but be."[104] Many of Hughes's songs *are* not, since they exist in multiple versions, though they *mean* with all the kitschy blatancy of cliché, and the versatility of Hughes's lyric procedure is such that they do so prolifically and elastically.

At first glance, Tiffany's analytic of poetic diction appears a useful means of approaching Hughes's elaborately versioned songwork. It's after all *diction*, as "verbal substance," that remains of a lyric when set to music. And it's diction, too, that does the work of hailing and securing an efficacious reception among distinct publics, since the difference between two versions of a song is often the difference between two lexical buckets. Thus the lever of diction allows Hughes to move across medial and generic boundaries. But in the case of a work like "Total War," it's hardly the lyric's vocabulary alone that instantiates its "semi-literacy," nor does diction fully account for its remarkable crudity. Much as it was necessary to supplement Tiffany's model of poetic kitsch with its twentieth-century apotheosis in midcentury pop, the stultifying diction of kitsch requires an accompanying concept in order to address the poem's bungled prosody—which, again, may only be "bad" to the extent that it indexes the absence of a lead sheet, of the melody that would justify its roughshod meter. For such a concept we can turn to the most exemplary, most extreme instance of songwork's semantic fungibility: the "dummy lyric."

The prosodic counterpart of counterfeit diction, a dummy lyric refers to the placeholding verse that a songwriter employs to retain the melody or the abstract rhythmic structure of the song lyric while working up the "real" words. Certain famous dummies dot the history of popular music as amusing trivia: sometimes an official lyric is revealed to have originated as a dummy, as in Vincent Youmans and Irving Caesar's "Tea for Two" (1924), or else the pop fan's knowledge of a dummy can puncture the seeming spontaneity and inevitability of a well-turned lyric, as in Paul McCartney's "Yesterday" (1965), known once by the title "Scrambled Eggs."[105] Though Hughes wrote his song verses in the absence of music, he was nonetheless sensitive to the compositional utility of the abstract musical or prosodic structures that a dummy lyric essentially delineates, and in fact marshaled dummies in his quest to write a wartime hit. Applying himself to the example of veritable hits, Hughes undertook studies of several popular standards, including Hoagy Carmichael and Stuart Gorrell's "Georgia on My Mind" (1930); Carmichael and Johnny Mercer's "Lazy Bones" (1933); and the war song "This Is Worth Fighting For," by Edgar De Lange and Sam H. Stept, which made *Your Hit Parade* in June 1942 (figs. 3.3 and 3.4).[106] Hughes produced detailed line-by-line analyses of the latter lyric's "metrical and musical construction," noting not only its foot and syllable arrangements but also the way linguistic *meter* interacts with extralinguistic *measure*, or the song's bar structure. The result is a genuinely intermedial prosody that opens its ears to both present and absent sounds. For easy visualization, Hughes then abstracts this information into patterns of repetition ("s" for "SAME") and variation ("V" for "VARIES"), a flagrant reminder that the most basic element of a certain type of pop song is the not-so-free play of identity and pre-calculated difference.[107]

In effect these exercises allowed Hughes to utilize preexisting hits *as* dummies—as purely formal skeletons for his own future songwork. Dummy lyrics speak to the primary importance of song verse prosody and, by extension, to the pop lyric's ultimate subservience to musical accompaniment. They also vividly travesty any notion of organic form; as these studies in songwork suggest, prosodic form becomes at once the sovereign feature of the lyric and also banally standardized, while the lyric's content is totally

```
                THIS IS WORTH FIGHTING FOR
                by DeLange & Stept (Harms, Inc.)

                I saw a peaceful old valley

                With a carpet of corn for a floor

                And I heard a voice within me whisper

                THIS IS WORTH FIGHTING FOR.

                I saw a little old cabin

                And the river that flowed by the door

                And I heard a voice within me whisper

                THIS IS WORTH FIGHTING FOR.

                Didn't I build that cabin?

                Didn't I plant that corn?

                Didn't my folks before me

                Fight for this country before I was born?

                I gathered my loved ones around me

                And I gazed at each face I adore

                Then I heard that voice within me thunder

                THIS IS WORTH FIGHTING FOR.

                                                              Feet----Sylables
                     Bars     **********                              Line--Bars
Same feet     2      I saw----a peace--+--ful old valley      3 ft.-8syl.4-4
       bars   2      With a car---pet of corn-+-for a floor   3       9    6-3
              2      And I heard---a voice-+--within---me whisper 4  10    5-5
              2      THIS----IS WORTH FIGHT-+--ING FOR.        3      6    4-2

              2      I saw----a lit---+--tle old cabin         3      8    4-4
  "      "    2      And the riv---ver that flowed-+--by the door 3  9    6-3
              2      And I heard a voice within me whisper     4     10    5-5
              2      THIS----IS WORTH FIGHT-+--ING FOR.        3      6    4-2

              2      Didn't----I build----that cabin?          3      6    4-2
   (R)        2      Didn't----I plant----that corn?           3      5    4-1
              2      Didn't----my folks---before me            3      6    4-2
              2      Fight---for this coun---try before---I was born 4 10  5-5

              2      I gathered---my loved-+-ones around me    3      9    5-4
  "      "    2      And I gazed-----at each face-+--I adore   3      9    6-3
              2      Then I heard---that voice-+--within---me thunder 4 10 5-5
              2      THIS----IS WORTH FIGHT-+--ING FOR.        3      6    4-2

CHORUS---Conventional four verses, 1st, 2nd, & 4th the same metre with the
title line repeated at end of each. The 3rd is the release, keeping the
same 3-4 XXXX feet to a line, but varying arrangement of third and fourth
lines.
```

Figure 3.3. Langston Hughes's prosodic analysis of Edgar De Lange and Sam H. Stept's "This Is Worth Fighting For," n.d. Box 511, Folder 12680, Langston Hughes Papers, James Weldon Johnson Memorial Collection, Beinecke Rare Book and Manuscript Library. Courtesy of the Langston Hughes Estate

THIS IS WORTH FIGHTING FOR

ANALYSIS: Conventional 32 bar chorus. Conventional 16 line
lyric, divided into 4 four-line verses: A, A, B, A,
with the title line repeated three times---at end of
1st, 2nd, and 4th verses. These three verses have
exactly the same metrical and musical construction.
The 3rd verse, B, the release varies but slightly
except in melody. A line by line analysis follows:

	BARS	SYLLABLES per bar:	SYLABLES per foot:	SYLABLES per line:	FEET	
(A)	2	4-4	2-2-4	8	3	
	2	6-3	3-3-3	9	3	SAME
	2	5-5	3-2-2-3	10	4	
	2	4-2	1-3-2	6	3	
(A)	2	4-4	2-2-4	8	3	
	2	6-3	3-3-3	9	3	SAME
	2	5-5	3-2-2-3	10	4	
	2	4-2	1-3-2	6	3	
(B)	2	4-2	1-2-3	6	3	
	2	4-1	1-2-2	5	3	
	2	4-2	1-2-3	6	3	VARIES
	2	5-5	1-3-3-3	10	4	
(A)	2	5-4	3-2-4	9	3	LINE VARIES
	2	6-3	3-3-3	9	3	
	2	5-5	3-2-2-3	10	4	SAME
	2	4-2	1-3-2	6	3	

VARIATIONS FROM A (1)

(A)	s	s-s	s-s-s	s	s	
	s	s-s	s-s-s	s	s	
	s	s-s	s-s-s-s	s	s	FIRST TWO VERSES
	s	s-s	s-s-s	s	s	EXACTLY THE
(A)	s	s-s	s-s-s	s	s	SAME
	s	s-s	s-s-s	s	s	
	s	s-s	s-s-s-s	s	s	
	s	s-s	s-s-s	s	s	
(B)	s	s-V	V-s-V	V	s	
	s	V-V	V-V-V	V	s	RELEASE VARIES
	s	V-V	V-s-V	V	V	
	s	V-V	s-s-V-V	V	V	
(A)	s	V-s	V-s-s	V	s	OFF 1 SYLABLE
	s	s-s	s-s-s-s	s	s	
	s	s-s	s-s-s	s	s	
	s	s-s	s-s-s	s	s	

Figure 3.4. Langston Hughes's schematic analysis of Edgar De Lange and Sam H. Stept's "This Is Worth Fighting For," n.d. Box 511, Folder 12680, Langston Hughes Papers, James Weldon Johnson Memorial Collection, Beinecke Rare Book and Manuscript Library. Courtesy of the Langston Hughes Estate

emptied of meaning, and thereby reinvested with the kind of protean polyvalence characterizing Hughes's endless versioning on single themes for multifarious audiences.

The dummy lyric, in a sense, writes large the enduring poetic logic of the nonsensical ballad refrain, reprise, or burden, which by repetition unwinds signified and signifier and establishes "the priority of lulling over meaning."[108] In Tiffany's account of poetic kitsch, this triumph "of sensibility over sense" that we find in the ballad refrain effects a "live burial of meaning" via reproduction that expressly "links the operation of the poetic refrain to mass culture, consumption, and to episodes of fleeting collectivity" galvanized in reception. The dummy lyric goes further, it might be said, by allowing us to glimpse the production side of this dynamic. It offers the utterly abstract (exchange) value of lyric—the very condition of its valuation vis-à-vis accompaniment, pluggability, remediation, and commodification. In this way, the dummy lyric forms the basic unit of songwork, and it maintains for that reason a singular relation to both media and genre. As a sensuous abstraction of sorts, the dummy presents an odd variant of the material condition I call lyric intermediality. One might do best to call dummies *pre-medial*, proper neither to print nor to any medium in particular but oriented nonetheless, as essentially phonic phenomena, toward that same tension between sonic and linguistic orders characteristic of lyric practice more generally. We can think of the dummy as lyric intermediality *in potentia*. In analogous fashion, dummies also exist curiously prior to genre. To the degree that Hughes's protean signaling of discrete and specific publics is understood to facilitate generic uptake, the dummy lyric itself—an artifact of pure pop—furnishes the condition for those social acts of recognition.

Divested of determined semantic content, stripped down to jingling kitsch, the dummy represents intermediality just as an exchange relation—the lyric text in its commodity form. Half stripped of its veil, the meaningless dummy-lyric-as-commodity-fetish lays bare the social relation of its collaborators in words and music up against the culture industry and up against the mass mobilization of the war effort. It's one compositional tool for working between media and between genre in the cultural arena we call songwork. It is what allows Hughes to craft a poem like "Freedom Road" for multiple audiences or to adapt a vision of "This Is Worth Fighting For" that might be of polemical use to the Double V campaign. It also allows Hughes to retain, on the side of production, a degree of the control he must cede when, as with Kingsley and "Hard Luck," the hard work of mediating a blues tradition in print is scuttled by white adaptors. Perhaps above all, it is a strategy for surviving the cultural marketplace.

As a heuristic for studying the proximity of song and poem in Hughes's life and work, songwork is a category that spans a wide variety of different media (magazines, books, songsheets, radio broadcasts, records, etc.) in order to regather critically what Hughes has, as a matter of pragmatic practice and his own livelihood, already brought together: lyric poetry and its paraliterary avatar, popular song lyrics written for a multitude of musical genres. Importantly, what lends a certain coherence to this category by putting lyrics in circulation across the boundaries of poetic and musical practices is also what stratifies and ruins it: the commodity form of the pop song. My goal has been to reconstruct something of this transgeneric, transmedial practice by appealing to the fragmented drafts of songs and poems that Hughes left behind—the visible ruins of songwork and the best evidence of a problematic (the proximity of lyrics to lyric) that for all its familiarity continues to render difficult the writing of lyric's media history in the twentieth century. Songwork implies the existence of the dummy lyric. For Hughes, it was a means of navigating an uneven market for song. For us, the dummy lyric is a theoretical tool for sounding out the cultural space of that idealized or nonexistent genre that we can't help projecting in songwork's wake, the spectral sum of lyric and lyrics.

4 "In Lieu of the Lyre"
Lyric Objects and the Audio Record

A Poem's Two Objects

The word "lyric," in this book, designates writing that is accompanied by its prosodic, performative, and ideological claims to sound. Whether these sounds are phonic, sonic, musical, bodily, ear-splittingly real, or wholly imaginary, lyric is that kind of writing historically invested in flouting the silence of manuscript, print, and digital script. Though it may prove reliably unsatisfactory as a term of art for poetry's style, form, size, rhetorical apparatus, subject, or affect, "lyric" continues to describe, minimally and accurately, poetry's primary concern with what Susan Howe names "the deepest mystery of writing": the certitude "that sound and sight are inextricably and mysteriously bound, and that every mark on paper is an acoustic mark."[1]

This intermedial condition is brought most plainly into view by meeting poems where they live—in historically specific media ecologies, among other technologies for the reproduction of speech and song. To that end this book lays a strong emphasis on lyric writing as *social practice*, taking some vital bearings from Raymond Williams, whose most important contribution to media studies may be an argument for the dissolution of the medium concept itself. We can study the technical means of cultural production, according to Williams, only in the solution of practice, as "work on a material for a specific purpose within certain necessary social conditions."[2] The contrary habit of isolating a given *medium* for analysis—be it oil painting, Appalachian folk singing, or printed poetry—represents the mystified yield of art's progressive spiritualization in the modern era, and it threatens to reify practice.[3] Thus, our search for the lyre in lyric aims at a horizon beyond the alienable text-object of New Critical and poststructuralist concern and beyond also the narrowly bibliographic artifacts of literary history, toward a conception of lyric writing as dynamic material practice whose constitutive ties to social life are often a matter of acoustic resonance: lyric forms are "material notations on paper" that index, evoke, and solicit a wide manner of acoustic phenomena, from the internalized reading voice to the commercial popular song.[4]

This chapter weighs a moment—the 1950s and early 1960s—when this methodological shift "from medium to social practice" was anticipated in the construction of poems and then worried over, processed, and dialectically reflected at the level of poetic theory. These years witness the dissolution not of any medium, per se, but of the poetic object itself. In the first half of the twentieth century, poets "took the side of things" ("le parti pris des choses"), in a phrase of Francis Ponge's, not only by praising the impassive allure of red wheelbarrows and Egyptian pulled glass bottles but by stressing the objective dimensions of the poem itself.[5] Take, for example, the referential confusion at the heart of Ezra Pound's imagism, where "direct treatment of the 'thing'" demands heroically ascetic attention to the sculpted *thingliness* of the poem, and not only to what it describes.[6] Modernist "poets of reality," in J. Hillis Miller's account, opened "a space in which things, the mind, and words coincide in closest intimacy."[7] On the side of criticism, lyric objectification won its corresponding apotheosis in New Critical descriptions of poetry as a well-wrought urn or verbal icon.[8] By the 1950s, of course, this broad commitment to the modernist object was forced to make room for an emerging "anti-artefactual aesthetics" consonant with developments across the arts—with visual art's embrace of the participatory "theatricality" famously scorned by Michael Fried, for instance, and with the more general "culture of spontaneity" finding expression everywhere from the happenings at Black Mountain College to the bebop of Charlie Parker.[9] This chapter greets its central figures, Marianne Moore and Russell Atkins, at just this flexion point in the career of modern poetry, when an emergent structure of feeling about what poems are begins to shift its priorities, from finished "thing" to participatory "process."[10] Inheriting to various degrees modernism's confidence in the poetic object, while also clearing ground for the naked, open, notational, and event-based forms like those gathered in Donald Allen's *The New American Poetry: 1945-1960* (1960), these poets reckon with and respond to the veering fortunes of lyric objecthood.[11]

This chapter traces the media-ecological foundations of this shift by examining how poets react to their soundscape's thoroughgoing technologization—to a situation of *ubiquitous recording*. "Ubiquitous" is a fuzzy term, of course. Douglas Kahn calls the mid-1920s and 1930s an era of "ubiquitous recording" to recognize that period's proliferating innovations in radio, sound film, phonography, and early television, whereas for others like Steven Connor it's the postwar introduction of magnetic tape, with its possibilities for personal recording and playback, that finally secures the ubiquity of technologized sound.[12] For our purposes I locate ubiquity's qualifying watershed in the promotion of recorded poetic performance—live or in studio—from the status of novelty to viable and visible mode for the circulation of poems. The 1950s sees the birth of literary audio labels like Caedmon Audio and the ascendance of text-based performance events—"poetry readings"—as a shaping influence on verse's printed forms and public life.[13] Poets in the Anglophone world had experimented with technologized sound since Robert Browning and Alfred, Lord Tennyson pressed their lips to Edison's bell at the end of the nineteenth century.[14] But only in the 1940s and 1950s, as poets begin to record and perform their poems as a matter of course, does the modern *desire* for poetic objecthood, and the *idea* of what a poem essentially is, come to reflect the full remedial influence of sound recording. Ubiquitous recording is a cultural situation in which lyric sound gains a future on phonograph records and tape at the price, paradoxically, of its own object permanence.

The words "desire" and "idea" imply a core supposition of the argument to follow: media history conditions poetic practice not only by setting the material limits of that practice, but also by setting terms on the no less operative plane of poetic ideology. Since this chapter endeavors to keep both poetry and poetics—practice and theory—in mutually illuminating play throughout, some conceptual modeling is in order. Let it be said that every poem comprises at minimum two distinct objects. We have at hand, most obviously, the historical artifact itself, a linguistic form manifest in ink on the page or else spoken out and captured as sonic signal on record, tape, or in digital code. But this artifact scarcely exhausts the poem's objecthood; circulating in contingent relation to this archivable entity, an *idea* of the poem-as-object, its hypothetical objectification in mind, just as significantly solicits our interpretive regard. I say every poem has *at least* two objects because any number of material texts may instantiate versions of a "single" poem, and any single poem may reflect or elicit different ideological claims to objecthood. The very same poetic text, for instance, may be regarded by its writer as the concrete index of "an intellectual and emotional complex," by a close-reading critic as an autonomous linguistic structure, and by the pragmatic anthologist as a collectible, alienable, "durable thing[] made of nothing but words."[15] Objecthood, in this sense, is a function of recognition and use.

These observations reflect bedrock principles in contemporary literary studies. Decades ago, textual scholars like Jerome McGann disabused critics of recourse to a poem's "Ideal Text," the abstract, hypostasized ur-object shared by New Critical and poststructuralist readers alike, in favor of attention to any number of bibliographical *texts*, instantiated in particular contexts.[16] Historically minded readers know well to heed the constitutive heterogeneity of any ostensibly single "work" by accounting for the many different material objects—the manuscript variants, the various modes and scenes of publication—nominally gathered under its title. We know, too, how to pierce the illusions of aesthetic objecthood, aura, totality, necessity, and talismanic identity that accrue to such ephemeral and contingent entities as poems. In other words, scholars understand that both ascriptions of objecthood—of the poem we hold in our archival gaze, and of the poem we summon and summarily bind whole in the critical imagination—are likewise fictions of a sort.

But this chapter will stress a point easily missed: *both* these senses of the lyric object are properly historical. The ideal entities we make of poems are no less historical artifacts than the onionskin page, little magazine, or deluxe edition on which that poem variously appears. As a consequence, not only a poem's material incarnation but its idealized projection must be grasped as a *media object* in its own right. The task of expounding this second claim will occupy the second half of this chapter. For now, I'll simply ask readers to recall Louis Zukofsky's investment in an objectified speech-music treated in chapter 1, and the concept of latent remediation formulated in chapter 2. In Zukofsky's case, I proposed the sound recording as a hidden referent for the poet's career-long endeavor to objectify lyric materials. Chapter 2 traced the influence of radio on page-bound lyric forms by partly dissolving the media-technological formation of 1930s radio into its constitutive social processes. In the present case, in not dissimilar fashion, we will discover how both the desire for lyric objects and their dissolution, however little these contrary tendencies have to do manifestly with technologized sound, are nonetheless motivated in real, specific, and consequential ways by a milieu of ubiquitous recording. I designate these phenomena "lyric" objects because it is lyric intermediality, the tension between writing and sound, that conditions the relation between the poem's material presence and the ideal object we inveterately mistake it for.

"In Lieu of the Lyre" 173

Recognizing that every poem has at least two lyric objects may occasion an adjustment in our thinking about *poiesis* itself—about how poems get made and then, in reading, recognized or made over. Modern poetry's media history teaches that when someone makes/reads a poem, they make/read it twice: once in fact, and once in mind. To be sure, this distinction evasively abstracts from complicated issues in philosophy and psychology, questions of intention and telos, the materiality of the imagination, and the interrelation of perceptual and cognitive processes that ties understanding of any poem to our sensuous experience of it. But the crude binary has the saving virtue of convening disparate notions of objecthood that in different ways underpin aesthetic discourse, and theories of the lyric in particular. Any poem's twice-madeness is evident, for instance, in the self-consciousness of poems *as* objects, the way poets may take their own poetic production as theme, "singing of singing" in the style of the troubadours.[17] In a different register, this distinction remarks the figural fold in reception between the poem as historical artifact and its intelligibility as an expressive utterance, a condition of possibility for the interpretive habit of "lyric reading" first sighted by Paul de Man and then elaborated by Yopie Prins, Virginia Jackson, and others.[18]

But this chapter's chosen guide to the poem's two objects is not a lyric theorist but a poet, Marianne Moore, who in marking out poetry's "raw" and "genuine" elements in her celebrated poem "Poetry" offers a neat shorthand for its material and ideal aspects, respectively.[19] Robert Lowell employed a similar-seeming dyad—"raw" versus "cooked"—to schematize the competing styles of postwar American poetry; shifting the analysis toward media and away from form, this chapter illustrates how a material drama of the raw and genuine transpired in this period at a level subtending—and not necessarily homologous to—contemporaneous debates between the academic formalists of an institutionalized modernism and the rejuvenated avant-gardes of the Beats, Black Mountain, and San Francisco Renaissance.[20] Moore, an elder poet preternaturally sensitive to the curious itineraries of the object world, opens a window onto this important chapter in lyric's media history.

We'll begin in the glow of Moore's late-career literary celebrity, at a midcentury moment when the poet emerges as a practical champion of the *displaced objecthood* refashioning lyric's "raw" materials. We then follow Moore to her encounter with the Cleveland poet and composer Russell Atkins, a lyric theorist with extraordinary ambitions toward the formulation of a "genuine" poetic object. The recovery of Atkins—a Moore mentee whose phenomenological theory of "psychovisualism" unfolds a wholly different program of objectification under the revisionary auspices of a Black modernist tradition—promises to demonstrate how lyric techniques on the page can disclose the latent influence of the midcentury audio record. Atkins's confounding interest in the primacy of *sight* seeds a critical meditation on what we might consider the degree zero of lyric intermediality, where the missing lyre meets the sensing body: the puzzling relations of eye to ear that render "every mark on paper . . . an acoustic mark."

Marianne Moore, Verbal Pilgrim

On May 11, 1965, a seventy-seven-year-old Marianne Moore visited Harvard at the invitation of the *Advocate*, the institution's long-running undergraduate literary magazine. Preserved on tape and digitized by the Woodberry Poetry Room, Moore's reading that evening in Sanders Theatre is one of the last extant recordings in the poet's sizeable audio archive.[21] We can hear the seasoned poet-entertainer still very much in top form, indulging the breezy, wry patter that was the métier of her late career celebrity. She opens the reading by recounting her receipt of a "gracious, unusual, original, eloquent" request by undergraduate Stuart Davis, editor of the *Advocate*, for an original poem to accompany her visit.[22] After extending the invitation—"I was delighted to come," she says—the savvy Davis had slipped in his request for a poem by way of an apparent afterthought, which Moore relishes in quoting for her audience: "And [then] he said, 'It just occurred to me, this moment, we'd be very grateful if you could give us something to print, something you had to spare.'" That this anecdote wins laughter from Moore's audience is partly a result of Davis's sophomore pluck and partly a response to what Moore calls the "formal-informal craftly rare" nature of his request: "And then he said, 'but if I shouldn't have mentioned it, pardon us, disregard it.'" Evidently charmed by words of strategic demurral that seem designed especially to gratify Moore's legendary predilection for "unpompous gusto," for the defense of deference, and for humility brandished like a shield, the poet calls this "the most eloquent form of persuasion that I've met with."[23]

Not surprisingly, then, Moore acquiesced to Davis's request. Though as she explained to her audience, her desire to submit to the *Advocate* not any "spare" poem but a new one, tailored to the occasion, presented her with a formidable obstacle. In 1958 she had signed a contract "not to offer any new verse that I write to any but one magazine." Those gathered in Sanders Theatre may have grasped this allusion to the *New Yorker*, which played an outsized role in the last decades of Moore's life, canonizing the poet (albeit in somewhat condescending fashion) as the chatty Dodgers fan of the tricorn hat and cape in their 1957 profile of Brooklyn's foremost "literary monument," and publishing the vast majority of her late poems.[24] If the poem she writes for the *Advocate* is to evade the letter of her first-readers agreement with the *New Yorker*—its prerogative to consider all her "new verse" for publication—the *poem* must not be *verse*. Davis's request put Moore "in a frenzy of concocting something" in "neither prose nor verse but [which] could pass for both." Moore often disarmed public audiences and reporters with the self-effacing quip that she called her writing *poetry* only "because there is no other category in which to put it."[25] In this case, however, she was legally bound to offer the *Advocate* something nominally uncategorizable, something "in lieu of the lyre," which is the felicitous phrase she adopts—"very accurately," she says—for her "poem's" title.[26]

"In Lieu of the Lyre," which Moore read for her Harvard audience, enacts in both its thematic and artifactual dimensions the vertiginous displacements of lyric objecthood at the center of Moore's late poetics and at the heart of this chapter's investigation of late modernist lyric media. What pretends to "poetry" in this work is dynamically, self-consciously challenged by printed prose, by spontaneous talk, and by technologized sound. This last is the catalyzing ingredient, the new raw material in the intermedial mix. All printed poems go "in lieu of the lyre," suspended as they are between writing and sound, but recording alters that relation. We saw in chapter 1 how phonography drives lyric intermediality into a crisis of fidelity. Once the recording of poetry achieves a measure of cultural ubiquity, we can begin to identify pivotal changes not only in the style and textual form of poems, all of which also latently remediate a technologized mediascape, but now in their shape and material status, too—in the very kind of objects we make of poems.

That such transformations appear especially legible in Moore's case has not escaped the attention of her critics, though emphasis here has fallen largely on *performance*, with varying degrees of attention to the conditioning role played by sonic technologies.[27] When Moore begins reading her poems publicly in the 1940s and 1950s, this "mid-career turn to performance" enjoins her to comprehend any poem "less in terms of a textual object, and more as an event of continual feedback between the author and the poem's various public appearances."[28] According to Peter Howarth, "the feedback loops that performance makes are ways for poems to be both things and events, or perhaps, to move beyond the categories themselves," challenging the notion, now in Christopher Grobe's phrase, that there "exist two such pristine and separate realms as print and performance."[29] In the interest of extending and substantiating these observations, I'll argue that we cannot entirely account for this new apprehension of the poem as a plural, dynamic, displaced loop of attention without reference to its sturdy, dialectical counterpart: that is, the reifying objecthood of the sound recording. To grasp the counterintuitive process by which sound's objectification contributes to the disruption of the lyric object, we must first remember that medial plurality—the intermedial difference that has long animated lyric practice on the page—*precedes* technologized sound. Poems were sundered from the first; ubiquitous recording amplifies that rift. In fact, Moore has a preferred wedge for prizing open the objects her poems supposedly are to the impress of ubiquitous recording and for exacerbating tensions between print and sound. That tool is *talk*.

At Sanders Theatre, Moore punctuates her reading of "In Lieu of the Lyre" with the humorous interjections and explanatory comments that characterized her celebrated effectiveness as a public speaker in the 1950s and 1960s. At the podium Moore would marry casual wit and self-effacing charm with her own "burning desire to be explicit," securing her audience's attention by explaining, like a human footnote, a line's meaning or lack thereof, deflating the impersonal authority of the modernist poem.[30] The asides and excursuses interrupting her recitation of poems ranged from comments compensating for the absence of the printed page—spelling clarifications ("'q-i-v-i-u-t, Eskimo word") and remarks on eye-rhyming prosody ("'dragon,' 'Solomon,' and 'phenomenon' are *supposed* to be rhymes")—to expositions of allusions ("if you don't attend the races—neither do I—I'd better say 'Bold Ruler' was a very promising horse"), anecdotes about *New Yorker* fact-checkers grousing over words like "pashm" they cannot locate in the dictionary ("well, it'll be there, presently, and I refuse to change it, because it rhymes with 'ram'"), and pointed echoes of current events ("Believe I'll read that again: 'They did not let self bar / their usefulness to others who were / different.' That has a very unemphasized bearing, I think, on our troubles there in Little Rock").[31] Given Moore's conversational style and the tempering flatness of her midwestern accent, those listeners unfamiliar with the work in print found it difficult to parse poetry from patter. Thom Gunn remembers how easy it was to assume "her explanations—which often came in the middle of the poem—were merely the portions of the poem she'd neglected to write."[32]

Moore often further effaced the slippery, ductile borders of the poem by cutting, rearranging, and amending her poems in the swift course of recitation. "I'm re-writing this as I read it," Moore told an amused audience in San Francisco as she worked an impromptu trimming of the "Labors of Hercules," respecting the poem's admonishment of "the bard with too elastic a selectiveness" by omitting that very line.[33]

Moore's habit of "re-writing" on stage is an extension of her lifelong commitment to revision in print. And yet, it's also clear that Moore's skill as a celebrity performer—one increasingly called upon to explain and entertain in print, on stages, on the airwaves, and even on television—developed in mutually enforcing relation with the more didactic and evidently more accessible poems of her final collections.[34] To the extent that Moore was "writing for a listening audience or for a specific publication or event" in her last decades, the figural "voice" of her poetry in silent print seems inevitably bound up with her *actual* performing voice, a point substantiated by scholars like Edward Allen and Allison Neal who have traced the elocutionary disciplining of Moore's voice for phonograph recording and its effect on her poems.[35] Such work hauls wisely into view the latent remediation of Moore's late poetry by sound recording—the sense in which the poems reflect, in their formal makeup on the page, the shaping presence of technologized sound.

Moore's talk attunes us to the complexities of this influence. Her habit of chattily interrupting herself on stage furnishes the obverse instance of what her *New Yorker* profiler, Winthrop Sargeant, describes as her ability to "talk literature," to let loose the quicksilver "flow of imagery and anecdote" that distinguished her conversations off stage as well.[36] Sargeant goes to extravagant lengths to capture Moore's "breakneck manner of expressing herself," using several column inches to quote one such "dizzying monologue" delivered "at an afternoon gathering in her apartment."[37] The set piece follows Moore's "tumbling associations" touched off by the label on a bottle of Harveys Shooting Sherry. What Elizabeth Bishop elsewhere describes as Moore's "long unnebulous train of words" in this case carries the latter's guests and by extension Sargeant's readers through subjects as far afield as bicarbonate of soda, Wallace Stevens, Goethe, a quip by Ruth Draper, her uneasiness writing prose, and her favorite musical instruments: "If I find that a man plays the trumpet I am immediately interested . . ."[38]

Just when readers may begin to wonder how Sargeant managed to record this verbatim effusion, Moore herself raises the question of fidelity and representation by swerving briefly to the topic of her relation to posterity: "Those readings of my verse I made for the phonograph—well, they're here forever, like the wheat in the pyramids."[39] This remark refers literally to the fixity and permanence of her voice on the commercial recordings she had recently cut for the Library of Congress's Recording Laboratory (1949/1953), Soundmark Records (1950), Caedmon Audio (1954, 1955), and the Yale Series of Recorded Poets (1951/1965). And yet the offhand comment arrives wrapped in several layers of suggestive irony. For one thing, Moore relies on a biblical idiom she knows to be counterfactual—the pyramids were not storehouses for grain but tombs, empty now in any case.[40] "In the pyramids" readers will find either an idiomatic figment or a product of historical surmise—precisely the opposite of a reliable artifact. In what sense, then, will these recordings be "here forever," if they never were there at all? These ironies are endorsed by the strange assertion's material status: we simply do not know whether Moore herself ever uttered it. It's scarcely clear how Sargeant was able to reproduce this "torrent of impressions" word for word. Did he use an early portable tape recorder? Was he scribbling in shorthand? Did Moore have a role in proofing her own monologue? Or is this comment regarding the endurance of Moore's voice a pure product of the profiler's own dubious imagination of her literary celebrity and her ability to "talk literature"?

Whoever is finally responsible for her voice on the glossy page, Moore's talk of phonography is patently mediated by print. Or, what amounts to the same, the meaning of these recorded performances remains inextricable from the print record, just as the odd idiom forces scare quotes down around "here forever." The dubious provenance of this comment on the fixity of sound recording permanently unfixes, or displaces, its source. Rather than read this moment prima facie as a performance of phonography's difference from print, we ought to register its broader emphasis on the complex relations between the protocols responsible for audio and printed archives. The displacement of lyric objects this chapter seeks to describe—and of which "In Lieu of the Lyre" will be our starkest example—proceeds not as a direct response to sound recording in any easily demonstrable sense but instead as the messier negotiation of just this mutually destabilizing relation between print and sound.

"In Lieu of the Lyre," a poem suffused with talk, confronts us with displacements before we even start reading, for the conventions of scholarly presentation require initial decisions as to what and where "In Lieu of the

Lyre" *is*. Is the poem in Moore's May 1965 manuscript, on the reel-to-reel tape digitized by Harvard's Woodberry Poetry Room, in the pages of the *Advocate*, or in one of its subsequent printings?[41] Reliance on any single material instance would be a mistake, of course, not because all these artifacts prove of equal interest or significance, nor because there exists some virtual "poem" abstractable from its various instantiations, but rather because each concrete artifact is shot through with—is immanently displaced by—other media and other mediations. In this critical pinch, I have reprinted "In Lieu of the Lyre" as it appeared in the *Advocate* and in Moore's *Tell Me, Tell Me* (1966), though I have also included transcriptions of the poet's interjections as heard on the 1965 recording, to which I urge readers to listen if they can (fig. 4.1).

Moore's first line announces that her efforts to evade the strictures of her *New Yorker* contract and write a poem for the *Advocate* will simultaneously rectify another institutional proscription: her debarment as a woman "from enrollment at Harvard." Bleeding the distinction between prose and verse, speech and literature, print and sound, Moore harnesses exactly these destabilizing mediations to effect a displaced lyric object, one that allows her to assert, vis-à-vis the Ivy League, just that measure of constructed, fluid, pluralistic authority that Cristanne Miller has identified at the heart of the poet's feminist achievement.[42] By exploiting lyric writing's intermedial condition, Moore challenges her own exclusion from Harvard by exhibiting full participation in its print-based networks of intellectual accreditation and prestige.

This reparative work transpires at the level of the poem's media unconscious. At the poem's surface, by flagrant contrast, we find Moore extravagantly deferring to Harvard's authority. "In Lieu of the Lyre" comprises a series of digressions, one atop the other, expressing gratitude to the *Advocate* for the invitation to read, and to Harvard more generally for its oblique promotion of her career. Though "debarred from enrollment" herself (Harvard and Radcliffe would commit to merging in 1969), Moore alludes to the significant relations she maintained with the university in her later years, summoning a complex social network of people and printed matter. Harry Levin, Harvard's Irving Babbitt Professor of Comparative Literature, had "invented" Moore's "French aspect" by detailing her affinity with the French classical tradition in a birthday *Festschrift* (published the previous year by Tamil poet Meary James Thurairajah Tambimuttu), though the reference also invokes Levin's direct aid in translating Moore's *The Fables of La Fontaine*

One debarred from enrollment at Harvard,   *"This is called, very*
may have seen towers and been shown the Yard—   *accurately, 'In Lieu of the*
animated by Madame de Boufflers' choice rhymes: *Lyre'—L-Y-R-E"*
Sentir avec ardeur: with fire; yes, with passion;
rime-prose revived also by word-wizard Achilles—[i]   [i] *"See note"*
    Chinese Dr. Fang.

The *Harvard Advocate*'s select formal-informal craftly rare
invitation to Harvard made grateful, Brooklyn's (or Mexico's)
        ineditos—   [ii] *"In T. Tambimuttu's curiosity*
one whose "French aspect" was invented by   *entitled* The Festschrift for
    Professor Levin,[ii]   Marianne Moore. Well..."
the too outspoken outraged refugee from clichés[iii] particularly,   [iii] *"I"*
    who was proffered redress
        by the Lowell House Press—[iv]   [iv] *"No, no"*
the Vermont Stinehour Press, rather. (No careless statements
to Kirkland House; least of all inexactness in quoting a fact.)[v]   [v] *"Hard little rhyme there"*

        To the *Advocate, gratia sum*
    unavoidably lame as I am, verbal pilgrim   [vi] *"'Pilgrim' is supposed to*
    like Thomas Bewick, drinking from his hat-brim,[vi]   *rhyme with 'brim,' and 'him'"*
    drops spilled from a waterfall, denominated later by him
        a crystalline Fons Bandusian miracle.

It occurs to the guest—if someone had confessed it in time—
that you might have preferred to the waterfall, the pilgrim and hat-brim,
        a nutritive axiom such as
"a force at rest is at rest because balanced by some other force,"[vii]   [vii] *"Comma"*
or "catenary and triangle together hold the span in place"
        (of a bridge),[viii]   [viii] *"Parenthesis"*

or a too often forgotten relevant thing, that Roebling cable
        was invented by William A. Roebling.[ix]   [ix] *Sic.* John A. Roebling

        These reflections, Mr. Davis,   [x] *"Then there are some notes,*
            in lieu of the lyre.[x]   *which we'll dispense with"*

Figure 4.1. Transcription of Marianne Moore's reading of "In Lieu of the Lyre" at Sanders Theatre, Harvard University, May 11, 1965.

(1954).[43] The "redress" provided by Lowell House Press/Vermont Stinehour Press refers to yet another printed item, the 1963 pamphlet publication of *Occasionem Cognosce*, a poem whose title adopts the motto of Harvard's Lowell House, where Moore was a frequent guest in the 1960s.[44] Even the ardent intellectual "passion" inspired by the sight of Harvard Yard, which Moore first glimpsed in 1941, finds expression only by way of the poet's debt to the institution; "Sentir avec ardeur" is a poem by the eighteenth-century aristocrat the Marquise de Boufflers, the existence of which Moore discovers, footnoted, in Harvard professor Achilles Fang's translation of Lu Chi's *Wên-Fu*, a third-century verse essay or "rhymeprose" on Chinese literature, itself published in the *Harvard Journal of Asiatic Studies*.[45] Given Moore's assignment—to write a poem *sans* verse—we can assume she took special delight in the "word-wizard" Fang's coinage-into-English of the German term for medieval rhymed prose, *Reimprosa*, which he applies in this case to ancient Chinese literature.

Whereas "rime-prose" reminds us of the poem's strategically uncertain generic status, the overall effect of all these registered debts is to assert that like so much of Moore's writing, "In Lieu of the Lyre" is constellated from *other* print media: in this case, from other books of verse, *Festschriften*, pamphlets, scholarly journals, memoirs, and cultural histories. In flashing the Harvard imprimatur of her print sources, Moore winkingly performs a nigh-philological "exactness" that defies her exclusion from "Kirkland" or from any of Harvard's residential houses. Consequently, Moore's expression of gratitude, delivered in a bewildering wash of printed matter, has its tongue firmly in cheek. Acknowledging as much amplifies the ironic overtones in Moore's ridiculous self-deprecation as "Brooklyn's (or Mexico's) *ineditos*" or "unpublished," as "unavoidably lame," and as a mere "verbal pilgrim." The circumstances of the poem's production dispel the strangeness of these self-effacements: Moore is "lamed" by her contract with the *New Yorker*, which prohibits her from offering new verse to the *Advocate*. The "waterfall, pilgrim, and hat-brim," meanwhile, are lifted from a wood engraving by Thomas Bewick (1753-1828). Moore implies that, like Bewick, she has taken what is modestly at hand—hat and waterfall, or the "fact[s]" and "reflections" sourced from her reading—and passed them off as poetry, as some "Fons Bandusian miracle," invoking Horace's ode on the Bandalusan spring, famous in Western literature for its "loquaces/lymphae," or "garrulous streams."[46] Moore is a "verbal pilgrim" at Harvard, reminiscent as Miller suggests of

Virginia Woolf, the trespasser at Oxbridge, but she is also a pilgrim—a foreigner, *peregrinus*—venturing beyond the pale of the poetic.⁴⁷

The poem concludes with one further deferral, though, for Moore fears even the prosy observations assembled from her reading and her history with Harvard may prove too "poetic" after all. She wonders whether her audience would have preferred, over her ekphrasis of Bewick, some "nutritive axiom" of science and industry—some "surely relevant thing" picked up, for instance, in Alan Trachtenberg's new book of cultural history, *Brooklyn Bridge: Fact and Symbol* (1965).⁴⁸ It's the distinction between fact and symbol that governs Moore's sense of what counts as poetry and what does not, as if to compose a poem of merely factual "reflections" were to write "in lieu of the lyre" and thereby evade the *New Yorker*'s proscriptions. When Moore submits her fair copy to the *Advocate*, she advises Davis not to print the poem should it seem "too flat . . . labored, and illiterate" ("I am depending on you to protect me").⁴⁹ But flat exertions straining against the silent letter are precisely what the situation's exclusionary conditions require: How else to enroll oneself in Harvard while gratuitously exaggerating one's debarment or write a poem while *not* writing poetry? By marshalling all manner of fact and printed prose into these lines "in lieu" of poetry, Moore installs an exemplary heteronomy at the very (displaced) center of her poem.

To be sure, the incorporation of extra-poetic matter has long been a key element in Moore's practice, and the tactical negation of literary convention is perhaps *the* strategy of the poetic avant-garde.⁵⁰ But the displacements at issue here run deeper than the avoidance of normative poetic subjects. At the Sanders Theatre, we find Moore working "in lieu of the lyre" in an altogether stranger and more material sense. Take, for instance, Moore's allusion to Dr. Fang's translation of Chinese "rime-prose," the source for "Madame de Boufflers's choice rhymes." In her notes to the poem, Moore quotes from the footnote in which Fang employs Mme de Boufflers's lines in a brief for paratextual economy: "As far as notes go, I am at one with a contemporary of Rousseau's: 'Il faut dire en deux mots / Ce qu'on veut dire; / Les long propos / sont sots. // Il ne faut pas toujours conter, / Citer / Dater / Mais écouter . . .' But I cannot claim 'J'ai réussi,' especially because I broke Mme. DE BOUFFLERS' injunction ('Il faut éviter l'emploi / Du moi, du moi')."⁵¹ "Say what you will in two / Words and get thru," directs Mme de Boufflers. Fang is apologizing for the length of his footnotes, and when Moore reproduces the former's remark in her *own* notes to the printed poem, she excises the second half of Mme de Boufflers's advice: "Long frilly / Palaver is silly . . .

You need not always narrate; cite; date,/But listen a while . . ."[52] It's ironic to spot this footnote-against-footnotes, which quotes only to contravene Mme de Boufflers's incitements to concision, itself quoted (and cited, and dated) in Moore's notes. It's doubly ironic when we consider how ambivalently Moore respects these notes in performance. She "dispense[s] with" the notes at the end of the poem, judging them not worth her audience's time, but mid-recitation she refers her listeners to just this piece of information regarding "Dr. Fang." "See note," she interjects, and the print-less Harvard audience laughs, amused to imagine themselves as readers of a virtual text. Ultimately this tangled intertextual web—yet another index of the poem's saturation in print sources—leads us to a simple observation: in the act of only partially quoting Fang's quotation of Mme de Boufflers, Moore suppresses the privilege that both these writers accord to *listening* ("écouter") as a salutary corrective to excessive speech and writing. But as indicated by her playful cultivation of a virtual textuality, for Moore, listening to a poet talk is *not* opposed to reading—it's not even opposed to *writing*. "In Lieu of the Lyre" so destabilizes the lyric object that listening emerges as a mode of composition, a reversal we shall shortly find elaborated more fully in the work of Russell Atkins, a Moore protégé. Poet and listener together create an object at the dynamic crux of intention and interpretation, set in permanent circulation between pages, stages, and records.

"In Lieu of the Lyre" is a strange kind of media object, one that is constitutively difficult to pin down, irreducible to any single object because that object, present and palpable though it is, will always also be materially elsewhere. Displacement, in this case, does not simply denote a plurality of material texts, as if critics need only adjudicate among different manuscript, print, and audio variants. Rather, lyric intermediality entails that the socially mediated tension between writing and sound is *internal* to the poem, present in any of its material forms. Even if we were to inhabit the perspective of Moore's Harvard audience and seal our analysis off from anything but the "audiotext," that analysis would necessarily involve the poem's print dimensions.[53] In the Sanders Theatre recording, we hear Moore enjoining her audience to create a virtual text-object, to effectively *write* the poem together, collectively evoking its print existence. Not only is "In Lieu of the Lyre" *about* networks of print culture; its audiotext refers to the absent page, enjoining the audience to "see note," for instance, and explaining eye-rhymes designed for visual appreciation: "'Pilgrim' is supposed to rhyme with 'brim,' and 'him.'"

Neither can we plausibly exclude the audiotext from study of the poem's printed versions, since Moore's visit to Harvard marked both the work's public debut *and* its occasion. Given that "In Lieu of the Lyre" centrally concerns Moore's debarred enrollment from the institution, the fact that the performance of her "enrollment" was captured on reeled tape now carefully preserved in Harvard's own Woodberry Poetry Room amounts to a crucial piece of interpretive context. In a broader sense, the special distinction of Moore's performance—with its implicit allegory for the displacement of the printed poem, and its showcase of the poet's remarkable ability to "talk literature"—lies in its vivid illumination of midcentury poetry's broader media condition. The disturbed and unstable objecthood of "In Lieu of the Lyre" exemplifies the complex, contingent interlacings of print and sound playing out across US literary cultures in the 1950s and 1960s, a moment when the public performance and recording of poetry were enjoying wide adoption. Though Moore's fame and her skill at the microphone made her a performer of unrivaled popularity, her situation is distinctive chiefly in degree, for these displacements index broader, historically specific developments in lyric's intermedial condition. Though lyric is always missing its lyre, the shape of that absence undergoes a transformation in the age of ubiquitous recording.

Moore's lifelong commitment to troubling the bibliographic integrity of her poems makes her an especially useful thinker with whom to examine this episode in modern literary history. According to Heather Cass White, editor of Moore's *New Collected Poems*, "the publication of a poem in a periodical, or the ordering of poems in a book, marked resting-places in her poetry's development, not its final form," and we have "good reason to think that her own death is, in effect, only a particularly protracted pause in a process of revision that would otherwise have had no end."[54] The most dramatic object lesson in Moore's revisionary practice is the perverse fate of "Poetry." Upon its initial 1919 appearance in *Others* magazine, the poem featured five long-line stanzas, but by the time "Poetry" was collected in *Complete Poems* of 1967, Moore had pruned it back to just its opening salvo:

> I, too, dislike it.
> > Reading it, however, with a perfect contempt for it, one discovers in
> > it, after all, a place for the genuine.[55]

Robin Schulze calls the poem "Moore's famous complaint against her own medium."[56] I hope the preceding discussion will license a misreading that

"In Lieu of the Lyre" 185

takes Schulze more literally than she intends: What if Moore's audaciously nebulous "it" refers precisely to the medium of poetry, that is, printed writing stressed by its sundry claims to sound? We might see Moore's insistent drumming on the opaque pronoun "it" as enacting some special pleading on behalf of a lyric object with material displacement and nonidentity at its core. However tendentious, this reading does conform with the conclusion to the version of the poem that appears in "Observations" (1924):

> if you demand on one hand,
> the raw material of poetry in
> all its rawness and
> that which is on the other hand
> genuine, then you are interested in poetry.[57]

Among the "raw material of poetry" circa 1919 Moore gathers "business documents and/school-books," but by midcentury, these materials also include *non*print media. The textual instabilities that mark Moore's poetics from the first now embrace technologized sound and cultures of performance, which further scramble and displace the lines between poetry and prose, talk and literature. In lieu of the lyre we find not only the nutritive axioms of schoolbooks but also the garrulous streams of sound and speech on tape and phonograph records.

But poetry's raw materials are only that. These famous lines announce *two* desiderata for poetry, and if an ascendant culture of ubiquitous recording displaces the lyric object by adding to the raw materials of lyric practice, then surely these media-ecological shifts can be expected to alter what Moore means by the "genuine." At the beginning of this chapter, I suggested we could adopt this passing distinction between the raw and the genuine to delineate the *two objects* composing any poem. I have focused thus far on the raw objects of poetic practice. Addressable by the disciplinary habits of literary history, these lyric objects are constituted by the archival gaze, furnishing matters of critical concern in textual editing, book history, and those fields of literary scholarship foregrounding social and historical context. The *genuine object*, by contrast, is a phenomenon proper less to literary criticism than to the discourse of poetics, the long history of thinking with and about poetry. In short, the genuine object signifies a poet or reader's overdetermined emphasis on the poem's aesthetic objecthood. This concept is open to immense historical and ideological variety, of course, since by "genuine object" we must include both the avant-garde imperative to work upon the "word as such" or "poem as thing"—that keynote of modern poetics reaching from symbolism through imagism and surrealism to late-century language writing and beyond—as well as the New Critical emphasis on the autonomous text-object.

What is the relation between these two senses of the lyric object? How do the materials of lyric practice, subject to the currents of media history, condition and contour our sense of what kind of object a poem is, what kind of phenomenal reality and objective force it possesses? It has long been apparent that ubiquitous recording alters the cultural ontology of poems. Building on the work of McGann and others, Charles Bernstein responded more than two decades ago to the burgeoning archive of recorded poetry by positing any poem's "fundamentally plural existence," replacing emphasis on the unitary text with a newfound regard for the multiple "textual *performances*" of any poem: "While performance emphasizes the material presence of the poem, and of the performer, it at the same time denies the unitary presence of the poem, which is to say its metaphysical unity."[58] Bernstein's revocation of "unitary presence" authorizes attention to elements of poetry's social practice—embodiment, ephemerality, aurality, and performance—long subordinated to the poem as an ideal "textual entity." All the same, dispensing too quickly with claims to "metaphysical unity" can inadvertently obscure important sites of materialist inquiry, for the long history of Western poetics teaches that not all presuppositions of unity are created equal. All, in any case, are created.

When we make a poem we make it twice, which is to say that poets and their readers fabricate from raw lyric objects a "place for the genuine"—an ideal objectification of lyric materials. This is a social process, and by tracing the variously historical character of these metaphysical unities, by tracking these desires for poetic objecthood to their sociohistorical sources, we attend to the ways that ideas about poetry—and maybe even that single ideal object, Poetry itself—are produced just *as* media artifacts. Moore has invited us to witness the displacements of the lyric object wrought by the age of ubiquitous recording. The rest of this chapter will explore how new conceptions of the genuine lyric object—new conceptions of *what poems are*—reflect the midcentury shake-up of its raw materials.

Russell Atkins, Literalist of the Imagination

In July 1951, Marianne Moore appeared on radio station WEVD's *The World in Books*, where interviewers asked the poet to recommend for listeners "two or three younger talents who perhaps are not known, but who, in your opinion, should be."⁵⁹ One of those writers was twenty-five-year-old Russell Atkins. "Atkins lives in Cleveland," explains Moore:

> He uses spaces, reiterated by characters and attenuating devices, staggering lines down the page somewhat in the manner of E. E. Cummings. He says, really quite touchingly, I think, "I never write typographically to hide or make up for anything." His sincerity and modesty go a long way with me, and so do his lines. These, for instance, about a locomotive:

>> Trainyard by Night
>>
>> Nn and UNandDER THandUNandDER
>> and you just know
>> its huge big bold blast black
>> hiss insists upon hissing insists
>> on insisting on hissing hiss
>> hiss s sss ss sss ssss s
>> ss sssss ssss
>> when whOOsh
>> the sharp scrap making his fourth lap
>> with a lot of rattletrap
>> and slap rap and crap—
>> I listen in time to hear coming on
>> the great Limited
>> it rolls scrolls of fold of fold
>> like some traditionally old—
>> (coldly meanwhile hiss
>> hiss insists upon hissing insists
>> on insisting on hissing hiss
>> hiss s ss sss sss ss⁶⁰

Reporting back to Atkins, with whom she had been corresponding for several months, Moore worries that despite carefully scoring her reading copy with cues for emphasis, tempo, and pronunciation (the opening syllable should be managed with a "bee-sound hum," for instance) she had botched the recitation.⁶¹ Regardless, Moore's on-air shout-out surely marked a highlight in the early career of this unusually precocious poet and composer.⁶² Perhaps the youngest late modernist running, Atkins for several years had been placing his work in avant-garde magazines such as Charles Henri Ford's

View and Marguerite Caetani's *Botteghe Oscure*. As a Black poet of the Midwest, whose "homegrown version of French Lettrism" stood out against the staid formalisms and speech-based naturalisms of the 1940s and early 1950s, Atkins was heralded by Langston Hughes and Arna Bontemps's *The Poetry of the Negro* anthology (1949) as a harbinger of the new Black poetry to come.[63] Hughes—like Moore another of Atkins's flattering fans and correspondents—would count him in the pages of *Phylon* and elsewhere as among the "three or four good young poets" yet to publish books.[64]

With hindsight, it's difficult not to regard Atkins as somehow also *too* young, his mature poetics of the 1950s emerging just one half step ahead of postwar literary developments that might have offered his poetry a more nurturing reception. Eight years older than Amiri Baraka, for instance, Atkins's exuberant dislocations of grammar and syntax, and his playful embrace of the material page, cleared aesthetic ground for the Black Arts Movement, and in the late 1960s, he would mentor the movement-affiliated Muntu Poets of Cleveland. At the same time, Atkins's stock in a familiarly modernist brand of aesthetic autonomy would lead him to insist on his own distance from art too narrowly "preoccupied with 'other people' or social problems," which tend to induce an "uncreative realism."[65] Aldon Lynn Nielsen, Atkins's most dedicated critic, frames the poet as a bridge figure linking Hughes with younger writers of the 1960s, a figure whose "deformative folk sound" and "experimentalist folk chirography" we ought to recognize as an early "form of black postmodernism."[66] Only in the second decade of the twenty-first century has a surge of scholarly interest in Atkins promised to enlarge upon Nielsen's insights and rescue Atkins from the literary-historical impression that he is stranded between poetic modernism and what comes after.[67] In the 1950s, by contrast, Atkins had to marshal, on his own, an intellectual context for his poems and essays. These appeared most frequently in *Free Lance*, the Cleveland-based little magazine he founded in 1950 with Helen Collins and Casper L. Jordan. Featuring white poets like Robert Creeley alongside Black writers like Baraka and Conrad Kent Rivers, *Free Lance* thrived under Atkins's editorship for an astonishing thirty-year run at a time when "black-bossed magazine[s]" were few and short lived.[68]

It was in the pages of *Free Lance* that Atkins exercised his editorial freedom to develop, between 1955 and 1958, his musical theory of "psycho-visual object-forms." This deeply idiosyncratic theory provides an essential critical lever for uncovering both the originality of Atkins's verse on the page and its media-historical significance, for Atkins's most distinctive poetic achievements have their conceptual root in his musical training and his pathbreaking musical thought. This book's first chapter drew to a close around the speculation that Zukofsky's Objectivist desire can be read as an indirect response to the auratic allure of the sound recording. Atkins, too, we find impelled by technologized sound to the objectification of music and poetry, and this impulse begins with a notably Zukofskian emphasis on the imbrication of sound and sight: music, Atkins will assert with rather extraordinary nerve, is a *visual* artform. And yet, the social meaning of Atkins's particular brand of objectification will carry us far afield of Shakespeare and anti-epistemological polemic. Psychovisualism is Atkins's bid for a "nondominant" aesthetic theory—a mode of *poiesis* disembarrassed of an otherwise debilitating reliance on exclusive forms of cultural knowledge and knowhow. I'll illustrate how this working theory, though formulated with regard to musical objects, all the same conditions the poet's subsequent work on the poetic page.[69] Eventually Atkins's poems will lead us back to Moore's distinction between the raw and the genuine, and to this chapter's central point of inquiry: the lyric object, in lieu of its lyre, formed and deformed and reformed again in the currents of media history.

Atkins's passion for music's *look*—its visual proportions—was touched off at an early age. In an autobiographical sketch recounting his childhood on Cleveland's East Side, he recurs frequently to his mother's "out-of-tune" player piano—"rolls and all"—on which she would delightedly "thump out" ("her term") "serviceable versions" of popular "warhorses" like Schubert's "Serenade" and Rimsky-Korsakov's "Song of India."[70] These early experiences engendered an absorbing concern with the materiality of musical compositions. The spectacle of sheet music "spread on the floor," the mechanical magic of the punched-out piano rolls, and even his mother's disharmonic "thumping," which registers the embodied thrill of music making at the same time as it ambiguously conflates playing the keyboard with pumping the player-piano's pneumatic pedals—these childhood details appoint the "out-of-tune" piano both a source and emblem of Atkins's radical inquiry into the objecthood of music.[71] This imperfect instrument is also the inception point of his compositional commitment to using "any old sound [he] wanted in any way [he] chose."[72] By his late teens, Atkins "was not paying that much attention to sounds"; he was "making the notes *do* things that *used* sounds . . . I went about playing this out-of-tune 'music' which I began to refer to as 'composition' and sound applied. I'm sure this was due to having written most of it on Mama's out-of-tune piano."[73] Given time to unfold its consequences, this decentering of sound—the formulation of a music literally "out-of-tune"—will emerge as the most audacious insight of Atkins's musical theory.

The generative disrepair of his mother's piano also affords a handy symbol for the uneven distribution of cultural resources conditioning Atkin's development as an artist. Coming of age in a city blooming with industrial prosperity, Atkins enjoyed degrees of access to an enviable array of arts institutions, from the Cleveland School of Art and the Music School Settlement to the Cleveland Orchestra and Cleveland Museum of Art, where, by age sixteen, special exhibitions of Picasso (1940) and Dalí (1942) had "clinched [his] determination to go experimental."[74] As an adult, though, Atkins would come to feel acutely the limits and perils of this access. In a 1963 essay for *Free Lance* entitled "The Invalidity of Dominant-Group 'Education' Forms for 'Progress' for Non-dominant Ethnic Groups as Americans," he formulates, in the thorny idiom of his "egocentrical phenomenalism," a fundamental principle of ideology critique: "that 'education' is often the mere objectification of another group's thinking processes; that words such as 'mind' 'intelligence' etc., do not represent incontestable properties of anything and need not be taken as seriously as they are taken in the dominant group's definition's context."[75] Atkins rails against the assumption, promulgated by a white bourgeois "cultural bureaucracy," that "creativity" is dependent upon "learning," the prior acquisition of culturally accredited knowledge, or, what amounts to the same, the transmission of cultural tradition.[76] "For a minority, this is dangerous reasoning for the future," since the universalizing pretensions of a bureaucratized "learning" so easily mistaken for "intelligence" are in truth contingent, structured in dominance, and profoundly limiting. Nondominant ethnic groups must instead nurture and protect what Atkins calls "THE FACTOR GENERATIVE OF TRUE FREEDOM: i.e., THE MARGIN OF 'MIND' INDEPENDENT OF LEARNING," that margin "THAT CREATES."[77] While "IT IS IMPOSSIBLE TO LEARN OF WHAT HAS NOT BEEN CREATED," "IT *IS* POSSIBLE TO CREATE WHAT HAS NOT BEEN LEARNED."[78] Under the rubric of a "non-dominant" mode of artistic practice, unprejudiced and uninhibited by prefabricated knowledge, all Atkins's aesthetic departures—from literary convention, from modernist precedent, from properly tuned pianos, and from harmonic technique in music—must be recognized as socially necessary innovations.

The most original consequence of Atkins's commitment to nondominant creativity is the theory of artistic production and reception he called psychovisualism. If Zukofsky's strategies of objectification reflect and reproduce aesthetic ideologies backed by patriarchal norms, psychovisualism promises to objectify lyric's raw materials in a concertedly nondominant manner. Beneath their elliptical stylings and abstract psychological framework, Atkins's essays on psychovisualism fortify the concept with pragmatic attention to the historical actualities of art production, from the embodied mind's capacities for making sense of music, to the realities of stratified access to musical instruction and musical materials. Put simply, the theory of lyric objecthood urged by psychovisualism affirms at its core the social mediation of lyric materiality.

This affirmation begins as an intuitive dissatisfaction with the reigning fashions of midcentury art music, and it was forged into positive shape through conversations in the early 1950s with Atkins's chief sparring partner, the composer Hale Smith, then a student at the Cleveland Institute of Music. As Atkins recalls, the friends disagreed "violently about Schönberg's twelve-tone technique."[79] Like other critics of serialism and of twentieth-century art music more broadly, Atkins argued that the "devices" of modern music do "not fulfill psychological functions for 'mind'"—do not bear expressive meaning—because their procedures are too arcane and their principles too abstruse for real-time listeners. But his true quarrel was not with the *difficulty* per se of cutting-edge techniques but with their claim to *inherent* significance—that is, the tendency to treat the tone rows of serialism, an arbitrary and unmotivated formal schema, as if they possessed some innate expressive value. Proponents of twelve-tone technique treat it like a language; it's not, Atkins advises.

Just here, his critique of serialism broadens, daringly, to impugn most tonal music as well. Atkins charges that composers routinely mistake the locus of musical meaning, endowing "sounds as found in nature" with an expressive content that it's the sole prerogative of composition to *invent*.[80] Hence the "error of music" secreted in the heart of the Western tradition.[81] Since the Greeks, "composition" has signified the organization of tone relationships derived from "a science of sound"—from physical facts of consonance and dissonance. But neither the twelve-tone scale nor any other preformulated arrangement of harmonic intervals bears "significantly expressive form meaning." Intervals—a perfect fifth, an augmented fourth, and so on—do not mean in and of themselves; scales furnish only "a collection of raw materials." The composer who regards sonic phenomena as anything more than raw material—anything other than stimuli to creation—abdicates the essential duty of composition: the task of creating what has not yet been learned.

By declaring that sounds, in themselves, possess no "semantical and syntactical structure," Atkins undermines the working assumptions of every composer who has relied on a ready-to-hand system of musical sounds, tonality included.[82] The only valid aim of composition is the expression of "form-meaning," understood as a constructed psychological state. When fully elaborated in his conversations with Smith, such a principle entails that "composing [is] but an extension of how we think, comprehension-as-composition."[83] " 'It's *psychovisual*,' [Atkins] told Smith. 'We see comprehension machinery at work by which the tones attach themselves although we fail to notice much of it because of the force of sound stimuli.' Composing is a *deconstruction* method that is fixed for the 'mind's eye.' " If composition is "but an extension of how we think," then the *object* of composition—the vehicle of expressive meaning—must pass before the "mind's eye," and also remain coextensive with mental processes. It must be both conceptual and visual—as much a thought as a thing. Where to look for such an object?

In the 1950s, Atkins spent long hours in the Philosophy and Religion Division of the Cleveland Public Library, where his most treasured resource was the library's "forty years of the *Journal of Experimental Psychology*," organ of the American Psychological Association. It was from Gestalt psychologists like Wolfgang Köhler and his American interpreters, and their studies of perception as a matter of closures and spatial relationships, that Atkins drew a theoretical foundation for the odd aesthetic entity he dubbed the "Object-Form."[84] Whereas a conventional composer begins with the harmonic palette of the twelve-tone scale, the psychovisualist starts with an Object-Form—a visual construct before the mind's eye—and then "deconstructs" this abstraction into the musical material of intervals, rhythms, dynamics, and texture.[85] At the scene of reception, the listener accordingly reassembles the Object-Form in the course of listening. Audiences are the real composers; writers merely deconstruct. What's more, making good on his early intuition that music is a matter not of sound but of what can be *done* with sounds, composition becomes "not a binaural but a VISUAL ART."[86] Any musical work is "only secondarily, a universe of sounds, but primarily a universe of objects."[87]

Though psychologists and composers have long noted the interactive relationship between aural and visual perception, Atkins's insistence on visuality as the *primary* dimension of musical form is unique.[88] We can dispel a measure of the theory's initial preposterousness by admitting the insight on which it's founded: the mind's use of spatial coordinates to distinguish and organize vibrational frequencies. Deep-seated convention leads listeners in many cultures around the world to identify sound waves vibrating more rapidly as "higher" than those vibrating at lower speeds, even though rate of vibration has nothing to do with vertical space as such.[89] According to Atkins, "The psychovisual composer accepts that 'high and low' is imposed by psychic phenomena on tones as a primary condition of meaningfulness."[90] Sound itself cannot be organized to express "form-meaning" until the mind takes interpretive hold, transposing sonic stimuli into a psychovisual field. Harnessing the language of Gestalt psychologists, for whom perception is a visualizing process comprising attention spans, closures, and liminal shapes, Atkins argues that the mind orders not only pitch but volume, duration, and timbre in spatial relation on a visual plane. The resulting psychovisual construct—the Object-Form—is simply what the mind *does* when it attends, for a prescribed temporal duration, to sound stimuli. The Object-Form is a perceptual field isomorphically related to patterns of actual brain activity—a cognitive record of musical perception, or "a great beautiful hallucination" in the words of Atkins's friend Jau Billera.[91] The afterimage of listening, the Object-Form flashes up as we harken to musical sounds organized to effect just that pattern.

The most extensive description of the psychovisual program, Atkins's 1958 essay "A Psychovisual Perspective for 'Musical' Composition," includes the depiction of one such Object-Form (fig. 4.2). We may be tempted to remark a similarity between this Object-Form and contemporaneous experiments in unorthodox graphic notation by avant-garde composers like Morton Feldman and Earle Brown, but Atkins's structures are significantly *not* scores.[92] The shape in figure 4.2 represents, in print, an idea or gestalt that the psychovisual composer "deconstructs" into musical "motions," or sequences of intervals, which can be indicated conventionally on a staff. Atkins calls the form a "blueprint": psychovisualism involves "deconstructing the blueprint of the Object as Idea to apply tones which listener comprehension re-constructs through successive hearings comprehension. They should be planned in a way that when comprehension inertia stops 'motion' an object appears."[93] Atkins goes to recondite lengths to explain the psychological dynamics by which, for instance, the "comprehension inertia" of a listening mind discovers the mental shape of a perfect circle in a series of musical notes. (Readers can listen to a performance of Atkins's *Objects for*

Figure 4.2. A psychovisual Object-Form as depicted in Russell Atkins's "A Psychovisual Perspective for 'Musical' Composition," 1955, p. 39, in *Russell Atkins: On the Life and Work of an American Master*, edited by Kevin Prufer and Michael Dumanis. Copyright © 2013 by Pleiades Press. Courtesy of the Russell Atkins Estate

Piano [1969] and assess for themselves the salience of the Object-Form.)[94] For our present purposes, we need only recognize the Object-Form as an *intentional* figure, perhaps even a figure for intentionality itself—for what is communicable, mind to mind, composer to listener, across socially instituted distances of time and space.

Though Atkins's essays nowhere make the point explicitly, psychovisualism is designed to redress uneven distributions of cultural knowledge. In modern art music, the ostensible complexity of compositional techniques drives a social wedge between production and reception, ensuring that appreciative listening remains contingent upon enormous outlays of cultural context and expertise—goods that are not held in common and that are reflective of "dominant" (white bourgeois) groups in any case. The psychovisual Object-Form, by contrast, requires from listeners no special familiarity with music theory or the history of Western music, nor indeed much prior "learning" at all. In other words, if psychovisual objectivity entails a strident idealism—and it certainly does—it's idealism of a strategically "nondominant" variety. By founding a compositional procedure not on tonal systems, or on any culturally specific aesthetic sensibility, but instead on innate perceptual processes identified by Gestalt psychologists as near universal, Atkins sets free both composers *and* their audiences to "create what has not been learned." Psychovisual compositions are unmediated by the prohibitive trammels of conventional technique, music theory, and cultural history—of all that is traditionally necessary for "music appreciation."

Owing to *Free Lance*'s relative obscurity and the slight number of compositions Atkins himself produced, psychovisual theory made few waves in the music world.[95] But his conceptual innovations were far from out of step with more prominent developments in the postwar avant-garde. Atkins's endeavor to distinguish "composition and sound applied" from Western music affiliates psychovisualism with the efforts of composers similarly compelled by critiques of "musicalization" (the process by which noise gains significance, and is thus domesticated, *as* music) and by bids to reenergize compositional procedures with new sounds and new modes of attention.[96] When confronted with Atkins's concept of "sound applied," critics have reached quite understandably for an exactly contemporaneous analogue in the musique concrète of Pierre Schaeffer, the French composer well known for applying recorded sound as compositional raw material.[97] Schaeffer's inaugural work in the concrete mode—1948's "Étude aux chemins de fer," or "Railway Study"—strikes an irresistible comparison with Atkins's early experiment in concrete verse, "Trainyard at Night," but closer inspection swiftly commutes this resemblance into a stark antithesis. Musique concrète relies on an orientation Schaeffer dubs "acousmatic perception," where sonic phenomena are strategically severed from their sources, such that the "dissociation of sight and hearing" attunes listeners to the "perceptive reality of sound, as such, as distinguished from its methods of production and transmission."[98] In Brian Kane's phrase, the Schaefferian tradition is distinguished by its emphasis on "de-visualized listening."[99] Atkins's "psychovisualism," meanwhile, prescribes the opposite, enjoining listeners to eschew sound qua sound in favor of its express visualization.

In truth, the antithesis of the de-visualizing Schaeffer and the psychovisualizing Atkins testifies to a homology running deeper than any to which superficial comparisons might attest. We can span the distance between Paris and Cleveland only by following Schaeffer beyond the concrete experiments of the late 1940s and 1950s to his theoretical masterwork, *Traité des objets musicaux*, or *Treatise on Musical Objects* (1966), wherein Schaeffer develops his account of the "objet sonore." The sound object is pointedly not the sounding body of an instrument, or the medium bearing sonic inscriptions, or even the physical sound waves themselves. Rather—and rather like its cousin the "Object-Form"—Schaeffer's object is "something perceptual."[100] "Coming from a world in which we can intervene" and yet *"entirely contained within our perceptual consciousness,"* the sound object is constructed in perception by the acousmatic posture Schaeffer calls "reduced listening," an attentional practice that brackets contextual meanings and isolates sound's "intrinsic qualities."[101] The reducing listener abstracts a knock on the door from any thought of knuckles and visitors and denudes a French horn's B-flat of the concepts "B-flat" and "French horn." In its complex makeup, the resulting object is hard to fathom; half world and half intention, it suggests an "objectivity bound to subjectivity."[102] Like Atkins's Object-Form, which reproduces an image of the mental processes spurred by sonic stimuli, the *objet sonore* presents an "objective correlate of the listener's perception."[103]

It's no historical accident that these two perceptual units, the sound object (c. 1966) and the Object-Form (c. 1958), share the quiddity of subjective intention. Their claims to ideality are in each case authorized by the materiality of sonic inscription. For his part, Schaeffer's debt to technologized sound could not be more obvious. When initially theorized in the 1950s, "acousmatic listening simply named a situation that had already become generalized through technological media of broadcast and of reproduction: radio transmission, magnetic tape, and phonography."[104] Splitting sounds from their worldly sources, "reducing" the scope of attention, these machines cultivate a listener's regard for sonic stimuli as such. And as the foci of perception shift from the sounding *matter* of the source to the durational *event*, perception is thrown back on itself, disclosing the objective-subjective hybrid Schaeffer terms the "sound object." Though such objects may be fetched just as surely by reduced listening to a live orchestra as to an MP3, the sound object can surface as an articulable ontological form only on the horizon of a world awash in the severed sounds of phonographs, radios, and tape players.[105]

Though somewhat less obvious, the latent remediation informing Atkins's psychovisual theory is similarly a matter of duration. Crucially, it *takes time* to recognize the Object-Form, this radically democratic vector for musical composition. In fact, "music takes years to see."[106] Whereas the eye can absorb relatively intricate forms in short order—imagine the first impression a Caravaggio or the *Taima Mandala* makes—the ear is less efficient at apprehending what Atkins calls "object-complexity density."[107] If such "densities could be compared," the same complexity yielded up by an object submitted to the "naked eye" for one minute might occupy the ear for an hour. It might take "a hundred years or so" for the listener to *recognize* that complexity as an enduring Object-Form. The relative impoverishment of the ear explains both why Atkins's few extant examples of the Object-Form (see fig. 4.2) are simple two-dimensional shapes—not photographs, for instance—and also why the only demonstrably psychovisual composers, according to Atkins, are the repertoire's "immortal" juggernauts like Beethoven: "To the psychovisualist, this is the heart of the matter of 'music' that is 'immortal' for it is the time that a work needs to develop what we will discuss presently: the 'illumination' by which it is seen psychovisually and is 'clear' or comprehended so definitely that performers must trace its exact 'form-qualities' or Object-Forms or suffer cruel critical attack. It is simply that the work has transmitted that 'agreement between sound and picture far greater than sheer probability would ever permit.'"[108] Since the Object-Form emerges only slowly from discursion—since it will take listeners "years of comprehension . . . to re-compose the deconstructed Ideal of the 'composer'"—any properly psychovisual piece "will necessitate hearing again and again."[109] Precisely here, in the requirement of repeated listening, we can tally the price paid for Atkins's nondominant compositional practice and its refusal to presuppose a level of culturally dominant know-how on the part of listeners. The Object-Form's long composure into meaning requires habitual listening, and as a result, pyschovisualism presupposes a specific material relation to music. Namely, one must possess the capacity to listen "again and again" to the same work. While an avid concertgoer, availing themselves of the full range of Cleveland's performing institutions, might be expected to hear a popular piece of music occasionally over several years, for most listeners the process of reconstructing a deconstructed Object-Form necessitates *recorded* music and the audile techniques of repeat listening.

When Atkins writes that "'new music' is impossible," then, he strictly means it; because it takes years for Object-Forms to assert themselves, by the time "new sounds" have been perceptively formalized into spatial dimensions, their newness will have expired. Lurking behind such claims is an unstated corollary, one so taken in stride by Atkins—a composer in the age of ubiquitous recording—that it's not worth spelling out: all modern compositions must be, as a matter of course, recorded, for how else will psychovisual forms reach their audiences? Technologized sound plays no explicit role in his elaboration of psychovisualism, but its latent presence empowers Atkins to wrest from dominant cultural practices a freer way of making music. With sound on records, Atkins can displace sound from music composition—and replace it with a visual object. Sonic inscription lies behind the ideal Object-Form, enabling its guarantee of equal access to cultural resources and to a nondominant creative practice.

While Schaeffer's sound object had the larger impact on avant-garde music, Atkins's Object-Form aspires to do more justice—its improbability notwithstanding. Just as phenomenologists must learn to bracket naturalized assumptions, reduced listeners of the Schaefferian variety must *practice*, as Michel Chion reminds us.[110] In other words, it takes training to disarticulate the foghorn from notions of fog, horn, ship, and harbor. The psychovisual attitude asks nothing of the sort. It requires only the ability to listen, over and over and over again, until the "scope of consciousness" grasps "tones as *stimuli*," and those stimuli stimulate the mind "*to complete what is not completed by the transmission's form itself*."[111] The listener need not understand anything about tonal structure, twelve-tone procedures, or the history of music to commune fully with the composer's intentions—indeed to "complete" those intentions. They only need time. And the frozen time of records.

In a 1970 special issue of *Free Lance* devoted to Atkins, the editors advise their readers to recall that Atkins's "'*Music*' space theory of eye and brain was prior to both space-age, Beatniks, 'consciousness expansion,' LSD, and a host of other thoughts that dominated the sixties. The entire rigmarole of 'multi-media' in its *recent* outlines dates back to and fades at Atkins's theory of psychovisualism."[112] Insofar as it promulgates a revised ratio of the senses in response to the social difference ubiquitous recording makes, psychovisualism is indeed a "multi-media" theory in the McLuhan vein.[113] Psychovisualism also counts as significant research and development into lyric intermediality, for at its core lies an objectifying idealization of lyric's raw materials—not a sublimated music of the spheres but a visible mental construct, an afterimage of perceptual processes that is distantly authorized by the stored time of sound recording. In the Object-Form we find, or fail to, another displaced lyre, one more recourse for processing the lyric tension between writing and sound. And like Zukofsky, whose investments in the signifier's material excess merges speech and music into tendentiously continuous practices, Atkins sets up psychovisualism as the conceptual spearpoint of an entire aesthetic theory, a "scientific aesthetic for technique in art" that will, by the end of the 1950s, direct the course of both his poetry and his music.[114]

It follows that Atkin's poems, like his musical compositions, also score objects.[115] The conceit of the poem-as-score had wide currency in the postwar period, when influential poets like Charles Olson and Amiri Baraka seized on the possibilities of a performative orality.[116] More fundamentally, in a manner this book has sought to vivify, the protocol of scored sound has been central to the long media history of lyric writing. Yet the scored poems of psychovisual theory are distinctive for two reasons. First, rather than scoring iterable performances, Atkins's poems orchestrate mental Object-Forms, structures seeded by the writer but fully shared by the reader. Atkins's linguistic structures refer to ideational entities that exist separately from the words on the page, just as the abstract Object-Forms of psychovisual music extrapolate from the possibilities of sound recording a concept of objectified sound that is a strange blend indeed of the ideal and the material: an object of perceptual intention that reflects social needs and the media conditions that make it possible. To wrest such claims into view, however, we must remark a second unconventional feature distinguishing Atkins's way of working with language. The ideal objects to which Atkins refers these poems are *not* what we would identify as signified content. The object (in all senses) of Atkins's lyric practice is what he terms "conspicuous technique."

As linguistic constructions, poems do not only score (phonetic) sound; unlike notes on the staff, words also *mean* semantically. Conspicuous technique is Atkins's conceptual tool for redressing the perversity of applying Object-Forms to lyric writing—for negotiating the fact that a putatively psychovisual poetics offers no account of signification. In Zukofsky's late Objectivism, semiosis evaporates in the aesthetic experience of love; for Atkins, by contrast, semantic meaning is reintegrated into a more general system of *expressive effect*. In the early 1950s, prior even to his development of psychovisualism, we find Marianne Moore incisively alive to this dimension of her protégé's poetics. Of his "Elegy to Hurt Bird That Died (buried in matchbox)," she writes in a letter: "This shows what you can do. The motion and mood are secure—eloquent. Only the words detract, (Excuse my rough and ready speech) [sic]."[117] Though Moore intends her judgment critically, her words echo those of her own champion, John Ashbery, who avers that Moore "communicate[s] [ideas] to us, in a definite way, by words which nevertheless do not express them."[118]

Both Moore and Atkins are masters, then, of "motion and mood," but only Atkins would parlay the "detraction" of word-sense and the promotion of "motion" into poetic fundamentals. Just as the subordination of sound (to sight) founds his theory of music, the subordination of meaning (to motion) furnishes the conceptual basis for conspicuous technique, a mode of writing that refunctions modernist poetic strategies by embedding them in Atkins's own phenomenological theory of creativity. Atkins premises this "non-representational" aesthetic he calls "egocentrical phenomenalism" on "an objective construct of properties to substantiate effect as object."[119] The psychovisual poet "*creates* an experience evolved independently of the stimuli that emanates from 'real' as experienced by himself." Put otherwise, signifiers and their real-world references are mere fodder for the creative process, raw material for fabricating the genuine. In the act of writing, the poet recomposes these stimuli into "motions," throwing off an "effect as object" independent of any predetermined "meaning"—that is, an effect that discourse will always fail to circumscribe. When the editor of *Accent* magazine rejected a poem on account of its "restless, perpetual ingenuity that acts to overwhelm rather than reveal," Atkins took it as confirmation that "my 'conspicuous technique' approach was working."[120]

Nowhere does Atkins imply that the experience of a poem corresponds to a particular mental shape of the sort that governs his musical works. Psychovisualism appears to insist on just this difference between music and poetry: the object of the latter is more an affective complex than a visible structure. In ironic fact, a psychovisual attitude toward poetry produces not visual figures but something closer to what music is commonly held to convey—something like "motion and mood." Together, psychovisual music and conspicuous poetic technique fulfill Atkins's avowed ambition to "'compose' like a painter and write poems like a composer."[121]

By privileging motion over meaning, and deactivating the distinction between form and content, conspicuous technique surpasses and skews all familiar aesthetic dicta. This is not the harmonious, equipoised union of

"In Lieu of the Lyre"   201

technique and meaning. "A technique should not serve meaning but rather meaning must not only *be* but SERVE technique," Atkins writes.[122] It's scarcely self-evident what it would mean for a poem's semantic meaning to *serve* technique, but Atkins insists, repeatedly, on the coextensiveness of creative practice and mental processes. The Object-Form is only an "objectification of mental life."[123] Atkins may be lending the word "technique" the amplitude of its Greek root, *technē*, which braids together the material labor involved in writing with the varieties of practical, "socially instituted know-how" that invest that labor with skill and transform it into a veritable craft, from the muscle memory necessary for moving a pencil to the history of poetic forms.[124] To the egocentrical phenomenalist, Atkins suggests, poets are only one class of "object-forming animal . . . e.g. birds, spiders, or beavers, etc."[125] Far from denigrating poetry, the conspicuous foregrounding of technique exalts the everyday maker excluded from the consecrated title of "author" by all aesthetic theories that persist in asking after the meanings of poems.[126] By equating technique with poetic effect, Atkins asserts the primacy of a *making* indifferent to all preformulated discursive meanings, a commitment that dovetails powerfully with his orientation to nondominant modes of creative practice.

Calculated effects of "mood and motion," Atkins's lyric entities are designed to be realized at the scene of reception, where readers join writers in the shared project of conspicuous technique—the creation of intentional, perceptual forms. "Lakefront, Cleveland," written in 1954 and published first in 1968's *Heretofore*, number seven in Paul Breman's Heritage series, is a privileged example:[127]

> The stretch cast out night's long,'d
> a hideous voyage of far
> under'd sepulchral sky,
> colossal as a grave's after. I
> stood by a monument of thrust rocks
> shouldered together, that tremendously
> vaulted and rent themselves over sea
>
> There was extremed
> wake of the city
> (a woman
>   somewhere having secreted her burden
>   cast in her toilet

a jellied foetus—
    a surgeon's blade
hysterically sharp)

waves slid away
a murmur of laps
    —there lo! saw I
it—pulp

There, as stretch cast out night's long,'d
and hideous voyage of far
under sepulchral sky
colossal as a grave's after,
this pulp came down
that wake of city—

Now, then, God, listen: I'd swear
I heard, heard low,
its sigh-sounds lapse
as from furious determination—
furious, horrid determination!
Though stretch cast out night's long;
though hideous voyaged afar;
though there was extremed
wake of the city;
though there's excruciation
under sepulchral sky;
though there is grave's after
and grave's before, I heard
I swear, some of furious determination—

heard go the sigh
before I swept it to muck
with a laugh of cry!

"Lakefront, Cleveland" was written during an especially repressive and deadly moment in the US history of abortion, when public-health and medical professionals were just beginning to convene in advocacy for the reform of laws criminalizing the practice.[128] In the longer view, this poem—like its gruesome counterpart, Atkins's poetic drama "The Abortionist" (1954)—partakes in a

"In Lieu of the Lyre" 203

lineage of male poets depicting abortions, a tradition to which Atkins contributes the note of ghastly callousness that exemplifies his urban gothic mode.[129] The poem is also characteristically Atkins for the manner in which the natural world throbs menacingly to the beat of human dramas—the "littleness" of the illegally aborted fetus "deified by the bigness against which it exerts itself," in Linda Myrsiades's phrase.[130] Telling details, from the word "pulp" to the hallucinated "sigh-sounds," indicate Atkins's familiarity with Gwendolyn Brooks's "the mother," the 1945 poetic landmark that begins, "Abortions will not let you forget."[131] I'll propose this grim lament is also in dialogue with another set of intertexts, two earlier poems that speak to the fortunes of lyric objecthood this chapter has sought to trace. I'm referring to Moore's two most famous seascapes, "The Fish" and "A Grave." In its orientation toward pure *effect*, Atkins's conspicuous technique ought by rights to hamper close reading, since literary interpretations tend to exposit technique, thereby subordinating it to meaning. But allegoresis is one way to slip this literary-critical double bind. Pressed into service as allegories of poetic objecthood, these poems by Moore together frame "Lakefront, Cleveland," despite the evident horror of its referent, as a meditation on the kind of lyric object that conspicuous technique can be.

In "The Fish" (1918), the speaker's attention fixes in the poem's latter half on a cliff face, a "defiant edifice" bearing "all the physical features of / ac- / cident—lack / of cornice, dynamite grooves, burns / and / hatchet strokes."[132] By calling attention to the poem's martial lexicon and restituting degrees of its "now-obscured social contingency," historicist critics like George Bornstein have encouraged us to peer behind decades of formalist reading and see the work for the "anti-military poem" it is.[133] But we needn't shut our ears to history or alienate the poem into a "commodified aesthetic object" to propose that this blasted cliff face may evoke the carnage of Verdun while *also* supplying a figure for the poem itself—an embattled and accident-prone but stubbornly transcendent object:

> Repeated
> > evidence has proved that it can
> > > live
> > > on what cannot revive
> > > its youth. The sea grows old in it.

The poem closes around the same nebulous "it" that nags the opening lines of "Poetry" ("I, too, dislike it"). Here, Moore transposes her riddling alloy of the raw and genuine into complexly temporal dimensions. Just as the battering sea surrounds but is also somehow "in" (filling the "chasm" of) the cliff face, which outlives it, the ideal objects that are poems survive with and against the "abuse[s]" of time and the "ac-/cident[s]" of history. We may even detect in these lines the voice of a weak idealism upbraiding any too crude historical-materialist: the raw "grows old" in the genuine; something just barely irreducible weathers history. But poetry's object remains fettered by time, all the same, because history "cannot revive its youth." Both the raw and the genuine objects deteriorate, both are subject to time; it's only that the former ages always a little faster. A slim fold in time appears to insulate the genuine object, setting it one half step in front (and thus outside) of history and soliciting any number of ideological names—from the aesthetic, to the moral, to the messianic. This mind-bending figure performs the difference that is the first step in marking off something genuine from something raw.

At least partly inspired by the same 1917 trip to Maine that gave rise to "The Fish," "A Grave" maps its allegory of lyric objecthood not onto the cliff's enduring solidity but onto the sea itself—a "well-executed grave"— and the human artifacts it swallows.[134] "[Y]ou cannot stand in the middle of this," writes Moore of the sea, and of death, for both demand the distance of the object, though it's easy to forget that both annihilate all signs of human presence from the objects they consume:

> the ocean, under the pulsation of lighthouse and noise of bell-buoys,
> advances as usual, looking as if it were not that ocean in which dropped things
>   are bound to sink—
> in which if they turn and twist, it is neither with volition nor consciousness.

Depth charges "dropped" from the land of the living to sound the limits of human agency, these indifferent, twisting "things" counter the ideal of poetic objecthood implied by "The Fish." No longer is the poem a "defiant edifice" enlivened by but importantly surviving its context; "A Grave" asserts the "things" we make are "bound to sink." If the former's outvying confidence supplies an emblem of consciousness, the latter portends the obliteration of mind and will. The privilege of both objects is *object permanence*, though one lives on because it just barely transcends its environs, the other because it's supremely material.

Side by side, the "defiant edifice" and the "dropped/things" raise the central question of lyric objecthood: by what means do poetic objects persist?

Is their continuance best understood as a function of volition, autotelic agency, and liveness or a function of a muter materiality, as of an object drowned in context and indifferent to human wishes? Through hard-squinting eyes, these two seaside allegories may snap onto our categories of the genuine and raw, but Moore's figures work properly to keep distinctions between the ideal and real dissolving before our eyes: the cliff claims transcendence only by the skin of its teeth, and the bone "A Grave" picks with existence is that the brutal sea (of history) responsible for those lifeless, serenely turning objects always "look[s] as if it were not."

Read in light of these intertexts, "Lakefront, Cleveland" seems to supercharge the dialectical subtlety of Moore's figures, throwing out in the process an alternative conception of lyric objecthood in league with Atkins's theory of conspicuous technique. The poem assails the ideological effort to lift a transcendently genuine object off the historical page, while at the same time illuminating the antisocial limits of a commitment to volition-less raw materials.

Whereas "A Grave" opens with a "[m]an looking into the sea," Atkins transfers Moore's oceanside existentialism to Lake Erie's "extremed" Rust Belt shoreline, where a "sepulchral sky / colossal as a grave's after" stretches to the horizon. As in "The Fish," the speaker meditates on a "monument of thrust rocks" that "tremendously vaulted and rent themselves over sea," but here the object most intensely at issue is not the cliff itself or any fish "wad[ing] / through black jade." As if to specify perversely the "dropped things" twisting and turning in the watery catacombs of "A Grave," Atkins describes an aborted "foetus" brought by sewage line to the lapping lake surf, "pulp" in the "wake of the city." And yet the qualities ascribed by the poem's anxious speaker to the bobbing fetus, refused by city and lake alike, invert those characterizing the "dropped things . . . bound to sink" in Moore's poem. The fetus's singular distinction is a "furious determination"—it appears to the speaker, unnerved by its "sigh-sounds," a terrifying specter of "volition and consciousness." The "determination" imputed to the fetus issues from its frightening habitation of the littoral space between "grave's after / and grave's before," death and life, the "excruciation" of nature and the "hysterically sharp" cruelties of a social order that alienates the mother to a parenthetical "(woman / somewhere . . .)." Seized between pity and horror, the speaker's own affective response gets choked in a vice of paradox, sending up "a laugh of cry."

But here allegoresis reaches an ethical limit. If "A Grave" and "The Fish" incline us to read the "dropped thing" in Atkins's poem as yet one more figure for the lyric object, we ought to feel keenly the *inadmissibility* of this interpretation. It's inappropriate to read the "jellied foetus" as a figure for anything, as any *object* at all, much less a figure for the object a poem is; the abortion—and the social order that renders the practice criminal and/or lethal and/or necessary—refuses the pat resignifications of allegory. Reading figural meaning into the fetus is wrong for how blindly such an interpretation would follow the speaker in relegating to the space of parentheses any and all social context not reducible to the encounter between speaker and fetus, and for how swiftly it bypasses the question of personhood at the agonized center of the poem. I want to tarry at this ethical limit, weighing the wrench it throws into our interpretive works, because the very *inappropriateness* of an allegorical reading illuminates the nature and significance of Atkins's lyric object.

The speaker's shocked, hallucinating interest fixates on an intolerable ambiguity issuing from the uncertain personhood of the fetus. The poem asks, can an "it" "sigh"? Can something never born be killed, and if so, is burial appropriate? In her extraordinary reading of Brooks's "a mother," Barbara Johnson has shown how grievous indecision around similar questions animates a structure of apostrophic address that is actually a double bind for the poem's speaking "mother." "If the fact that the speaker addresses the children at all makes them human, then she must pronounce herself guilty of murder—but only if she discontinues the apostrophe. As long as she addresses the children, she can keep them alive . . . It begins to be clear that the speaker has written herself into a poem she cannot get out of without violence."[135] Though Atkins's speaker is not a mother existentially locked into a specular situation of address, "Lakeside, Cleveland" also reacts to the fetus's tragically ambiguous address with a gesture of violence: "I swept it to muck." The speaker refuses to entertain the fantasy of personhood by listening any longer to its hallucinated speech-sounds. So unbearable is the speaker's own crisis of responsibility in the face of this "extremed" horror that they refuse even to let the fetus persist as an object in the world.

Returning to the ostensible difficulty of "reading" conspicuous technique, we can now trace a series of echoing correspondences between the terrible action of the poem, the scene of its interpretation, and the broader context of Atkins's poetics. The speaker's resistance to any meaning the fetus might possess rehearses how readers of this poem bear a responsibility to refrain from ascribing to the "jellied foetus" any figural significance that would detract from its literal force. Not only would it be wrong to call the fetus an allegorical figure for the poetic object; readers are not entitled to any code of meaning that would prove sufficient to the crisis of the person represented in this poem. But just here, just as we move to admit this latter claim, we discover in what sense the poem *does* proffer an allegory for Atkins's lyric objects: his poems are linguistic structures that constitutively parry representation.

Atkins's objects, we remember, have no truck with representation and the distinctions between form and content underwriting the concept. They aspire to present experience just *as* technique, and in this way to create with a reader that which has not been pre-patterned by dominant codes of meaning. "Lakeside, Cleveland" suggests that representation would fail an existential problem of this magnitude, one that implicates not only the social arrangements precipitating the secret abortion and the speaker who refuses to decide its personhood but also the readers who must repeatedly reassume the ethical burden of that decision. The reverberating violence that wrenches Atkins's syntax and radically destabilizes judgment can only be shared—an enigma to be lived with and through. "Technique," in Atkins's case, is a synonym for thinking and feeling together.

A lyric object so resolutely on the side of technique must abjure altogether the categories of the raw and the genuine. "Lakeside, Cleveland" aspires to be neither subject to history nor immune from it, neither shot through and shattered by material contingency nor a transcendent vessel of subjectivity. In refusing one and the other, in fact, it entails both possibilities. Atkins's phrase "furious determination" marks the collapse of the raw and genuine, for his lyric object is full of furious determination—like the defiant edifice, a monument to consciousness—and also furiously determined by material context. The objective effects that orient Atkins's mature poetics do not simply transcend their milieu as autonomous expressions, nor are they reducible to the negativities of external pressures. One might say they are designed in response to social needs, as objects of mutual attention and concern, but Atkins's robust conception of a nondominant creativity may lead us to conclude these lyric objects simply *are* social needs.

It only remains to be observed that both Moore's lyric objects—the genuine edifice and the volition-less raw thing—have direct referents in the protocols of technical sound reproduction. Moore's defiant edifice reminds us of her quip in the *New Yorker*, where her phonographic voice is wheat in the pyramids. Back of the genuine object we find a desire for the preservation of lyric speech against time and history. Meanwhile, the silent indifference of the raw object, utterly subject to time and history, indexes the posthuman materiality of sonic inscription. Atkins invites us to understand that *neither* vision accurately reflects the way lyric objects respond to a literary culture of ubiquitous recording. As his psychovisual Object-Forms readily demonstrate, the historical relation between technologized sound and lyric practice, and the opposition between the raw and the genuine that has been this chapter's map to the cultural influence of recording, are furiously mediated by the determinations of social processes.

Coda
Helen Adam and the Invention of Poetry

> If there were no poetry on any day in the world, it would have to be invented.
>
> —Muriel Rukeyser, *The Life of Poetry*

At "Limbo Gate"

The question of how literary scholars of the future resolve to study lyric writing will rest, ultimately, on their conception of lyric's past. At present, our sense of this past seems more unsettled than ever—or at least more unsettled than past scholars generally have had cause to admit. Indexing this generative derangement is a recent shift in usage of the word "lyric" itself. Once a term of art cherished by post-Romantic literary cultures or openly resisted by experimental poets, "lyric" now increasingly designates an interpretive protocol, a way of reading poems *as* lyric, that is, as expressions of interior thought and feeling that a reader can access and understand, via the mediation of print, without recourse to a great deal of social and historical context. While this book has tried to promote its own media-theoretical definition of lyric, and has in that sense contravened the tide of scholars keen to regard lyric as a literary critical operation, this venture has been possible only because these same scholars have so productively loosened lyric's idealized moorings and enjoined us to historicist study.

Sometimes dubbed the new lyric studies, this latter vein of scholarship is most vividly encapsulated by the opening gambit of Virginia Jackson's *Dickinson's Misery: A Theory of Lyric Reading*.[1] Jackson poses an enchanting question: If you walked into Emily Dickinson's bedroom a week after her death in 1896 and discovered a portion of her verses copied onto the folded and stitched bundles we now call her fascicles, and the rest of her now-monumental literary inheritance scribbled on letters, loose pages, scraps, memos, and envelopes, *would you be right to call these lyric poems?* Jackson designs the question to bring us up short, for we cannot "return to a moment before Dickinson's work became literature, to discover within the everyday remnants of a literate life the destiny of print."[2] Instead, we can wonder at what has transpired since then to make the conclusion that Dickinson wrote lyric poems appear both wholly inevitable and also naggingly insufficient. We can chart the consolidation around the printed lyric of the interpretive horizon that has so exclusively organized all that we have been able to say to one another—as readers and as literary historians—about poetry. By asking the impossible question, "What if Dickinson did not write lyric poems?" Jackson drives home that modern verse is always already read (printed, framed, interpreted) as lyric, which is to say, as that kind of poem "thought to require as its content only the occasion of its reading."[3] With the new lyric studies, this history floods back into relevance, beginning with such contextual artifacts as the desiccated insect that Dickinson mailed to a correspondent alongside her poem "My Cricket."[4]

This coda proposes a media-theoretical recension of the new lyric studies. Such a project can begin with the observation that Jackson's moment of definition (*this Dickinson verse is lyric poetry*) rests on a first-order perception easily rushed past on the way to generic identification. In short, this act of lyric reading depends upon Dickinson's inheritor holding up these nineteenth-century pages and seeing something *other* than writing. All claims to lyric as an alienable genre proceed from the acknowledgment that poems written in hymn meter comprise the special kind of inscription I have termed throughout this book *intermedial writing*.

To be clear, I don't believe Dickinson's writing could be mistaken for anything else. I'm suggesting, rather, that the decision to call a piece of text "lyric poetry" presumes an encounter with intermedial writing—an encounter passed through so rapidly, and taken in such easy course, that it may only occur at a level we could call poetry's material unconscious. These final pages aim to freeze and dilate the moment when a reader decides, in a swift and non-acknowledged realization, that intermedial writing is different from writing's other kinds. Inspired by Jackson, I use another thought experiment to link this automatic recognition of a material difference to two issues of far-reaching (and related) importance; namely, lyric's interdisciplinary appeal to media scholars and its vaunted claims to time-traveling autonomy. The first of these has been a central concern of this book; the other has been conspicuously absent.

Coda

Imagine, if you will, a scholar of media history. Let's say this hypothetical academic, a speaker and reader of English, is professionally credentialed by a department of media studies in the United States. For the sake of specificity, though we might just as easily cast a specialist in eighteenth-century French bookkeeping, decolonial soundscapes of the Caribbean, the Astrodome, queer digital counterpublics, or ancient technologies of the body, let's say this person is presently at work on a book about experimental postwar German film. Deeply read in history, philosophy, and the arts, this scholar is thoroughly at home in the aesthetic humanities. Except in one respect.

It so happens that by some providential fluke, some pleat or tear in the universe, *this person has never heard of poetry*. Their ignorance of poets and poems is unremittingly total. They learned no nursery rhymes as a child, missed every day of school when poetry was discussed, journaled no angsty confessions as a teen, never melted to the words of any pop song, sang no hymns, psalms, chants, or sacred songs of any kind, cherished no Valentine love poems. They've certainly attended exactly zero lectures on the subject. The word itself—"poetry"—means nothing to this scholar otherwise unimpeachably versed in bookish things. By a miracle of cosmic censorship scrubbing every potential referent from this person's ken, these three quick syllables are emptied of meaning. All the sublimely variegated linguistic practices to have ever gone by the name "poetry" are simply missing, blank spaces in the tapestry of global cultural history.

With this outlandish fantasy I mean to set the stage for the entrance of poetry into this scholar's perversely omissive lifeworld. Imagine that while combing the archives for their chapter on the German filmmaker Rosa von Praunheim, they stumble upon an unmarked envelope, and out of this unassuming portal between worlds—the one devoid of poems, the other replete with them—they shake out three items: a sheet of paper with printed text, a page of sheet music, and a cassette tape. All three artifacts are labeled "Limbo Gate" (1974) and attributed to someone named Helen Adam (1909-93). A quick Google search reveals that Adam is a Scottish-American "poet," twentieth-century champion of the literary "ballad," doyenne of something called the "San Francisco Renaissance," and—as it happens—an occasional actor, later in life, in films by von Praunheim.[5]

Curious, the scholar peruses the printed page, its text oddly huddled in the left margin. They apprehend the work clearly enough as a Cold War artifact, a short fiction transporting readers to the threshold of an afterlife teeming with victims of nuclear holocaust. Here Limbo seems a bizarro Eden, the elaborate extrapolation of a simple pun on Adam and "atom," where the biblical forefather is once again responsible for universal ruination:

> They were all in the Limbo Gate
> When Adam ended mortal fate,
> The naked, the naked,
> The hosts of whirling clay;
> All the awful swarms of man
> That trod on Earth since time began,
> They were in the Limbo Gate
> On Halleluiah Day.[6]

In Helen Adam's account of the apocalypse, *everyone* ends up in Limbo, regardless of their moral comportment in life. "Some were peaceful from their birth. / Some with blood had shrouded Earth." Jesus, Napoleon, Saul, St. Francis, Blake, King David, lovers and murderers both—"We are one"—are present for the roll call, and rabbit, fish, and stag, too: "All poor beasts that felt his hate / And shared in Man's disastrous fate. / I am in the Limbo Gate, / Glad at last, and free." Beyond good and evil, the world's end is a joyous affair, and no place for retribution or repentance. Adam's speaker eschews the trappings of any particular eschatological tradition, East or West, for a Bacchic, sun-dashed peace.

> Loosened now from time and space,
> The multitudes through Limbo race,
> The naked, the naked
> On Halleluiah Day.
> Immortal spirits dearly bought,
> Or but the flickering of a thought,
> A dream of Krishna's meaning nought?
> Green waves, and golden spray.

Only "gigantic Adam" is discomfited with his state. He sits "Silent and alone," abandoned by Eve, herself transformed into the planet Venus:

> Adam on his nuclear bed,
> Venus blazing at his head.
> Woman once, now star to shed
> Light along his clay.
> Adam sprawls in Limbo shade

Coda 213

> By a blazing star betrayed.
> Raging powers with which he played
> Toss his world away.
>
> Towers of atoms fall and rise
> Where gigantic Adam lies.
> Adam lies in Limbo Gate
> Dwarfing night and day.
> Eden's lark beside him sings.
> No tomorrow lifts her wings.
> Silence takes all living things.
> Green ways, and golden spray.

Here a dark Romantic feminism recasts nuclear physics as the "[r]aging powers" of Eve/Venus, fission as the original sin, and planetary extinction as a liberation into eternity. Because he "played" with Eve, the biblical Adam lost his world—again—but those atom-splitting powers return all souls to Eden. This particular Limbo marries Heaven and Hell in the Blakean sense, where "Energy is Eternal Delight" and the "Eternal Female groan[s]" to silence forever "all living things."[7] The new denizens of Limbo await the judgment of the Eternal Female, but the decree is a jubilant one, always and forever. "[E]very thing that lives is holy," pronounces Blake, and everything that dies, says Helen Adam, is a holy innocent.[8]

These Blakean echoes would be lost on our media scholar, of course, as would the reference to Rudyard Kipling's poem "Helen All Alone," from which Adam lifts her titular phrase.[9] One wonders, moreover, what ears and eyes innocent of poetry would make of the ballad's strong-stress rhythm, chiming euphony, and the patterned shapeliness of its broken lines. What might our scholar infer about the meaning and function of the formal peculiarities that distinguish this text from prose? If our imaginary reader is to any degree stirred, seduced, enchanted, frightened, or awed by Adam's language, how might they contextualize their response to the "[r]aging powers" of the poem's rhetorical and prosodic machinery, lacking as they do any sense of the printed object's live negotiation with precedent and tradition? Lying before them is a fiction on apocalypse, conveyed in a manner both elliptical and repetitious, unfolding in a spare language bent to alien rules, alluringly charged by some discipline unknown to the codes of prose. What *is* this thing?

Before they have a chance to solicit the opinion of anyone less blinkered, the scholar remembers that the envelope contained two further items. The cassette tape preserves a recording of Adam's performance of "Limbo Gate" at a 1976 reading alongside Robert Duncan and Allen Ginsberg at the Naropa Institute.[10] On the recording, Adam's lilting brogue—more song than speech—intones a strong-stress, up-tempo rendition, to a melody "straight out of the Scottish folk ballad tradition."[11] The sheet music, meanwhile, features the setting of "Limbo Gate" that appeared in *Songs with Music*, a 1982 collection of Adam's ballads with transcriptions by Carl Grundberg of the traditional tunes to which she wrote and performed them (fig. C.1). In that volume's introduction, Grundberg informs readers that he crafted this score on the basis of tape recordings from Adam's 1976 Naropa performance.[12] These three artifacts—the printed text, the score, and the cassette—all bear the title "Limbo Gate" and the authorial name "Helen Adam." What would our scholar make of the relation each bears to its others? If these works are versions of the same, is one a truer, more complete, more authoritative instance of the work? Unacquainted with the normative priority of print media in modern poetic history, our scholar would be hard pressed—assuming the objects are undated—to assign interpretive priority to any one incarnation of "Limbo Gate." It would be entirely possible, for instance, to ascribe precedence to the cassette tape; thinking like an ethnographer, our scholar might assume that not only the musical score but also the printed writing are derivative entextualizations of an oral performance. Then again, since sheet music is conventionally understood to score performances, it would be far from obvious that "Limbo Gate" was not originally a piece of music—a veritable "*song with music*" best represented by the notation of words and music—and its merely textual form just a partial record, less lyric than lyrics. That printed words alone would represent a standard, sufficient, or even representative instance of the work may strike our scholar of media as far and away the unlikeliest scenario.

I hope readers of this book are willing, by now, to grant that the oddly distributed object we identify as "Limbo Gate" would pique a media scholar's interest—that such a scholar would understand instantly how the riddle of its distribution calls directly upon their field of expertise. Our thought experiment obliterates the cultural memory of poetry so as to unclutter the issue of its material condition, putting into stark relief the intermedial dynamics of poetic practice. But let's be clear: it's not simply the *existence* of these three discrete mediations that secures our scholar's attention. What could be more run of the mill, for a theorist of media, than episodes of

Figure C.1. Score for Helen Adam's "Limbo Gate" as transcribed in Carl Grundberg's *Songs with Music: Helen Adam*, 1982, p. 41. Courtesy of Carl Grundberg

remediation—than novels become films become video games become Broadway shows? What makes "Limbo Gate" a special case, what solicits the full bore of a media-theoretical gaze, is rather the work's *intrinsic* intermediality. Each of these three artifacts strongly entails its others, presupposing their existence, and this degree of mutual implication ought to confront a scholar trained in medium-specific analysis as a fairly singular phenomenon.

Consider that *Harry Potter and the Sorcerer's Stone* (1997), the book, does not innately necessitate the Wizarding World of Harry Potter, the theme park, nor do even the most apparently cinematic of novels depend for their existence on realization in film. By contrast, the tension between sound and writing at the heart of lyric practice locks each instance of "Limbo Gate"— text, score, and tape—into a simple but ineluctable network of supplementary relations. The tape recording is both the basis for textual and musical transcription *and* the record of one specific performance in 1976. The musical notation scores such performances, though it may also affix the melodic shape of previous ones. Meanwhile, as a score and a record itself, the written ballad is phenomenally contoured by a missing sonorous share ("I can't write anything unless I'm writing it to a tune," admits Adam) that may leave even the poet's most sympathetic readers dissatisfied with the poems on the silent page.[13] Without the interpretive protocols that have grown up among communities of print readers, our scholar lacks ideological access to an experience of the written poem as a potentially self-sufficient object. This is importantly *not* the same species of decontextualized self-sufficiency enforced by the idealizing maneuvers of "lyric reading," the focus of critical interest for Jackson and others. The claim to autonomy at issue here is more fundamental: it's the ability to look at writing that clamors for sound and not look elsewhere for its music. Unschooled in the reception of printed sound, our scholar doesn't know how to read this writing *as* lyric—the kind of writing that misses its lyre—and to fold lyric's historical, prosodic, and ideological claims to sound into the silence of the page.

To linger one beat longer with our thought experiment's perverse reduction of poetry to just the technologies supporting lyric practice, it's worth considering how far an account of modern poetry departs from conventional literary histories when reconstructed with the benefit of our media scholar's ludicrous naivety. First, the de-privileging of printed poetry elevates lyric productions in other media, including an extraordinarily various array of musical practices, oral poetries, habits of recitation, poetic performance cultures, and above all—in the modern period especially—popular song, from folk tunes to hip-hop, effectively healing the rift we plumbed at the heart of Langston Hughes's songwork. The range of works assembled in this book is designed to herald proleptically this expanded archive. Second, such an account recognizes the generative absence of the lyre in poetry's history, a history that is in large measure an endlessly renewed *negotiation* of that missing lyre. To identify an ever-missing lyre is to describe the tension between modes of writing and modes of sounding as a motor force in literary history, a force inextricable from the social history of other media technologies.

This brings us to a third major frameshift: the record of specific poetic genres, forms, movements, fashions, techniques, and institutions—all the agents and indicators of poetry's past—are now relativized access points or nodes in a larger media history. Such an outlook has well-established precedent in literary scholarship on the ballad. To comprehend the long cultural processes that eventuate in the artifactualization of the ballad as a printed form, scholars attend not only to the emergence of modern print culture but also to the live traditions of musical practice and sonorous reception that inform the ostensibly silent circulation of ballads in books and on broadsides. To this end, ballad scholars listen through the archives for the tunes that once animated the composition and the social life of printed poems like "The Flowers of the Forest," discussed in this book's introduction. And they trace these intermedial dynamics as they fluoresce in the twentieth century, under new media conditions, into what I have termed *songwork*—lyric practice under the cultural-industrial auspices of the commodity form. In the transvalued version of poetry's history conjured from the media-minded ignorance of our scholar, the ballad and the pop lyric—as opposed to the Keatsian ode, say, or the broken linguistic crystals of Paul Celan—would occupy the very center of modern poetic history, and not its vernacular, popularized, attenuated, or suborned outer limits.

If I have tested readers' patience by laboring this gimmicky thought experiment, I wager they will forgive its tendentious fancy precisely to the degree it affords a glimpse of a literary history in which ballads do not need to be levered up to literary-critical interest by the consecration of print. By the lights of media history, lyric works with perturbed objecthoods (Partch's speech-music, Hughes's blues poems, Moore's talky recitations, and indeed ballads themselves) are not special cases but lyric artifacts par excellence.

Lyric Suspension

Given that the ballad is an exemplary intermedial form, one might object that I have prejudiced our media scholar's first encounter with poetry by selecting "Limbo Gate" in particular. Surely not all modern poems are sung, and still fewer are, like Adam's ballad, written to preexisting tunes. And if a poem lacks close association with melodic content in its composition, performance, or reception, it would be hard indeed to convince someone that a musical scoring of that work possesses significance equal to that of a printed text or recording. And yet, the example of Partch's speech-musical methods suggests that all poems *do* possess an immanent tune; lyric intermediality's overinvestment in alphabetic sound guarantees that poems are always accompanied, at least potentially, by sonorous phenomena. Partch's microtonal scales and instruments—his recording apparatuses—could find in any poem a melody, and so can the phonograph, tape recorder, and the iPhone's Voice Memos app. While it's unlikely that these speech-musical tunes will conform to tonal systems, and while each reader will probably hear their own unique music, these distinctions scarcely matter from our medially reductive perspective. In this sense, any poem would have served our experiment.

My specific choice of "Limbo Gate" only begins to matter when we disassemble our experimental blinders and begin reading the ballad not merely as a media artifact but as a *poem* after all. Though any lyric work—any piece of writing so constitutively riven between sound and print—would have secured the interest of our media scholar, Adam's ballad was, I admit, a motivated selection. We can unfold the significance of this selection—and thus of the ballad's special relationship to media history—by following our thought experiment around a turning point. I have suggested that the bare materiality of the displaced lyric objects we call "Limbo Gate" will intrigue and compel our scholar, confronting them with a material problematic they are uniquely equipped, as a thinker of media, to ponder. But all I've *really* said, thus far, is that medial analysis can shake up literary history and lyric theory. We have not yet engaged the question of what a media scholar has to *gain* from reading lyric writing—that is, what lyric history can offer media studies.

We must first invent poetry. Let's imagine our scholar knocks on the office door of any colleague in a department of literature, brandishes the materials of "Limbo Gate" with a querulous shoulder shrug, and gets promptly initiated into the history of poetry. To drape tinsel on an already kitschy set-up: in the Hollywood version of this interdisciplinary tutorial a time-lapse camera logs the sun's repeated rising and setting over a university office building, as a voice-over stitches together verses from Sappho, Sidney, Seferis, Sanchez, and so on. Who knows how long such a conversation would take, or where it would begin, or what incomprehensions would crop up along the way? I have already proposed that one so miraculously unhabituated to the history of poetry might be forgiven for wondering at the dominance of *print* in the history of modern poetry—for asking why cultures of performance and multi-medial artifacts do not play a larger role in the authorized history of a writing practice constitutively engaged with musical and linguistic sound—and so our scholar (exhausted, surely, by the enormity of their oversight) may enjoy a moment of satisfied expectation when the lesson turns to the ballad tradition.

Since lyric's generative tension between writing and sound accounts for much of what is enduringly fascinating about both the material history and the ideological substance of the literary ballad and its antecedents, the subject of the ballad forces into the open intermedial dynamics otherwise obscured by print. It was only by taking the ballad tradition firmly in hand that British and European poets of the eighteenth and nineteenth century could assert a live connection with ancient bards, oral cultures, and the contemporary lifeworlds of rural "folk," and if that period's "romance of orality" appears to be a complexly overdetermined structure of feeling, reflecting an array of cultural-national and aesthetic interests, its first condition of possibility is plainly the fact that even a printed ballad *sounds*.[14]

The ballad's "romance of orality" is also a romance with history. Susan Stewart has taught us that literary ballads emerge from complex structures of historicizing desire. "Distressed forms" like the forged and imitated ballads of the eighteenth century, distant ancestors of "Limbo Gate," evince the artifactual logic of faux-antique furniture or pre-ripped jeans.[15] Over and above their content, such forms are designed to *"mean* historically" through processes of "separation and manipulation," artificially removed from history so they can be reinserted more conspicuously therein.[16] Modern balladeers aspire to negate "the contingencies of their immediate history" by composing "new antique[s]" that will necessitate, on the part of readers, new feats of recontextualization: "an invention of a version of the past that could only arise from such contingencies."[17] According to Stewart, "We see this structure of desire as the structure of nostalgia—that is, the desire for desire in which objects are the means of generation and not the ends." We might recall the example of "The Flowers of the Forest," wherein Jean Elliot's attempt to reconstruct an early modern ballad by putting new words to an old tune results in a song that flagrantly flouts its own time stamp. Insofar as the ballad records the scene of its own oral transmission, its *content* is also its own (nostalgically projected) *context*. To lilt in the present about "Lasses a' lilting" in the past is to reroute an oral tradition through the silence of print. When a distressed artifact turns context into content, it arrests itself, effecting an aspirational suspension of historical time. Coleridge describes the "poetic faith" solicited by the supernaturalism of the literary ballad as a "suspension of disbelief."[18] The bid of a distressed poem for faith in a live poetic *tradition*, and the suspension of history that bid implies, is no less a matter of imagination and ideology. And yet insofar as this latter brand of suspension is authorized by the medial dynamics of lyric writing, it also supplies, or so I suggest, the material ground for claims to lyric autonomy.

Few poets have insisted on historical suspension as intensely as Helen Adam, who managed nothing short of a one-woman ballad revival in postwar American poetry. "She was a poet out of time, stretched," as Maureen Owen recalls.[19] Adam began writing ballads and other rhyming verse as a child in Scotland, publishing her first collection—*The Elfin Pedlar and Tales Told by Pixy Pool* (1923)—at age fourteen. By 1939, the year she arrived in the United States, the thirty-year-old émigrée had disavowed this "dreadful doggerel" but not its metrical structures and supernatural pulse.[20] When Robert Duncan met Adam fifteen years later in San Francisco, at just the moment the city's poetic renaissance was reaching fever pitch, she appeared a veritable time machine. "If she was an original in her revival of the English ballad tradition," writes Michael Davidson, "she was an absolute incarnation of that spirit in her performances, singing or chanting her songs in a high, strong voice that had retained its Gaelic lilt."[21] Duncan recalls an early, revelatory encounter in his workshop at the Poetry Center in 1954: "Helen Adam brought Blake's introductory song from *Songs of Experience* as an example of great poetry to a workshop and read it in a sublime and visionary manner, as if what was important was not the accomplishment of the poem but the wonder of the world of the poem itself, breaking the husk of my modernist pride and shame, my conviction that what mattered was the literary or artistic achievement."[22] Duncan's "modernist pride" is punctured not by the poem but by its "world," a distressed context conjured up by the "sublime and visionary manner" of Adam's recitation. Duncan learned from Adam "that there is a continuity of what is at work in Coleridge or Blake, and a life of poetry kept outside the fashions of the day—of a fireside art of wonder making."[23] In Donald Allen's epochal anthology *The New American Poetry*— where Adam, one of three women, is the only regular rhymer—Duncan calls her ballads "the missing link to the tradition" of the "forbidden romantics": "They fascinated; they seemed entirely anachronistic. There was the mere lapse of time through which they had traveled."[24] This "lapse of time" inheres in the ballads formally, in their recurring rhymes and refrains, as well as at the level of their supernatural content, where feminized power triumphs, again and again, over mortal affairs—where the bright tresses of a dead wife return from the grave to strangle a murderous husband, for instance, or when the "witch Queen" Anne Boleyn dictates the terms of her own execution: "A sword that will waft me/As if upon wings/To a world I remember/Unshadowed by kings."[25] As Norman Finkelstein has argued, Adam's notion that "the cycle of human life" is an "endless round of unfulfilled desire," and her resistance to progressive models of history, neatly repeats her commitment to the ballad, wherein—like Elliot in "The Flowers of the Forest"—she often "recapitulate[s] the tradition out of which she writes."[26] "Strangely enough, Adam's personal belief in reincarnation strikes us as appropriate," writes Finkelstein, since "many of her poems are themselves re-embodiments of Romantic masterpieces, transformed by a sensibility derived from the nineteenth-century literary tradition and tempered by intervening historical circumstances."[27]

It hardly bears mentioning that poems do not actually suspend time, for that would mean escaping, defeating, evading, or mitigating it. Scholars are right to look suspiciously at any claims to the contrary, not only where they most obviously appear, as cultural cliché or as glittering remnant of reified Romantic thought, but elsewhere too, in our finer habits of critical abstraction.[28] As theorists of lyric like Jackson remind us, the popular notion that a lyric poem is a "message in a bottle" from ages past, or a time-obliterating opportunity for readers "to step into the writer's shoes and speak the lines ... as though they were [their] own," are only the most overt specimens of aesthetic ideologies and practices of "lyric reading" that underpin, in subtler ways, modern literary criticism itself.[29] Sprinting past difference for identity, the idea that poems possess a specially flexible relation to time conspires with the imposition onto poems of an abstract lyric "I" that reliably mistakes its own constitutive whiteness, and its unmarked privilege more broadly, for universality.[30] In short, to identify anything extraordinary in the historicity of lyric writing is to court all kinds of bad transcendence.

Stewart tells us that distressed ballads do not escape time—they take us deeper *into* history, or someone's interested sense thereof. A ballad that wields its hypostasized context as content presents us not with suspension in its temporal sense—as adjournment or interruption—but with the kind of suspension figured by a cable bridge or a musical passage in which one note carries over, blending, into successive chords.[31] Richer than nostalgia in its possibilities for unsettling the historical present, this latter mode of suspension lights up a "constellation" in the Benjaminian sense, one "saturated with tensions," a dialectical convoking of "what-has-been" alongside "the now"—in the case of Adam, British Romanticism alongside the renaissance afoot in hipster San Francisco circa 1954.[32]

In Adam's case especially, the ballad's time-lapse suspensions are laced with affirmative possibilities. Combining Blakean visions, the queer supernaturalism of the Scottish border ballads, and an incipient second-wave feminism, Adam spins "dramas of female independence" from an interested recovery of the Romantic tradition, staking out a position of "gynocentric authority" among her male peers in the Bay Area.[33] In her ballads' "most common narrative," explains Davidson, "an earnest but naive male asserts his right to 'possess' love, discovering too late that he becomes possessed *by* it. The allegory is a Gnostic retelling of the Fall in which Eve, by eating the fruit of knowledge, is possessed of a divine wisdom about the unity of all nature."[34] Such is the drama unfolded in "Limbo Gate," where nuclear apocalypse raises the theme of suspension to metaphysical proportions, and where Adam redeems those cast between realms in what is not purgatory but permanent bliss.

In charting a correspondence between the eternal suspension celebrated in "Limbo Gate" and the distressed form of the ballad, I propose that the poem allegorizes poetic tradition itself as a realm "loosened . . . from time and space," where "[n]o tomorrow lifts her wings." Won through an assertion of female power, this transcendent limbo does after all image a radical homeland for poetry, a place where the likes of Blake and King David, who "loved his harp to play," join the ranks of the world's acknowledged legislators like Saul, St. Francis, Jesus, and Napoleon. In Andrew Marvell's "The Garden," yet another source for Adam's verse apocalypse, the speaker descries the supremely powerful poetic mind, an "ocean" of creative power, "[a]nnihilating all that's made / To a green thought in a green shade."³⁵ Raising the possibility that all reality is "but the flickering of a thought," Adam transmutes Marvell's image, heightening its splendor to an everlasting seascape of "Green waves, and golden spray." And yet she does so by turning the English poet's "garden-state" inside out; whereas in Marvell's case it's Eve's injection of human community and sexual delight that spoils this "place so pure, and sweet," for Helen Adam, it's Eve who wins back paradise, from the first man and for us all.

Marvell's high pastoral would seem an odd reference point for the low country of the ballad did "Limbo Gate" not so evidently call to mind the literary-traditional pairing of ersatz Edens and lyric suspension. Patricia Parker has carefully tracked this tradition, from the lament of the exiled Judeans who "hung [their] harps upon the willows" in protest against their Babylonian captors in Psalm 137, to Pan's pipes suspended in Jacopo Sannazaro's *Arcadia* (1504), to demonstrate its convergence in the Bower of Bliss episode of Edmund Spenser's *The Faerie Queene* (1590).³⁶ There we find Acrasia, the Bower's Enchantress, having so charmed and corrupted the knight Verdant to "lewd loues and wastfull luxuree" that the very sign of his impotence is his "warlike armes, the idle instruments / Of sleeping praise . . . hong vpon a tree."³⁷ Though Adam has fully transvalued it from a den of iniquity to an eternal paradise, her Limbo, ruled over by Eve/Venus, is a Bower of Bliss in just the terms Parker lays out: "The sense not just of lyric but of sexual contest within the stanza of Verdant's suspended instruments evokes a recall not only of Mars and Venus but of a whole series of subject males and dominating female figures, from Hercules and Omphale to Samson reclining in the lap of that Delilah who deprives him of his strength."³⁸ Often these episodes of lyric suspension and the "male subjection" they signify are but "one moment in a larger progression," typically resulting in reassertions of patriarchal dominance: Samson brings down the temple, the good knight Guyon destroys the Bower.³⁹ But no such backlash or last word arrives in "Limbo Gate," where "[n]o tomorrow lifts her wings" because the angel of history is forever done with flight. Acrasia's garden is a "threatening female space" not only because it enervates masculine power but also because it "arouses hopes of gratification" only to leave them unrealized or to "fulfill them only in an illusory or compromising way."⁴⁰ "Limbo Gate," by contrast, is all fulfillment, all the time.

Adam's ballad figures lyric suspension not by picturing enervation (Verdant's suspended instruments) or abjuration (the hung harps of the Israelites) but by instituting a particular type of powerful song. This is the kind of song that endures its own impossibility, just as, in the ballad's conclusion, "Eden's lark" continues to sing as "[s]ilence takes all living things." Though "Limbo Gate" makes no mention of suspended lyres, my interpretation rests on its *being* one: a song in silence, out of time, the music that persists when the harp is laid aside. Or—as I hazard we're now prepared to see—when that harp goes missing. The hermeneutic of lyric suspension I've been seeking to describe is, after all, but one expression of lyric's intermedial condition. Adam's time-lapsing performance of a late Romanticism and the apparently "timeless" currency of the distressed ballad are both authorized, in the last analysis, by that lyric tension between sound and writing coeval with the disappearance of the lyre. In its apocalyptic content "Limbo Gate" seems to gather in an echo of these intermedial dynamics, as if to suggest the end of the world is also the silence of print.

By yoking together the ideas of suspension we've developed with the help of Stewart and Parker, on the one hand, and the absent lyre of literary history, on the other, we arrive at this coda's central argument: *lyric intermediality suspends media history*. To repeat, I don't mean that lyric poems *stop* media history. Rather, they bridge, blur, and short-circuit historical moments by refusing—or at least ignoring—linear-progressive conceptions of history and moving at an oblique relation to developments in material culture. The point is not simply that readers of poetry in English still read Caedmon alongside George Herbert alongside Phillis Wheatley Peters alongside Marianne Moore alongside Joy Harjo. It's that all these poets, different in so many respects, nonetheless make the relation of writing to sound the cardinal object of their literary practice. While that object gives rise to an immense variety of linguistic shapes, one of its chief affordances is a relative *flexibility* and *freedom* from the determinations of media history.

That may seem a strange admission to find at the end of a book about the way historically specific technologies have conditioned poetic practice, but poetry's openness to the impress of media history is the result precisely of its freedom from it. Lyric intermediality means that poems can move from print to bodies to machines and back—in short, that poems can *travel*. Though we owe the portability of poems, certainly, to their size—their "trademark compression"—and to their fabrication of "memorable speech," we owe an even more fundamental debt to their intermedial privilege.[41] Already suspended between writing and sound, printed poems slip with relative ease across media boundaries. As a consequence, different reading

Coda

cultures can distribute poems across diverse material forms, and as these forms circulate, so are they distributed across time as well. Lyric's suspensive potential thus explains its persistence as a "new-old" medium, an increasingly *residual* literary practice with an ever-more attenuated relationship, in the twenty-first century, to the dominant media technologies of its cultural milieu. Even as poems circulate with unprecedented range and rapidity on digital platforms infrastructurally written by machine code, alphabetic lyric writing remains a practice in the kind of script that sounds and is thus wedded to pages and page-like screens.

We have already seen how "Limbo Gate" is materially suspended between print, score, tape, and sounding body. This rich distension is responsible as well, though in a more figurative sense, for the ballad form's apparent suspension across centuries and for the way Helen Adam the post-Romantic poet appears "out of time, stretched." Recalling our media scholar's precipitous baptism into poetry, here lies the point of sharpest critical interest for a thinker trained to analyze "materialities of communication."[42] Though any individual ballad is stamped by historically specific social processes, the form itself moves through modern print history with a relative degree of freedom from the determinations of material culture. In "the mere lapse of time through which they had traveled," ballads convene together disparate moments in media history. A poem like "Limbo Gate" operates like a slipknot of sorts, tying fast oral cultures, worlds of print, and the age of ubiquitous recording. Because the absence of the lyre means that lyric writing readily traverses media boundaries, the ballad can be sung, printed, transcribed, recorded, remediated, reinvented, and *remain* a ballad—that is, one particular literary practice in which writing gets made conspicuously to sound.

By contrast, the example of lyric writing brings home to our scholar that other cultural artifacts with less unstable media conditions lack this unique passport across time. Compare the case of theater, another decidedly intermedial aesthetic practice. The difference is available intuitively. *Hamlet* the Globe performance in 1601, *Hamlet* the Orson Welles broadcast in 1936, and *Hamlet* the 1996 film adaptation are three different mediations of a play by William Shakespeare. Each bears its own relation to the text. But if you were to line up an eighteenth-century broadside of an early modern ballad, a nineteenth-century transcription of its melody, and a twentieth-century recording of that ballad in performance, you would be left with the impression not of three distinct works but a single work in three guises, each disclosing various aspects of the same. What accounts for this contrast? It would appear that the confluence of dramaturgy, acting, music, dance, and set design in a theatrical work are not contoured by an insistent lack, pulled apart and together, in the manner that sound and writing, in lyric, so generatively rive literary forms. While a theatrical work does score future events, unlike a poem it doesn't also pretend to transcribe prior, now-absent phenomena. Future performances of the play are not implicitly linked by the manner in which each promises to answer a fundamental lack in the original.

Lyric intermediality distributes a poem across space and time, such that any particular poem is both a networked repository of material culture and a promissory note on the future. Stewart anticipates this latter point in *Poetry and the Fate of the Senses* (2002) when she proposes that "the sound of poetry is heard in the way a promise is heard."[43] Stewart means to emphasize that lyric sound is performatively efficacious. It's an "action made in speech" that "create[s] an expectation, an obligation, and a necessary condition for closure." While lyric sound is never fully present—always "recalled"—it's the commissive guarantee of that "good faith in intelligibility" that is the precondition for poetic address.[44] What endorses this promise, though, what makes it felicitous across time, is poetry's intermedial capacity. Because lyric exists between writing and sound, this promise can be constantly renewed; in other words, it "can be called on, called to mind, in the ongoing present," as Stewart writes, only insofar as poems can circulate among the historical bodies and material media that compose this ongoing present—not out of time but *always in* history.

In marveling at this consequence of lyric intermediality, our media scholar has discovered nothing less than a kind of poetic autonomy. Indeed, a qualified autonomy seems the marvelous upshot of lyric intermediality. Lyric's residual media-ecological position grants to poetic practice a degree of freedom from the technologies and social protocols that also, at every historical moment, underwrite it. One principal goal of this book has been to resolve the contradiction haunting my previous sentence. I'll adduce one last poetic example to show how much more swiftly Helen Adam settles it.

"The Birkenshaw" (1974) is inspired by the Scottish ballad "Tam Lin," a favorite of Adam's for its "mysteries of sex and the supernatural."[45] As in "Tam Lin," the former poem's antagonist is the murderous Elf Queen, who together with her "hunting maidens" rules the "Birkenshaw" forest, "riding the young men down" who trespass into her "green uncanny land" and sparing only those handsome ones they take to bed. But when Robin o'Leith, erstwhile bard of Atlantis, enters the forest with "his great harp," the Elf Queen declares the law of her land inhospitable to o'Leith and, more pointedly, to his claims for poetic autonomy:

> "What are ye doing, human harper,
> Breaking the Elf Queen's law,
> Playing your harp up Badenoch
> In the magic Birkenshaw."
>
> "I pluck my harp in the Birkenshaw
> Where the silver branches blow,
> For they mind me of the running waves
> When I harped Atlantis low."
>
> "I'll gae ye gold tae spend, Robin,
> And a gold crown for your head,
> If ye will enter my hollow mountain,
> And harp whaur my feast is spread."
>
> "I will na' harp for your gold, Lady,
> Nor yet for a kingly crown.
> The speech o' my harp can never be bought
> Though the hollow hills tumble down."[46]

Robin's refusal to sell his talents sets off a Bacchic chase. "I'll hae the head o' that proud poet," asserts the Elf Queen, spurring her fairy maidens to frenzied pursuit. The bard fleeing the fairy maidens recalls Orpheus's flight from the women of Thrace, but o'Leith is equal parts Actaeon and Proteus, for as the maidens hound him through the Birkenshaw, "this cunning poet" and "shape changer" transforms repeatedly: first "tae a white owl flying," then "tae a tom cat cursing," "a circling eagle," and a "league-long swimming seal." Adam puts her own new twist on the familiar folk trope of the transformation flight when o'Leith is finally run down in the Birkenshaw.[47] The Elf Queen advises the poet to destroy his own harp, for she intends to keep him "a thousand years / In [her] silent fair land." He responds defiantly:

> "The strings o' my harp are strong, Lady.
> The strings o' my harp are strong.
> My harp has ridden a doomsday wave
> Wi' a mane a rainbow long.

> When I have broken its strings, Lady,
> The floods o' my heart will flow,
> As once they flowed for the truth o' love
> When I harped Atlantis low."
>
> He's broken those harp strings clean and quick.
> The moon shone broad and chill
> As the great harper the Queen had got
> Strode intae the hollow hill.
>
> She's seized on him wi' her arms sae cauld
> But he melted frae her clutch.
> He's changed his shape tae the holy harp,
> And that she daur na' touch.
>
> His sangs flew up like birds about her
> And blinded her wi' their wings,
> Till his banes became the base o' the harp
> And his hert became its strings.
>
> The harp stands in her hollow mountain,
> And whiles the harp will sing.
> Pure and strong is the harps' voice
> Wi' nane tae pluck a string.
>
> The harp utters the truth o' love,
> And tae a' the host that hears
> A thousand years are but as a day,
> And a day a thousand years.

In this final shape-shift, in which o'Leith self-destructively *becomes* a harp, securing for lyric practice a kind of monumental afterlife, Adam implies that her poet's autonomous relation to Birkenshaw's sociopolitical order was fundamentally *a matter of time*. It is not by resisting the Elf Queen on the plane of action where she reigns supreme that o'Leith defeats the Elf Queen but rather by establishing the conditions for lasting song.

By transforming himself into an aeolian harp, o'Leith has not escaped the punishment meted out by his enemy. He has merely, if crucially, changed its terms. He'll languish in "fairy land" "a thousand years," but he won't be "silent." He trades a bad eternity for a good one, and anyone who hears this song is likewise suspended in historical time-lapse. All this requires is o'Leith to forfeit his life; as in "Limbo Gate," eternity is won only at the cost of a violence that we're now prepared to grasp as nothing other than the violence of history itself. Finkelstein is right to locate a wise contradiction in this image of the flesh-and-bone harp, which sings "against the demands of time" even as it's "still made—quite materially—from the individual poet, who must always exist within time, vulnerable to its tyrannical authority."[48] But the figure of contradiction is not quite right, since for Adam, and for the ballad tradition at large, art and time are not hostile concepts.

In this poem, lyric's missing lyre steps forth to figure art and time, like the force exerted by the two brackets of a lyre string, in resonant tension. Lyric suspension is a way of refusing the distinction between poetry and history while also retaining for the former a degree of motility, the ability to syncopate or sound off. Adam's hero harps against the law of the land, but by fully dissolving into his medium, by giving himself over to the suspensive properties of lyric materials, he does so always *in* time. Whatever it is, the "truth of love" is *in* history.

Acknowledgments

This book began with the intuition that on the subject of modern poetry's relation to music, scholars could stand to take the claims of poets and composers more seriously than has often been the case. The project has survived because more than a few people took *me* more seriously than I have perhaps deserved. I need to thank, first of all, Jeremy Braddock. His critical acumen and unflagging encouragement have nurtured this book from its inception, and his own sense of the project's furthest horizons and deepest stakes has been an indispensable guide. I'm grateful also to Rayna Kalas, who set this study on course by urging me, early on, to embrace my hunches (why not just call it "the lyre"?), and whose galvanizing insights—arriving like Vico's lightning bolts—have been this project's goads and guardrails. Finally, thank you to Tom McEnaney, who showed me, before I had even the slightest inkling, what poetry has to do with media, and who taught me how to listen. All the missteps are mine, of course, but my highest hope is that this book will warrant, even just by half, their outlays of generosity, professional grace, and enthusiasm.

My life to date is one long schoolroom, and I'm grateful for early and formative visits from some remarkable teachers. At Oberlin, Desales Harrison, David Walker, and the late Jed Deppman lit the fire. Bruce Weigl arrived at a pivotal moment to show me what saves us. Wendy Flory, Marianne Boruch, and Donald Platt, at Purdue, modeled a life in poetry, and Mary Leader, master spinner, taught me what to make of the threads I had to follow. I hope she sees her handiwork work here.

Ithaca was a wonderful place to weather the PhD, and I'd like to thank my teachers at Cornell: Paul Sawyer, Cynthia Chase, Jonathan Culler, Tracy McNulty, Patrizia C. McBride, Laurent Dubreuil, Timothy Murray, Roger Gilbert, and Greg Londe, inspiring examples all. My abiding gratitude, as well, to Liz Anker, Shirley Samuels, and Kara Peet, for their essential help in navigating the job market, and to Caroline Levine, for some wise counsel in a pinch. Stuart Davis shared with me his sense of the past, and my chapter on Marianne Moore is stronger for it. And thank you to Joe Martino, Cornell class of '53, for funding the Joseph F. Martino Lectureship in Undergraduate Teaching, an appointment that gave me breathing room to redraft this book. Finally, with heartfelt gratitude for their friendship, I thank Chris Berardino, Marquis Bey, Ben Fried, and Sasha Anemone, who showed me what this Ithaka means.

I had the extraordinary good fortune to wind up at Notre Dame, in a department bursting with serious and seriously kind scholar-teachers. Thank you to Romana Huk, the spirit of poetics at ND, for her generous mentorship, and to my colleagues Laura Knoppers, Laura Betz, Steve Fredman, Joyelle McSweeney, Johannes Göransson, Barbara Green, Kate Marshall, Barry McCrea, Nan Da, Sandra Gustafson, John Duffy, Stephen Fallon, Susan Cannon Harris, Essaka Joshua, Jesse Lander, Tim Machan, Sara Maurer, Ian Newman, Susannah Monta, Mark Sanders, Roy Scranton, Yasmin Solomonescu, David Thomas, Orlando Menes, Elliott Visconsi, Alissa Doroh, Lynn McCormack, and Kelly Huth for going out of their way to show me the ropes. It is a pleasure—and in these dire days of the profession, an astounding privilege—to work among an electric group of junior faculty: thank you to Francisco Robles, Brandon Menke, Ranjodh Singh Dhaliwal, Sara Marcus, Chanté Mouton Kinyon, Dionne Bremyer, Xavier Navarro Aquino, and Katie Walden. Within the department, my thinking on verse has been refreshed by the stalwart Poetics crew: thank you to Sara Judy, Jake Schepers, Sally Hansen, Spencer French, Nidhi Surendranath, Jenkin Benson, and Kyler Schubkegel. Beyond the English Department, my thinking on technology has been vitalized by the Navari Family Center for Digital Scholarship and the Lucy Family Institute for Data & Society: thank you to Daniel Johnson, Julie Vecchio, Katie Liu, and Nitesh Chawla for their intellectual hospitality and institutional support.

This book has received generous financial support along the way from the Institute for the Scholarship in the Liberal Arts, College of Arts and Letters, University of Notre Dame; the American Council of Learned Societies; the Beinecke Library at Yale University; the Northeast Modern Language Association; the Cornell Society of the Humanities; and the Cornell English Department. Since many of the sharpest turning points in this book's genesis occurred in archives and reading rooms, I am beholden to Scott Schwartz

Acknowledgments   233

at the Sousa Archives and Center for American Music, André Bernard at the John Simon Guggenheim Memorial Foundation, Heather Cole and the special collections team at Brown's John Hay Library, Alison Fraser at University of Buffalo's Rare and Special Books Collection, and staff members at the Harry Ransom Center, the Thousand Oaks Library, the Department of Special Research Collections at the University of California Santa Barbara Library, the Special Collections Research Center at Syracuse University, the Rosenbach Museum and Library, the National Library of Ireland, the National Archaeological Museum in Athens, to Jeffrey Twitchell-Waas, and Christopher Husted, as well as to Kevin Prufer, Robert McDonough, Diane Kendig, P. K. Saha, and Aldon Lynn Nielsen, for sharing information and materials regarding Russell Atkins—and to Russell Atkins himself, for his blazing thought and music. For helping me to hear this music, thank you to Dan Sedgwick, Eric Sedgwick, and Kelly Guerra.

A section of chapter 1 appeared in *PMLA* 135, no. 3 (2020), and a section of chapter 2 in *Journal of Modern Literature* 43, no. 1 (2019). I reprint them here with the gracious permission of the Modern Language Association of America and Indiana University Press. Another section of chapter 2 appeared in *The American Sonnet: An Anthology of Poems and Essays*, edited by Dora Malech and Laura Smith, © 2022 University of Iowa Press, and is used with the press's permission. Thank you to Dora, Laura, and all the reviewers and editors who have improved these pieces along the way.

I couldn't be happier that this book has found a home in the Hopkins Studies in Modernism series at Johns Hopkins University Press. Time and again, I've been stunned by the generosity and acumen of my series editor, Doug Mao, whose record of scholarship continues to offer inspiring examples. The fastidious and authoritative insight of the manuscript's reviewers have drawn from me a much better book, and Catherine Goldstead was an unfailingly responsive and conscientious editor. Thank you to Matthew McAdam and Adriahna Conway for seeing this book into print, to Carrie Watterson, for lending my prose the clear light of her facility with the language, and to John Grennan, for the first-rate index.

Though I won't enumerate them here, this book owes debts more profound than anyone knows to some extraordinary friends: Terrance Manning Jr., Mike Campbell, Mark Mengel, Asha Tamirisa, Bryce Roe, and Dan D'Amore, a reader I'll always have first in mind. And a special heart-brimming thanks goes to Mark Feldman and Eileen Bringman, oldest buddies, for making home always feel like home.

For safe harbor and sense of deepest purpose, thank you to my family: to Ryan, Amy, Kelsey, John, and Brennan, to Nani and in memory of Papa. A big high-five to my pal Reese, who is about as old as this project but already wiser and way more fun.

The book is dedicated to my parents, Mike and Karen, who have always asked me to do what I love and have never stopped teaching me how.

And my final debt is to Sarah, whose brilliant love, kindness, and appetite for joy have made the long years of this book's making into something of a dream. The next book is for you. Everything is.

* * *

I gratefully acknowledge permission to reprint excerpts from the following works:

Helen Adam, "Limbo Gate" and "The Birkenshaw," from *A Helen Adam Reader*, edited by Kristin Prevallet. Copyright © 2007 by the National Poetry Foundation. Reprinted with permission of the Poetry Collection, University Library, the University at Buffalo. "Limbo Gate," from *Songs with Music: Helen Adam*, transcribed and edited by Carl Grundberg. Copyright © 1982 by Aleph Press. Reprinted courtesy of Carl Grundberg.

Russell Atkins, "Lakefront, Cleveland" and "Trainyard at Night," from *World'd Too Much: The Selected Poetry of Russell Atkins*, edited by Kevin Prufer and Robert E. McDonough. Copyright © 2019 by Russell Atkins. Reprinted with the permission of The Permissions Company, LLC, on behalf of Cleveland State University Poetry Center, csupoetrycenter.com.

Sterling A. Brown, "Southern Road" and "Memphis Blues," from *The Collected Poems of Sterling A. Brown, Selected by Michael S. Harper*. Copyright © 1980 by Sterling A. Brown. Reprinted by permission of the John L. Dennis Revocable Trust.

"Caedmon's Hymn," translated by John C. Pope, from *Norton Anthology of Poetry: Sixth Edition*, edited by Margaret Ferguson, Tim Kendall, and Mary Jo Salter. Copyright © 2018, 2005, 1996, 1983, 1975, 1970 by W. W. Norton & Company, Inc. Used by permission of W. W. Norton & Company, Inc.

Jean Elliot, "The Flowers of the Forest," *Halfpenny Lyre*, reproduced under a Creative Commons Attribution 4.0 International (CC-BY) license with the

Acknowledgments

permission of the National Library of Scotland. https://creativecommons.org/licenses/by/4.0/.

Langston Hughes, "Notes on Commercial Theatre," "Red Cross," "Fragments," "Burden," and "Total War" from *The Collected Poems of Langston Hughes*, edited by Arnold Rampersad with David Roessel. Copyright © 1994 by the Estate of Langston Hughes. Used by permission of Alfred A. Knopf, an imprint of the Knopf Doubleday Publishing Group, a division of Penguin Random House LLC. All rights reserved.

Reprinted by permission of Harold Ober Associates/International Literary Properties: "Fragments," "Burden," "Hard Luck," "Total War," "Note on Commercial Theatre," and "Red Cross" from *The Collected Poems of Langston Hughes*, copyright © 1951 by the Langston Hughes Estate; "Democracy, Negroes, and Writers" and "Negro Writers and the War" from *The Collected Works of Langston Hughes*, vol. 9: *Essays on Art, Race, Politics and World Affairs*, copyright © 2002 by the Langston Hughes Estate.

All other material used by permission of International Literary Properties/the Langston Hughes Estate.

Marianne Moore, "In Lieu of the Lyre," copyright © 1965 by Marianne Moore, renewed; from *The Complete Poems of Marianne Moore*. Used by permission of Viking Books, an imprint of Penguin Publishing Group, a division of Penguin Random House LLC. All rights reserved.

"In Lieu of the Lyre" and excerpts from "Poetry," "The Fish," and "A Grave" from *New Collected Poems* by Marianne Moore, edited by Heather Cass White. Copyright © 2017 by The Estate of Marianne Moore. Reprinted by permission of Farrar, Straus and Giroux. All rights reserved.

"The Fish" and "In Lieu of the Lyre," from *The New Collected Poems of Marianne Moore*. Used with permission of Faber and Faber Ltd.

"The Fish," "A Grave," and "Poetry," From *The Collected Poems of Marianne Moore*. Copyright © 1935 by Marianne Moore, renewed 1963 by Marianne Moore. Reprinted with the permission of Scribner, a division of Simon & Schuster, Inc. All rights reserved.

Lorine Niedecker, "for sun and moon and radio," "Now in one year," "The museum man!," and "Remember my little granite pail," from *Lorine Niedecker: Collected Works*, edited by Jenny Penburthy. Copyright © 2004. Courtesy of University of California Press Books.

Harry Partch, *Bitter Music: Collected Journals, Essays, Introductions, and Librettos*. Copyright © 1991 Board of Trustees (compilation, scholarly apparatus). Used with permission of the University of Illinois Press on behalf of the Harry Partch Foundation. All other material reproduced courtesy of the Harry Partch Estate, Danlee Mitchell, Executor, and the Sousa Archives and Center for American Music, University of Illinois at Urbana-Champaign.

Sappho, "118 [yes! radiant lyre speak to me]," from *If Not, Winter: Fragments of Sappho*, translated by Anne Carson. Copyright © 2002 by Anne Carson. Used by permission of Alfred A. Knopf, an imprint of the Knopf Doubleday Publishing Group, a division of Penguin Random House LLC. All rights reserved.

May Sarton, "From Men Who Died Deluded." Copyright © 1939 by May Sarton, from *Collected Poems 1930-1993*. Used by permission of W. W. Norton & Company, Inc. and the May Sarton Estate.

John Wheelwright, "Train Ride" and "The Word is Deed," from *Collected Poems of John Wheelwright*. Copyright © 1971 by Louise Wheelwright Damon. Reprinted by permission of New Directions Publishing Corp.

All Louis Zukofsky and Celia Zukofsky materials copyright © Musical Observations, Inc. Used by permission.

Notes

Introduction

1. Pound, *ABC of Reading*, 46.
2. Pinsky, *The Sounds of Poetry*, 8. For poetry in DNA, see Bök, *The Xenotext*, 150. See also Brian Reed's provocative unpacking of the same question in "Visual Experiment and Oral Performance," 270-84.
3. Jakobson, *Language in Literature*, 69.
4. Jakobson, *Language in Literature*, 70, 71.
5. Guillory, "The Genesis of the Media Concept," 352; Jakobson, *Language in Literature*, 508; McLuhan, *Understanding Media*, 23.
6. Guillory, "The Genesis of the Media Concept," 352. Of the six functions of language identified by Jakobson, "phatic" messages are those "primarily serving to establish, to prolong, or to discontinue communication, to check whether the channel works . . . to attract the attention of the interlocutor or to confirm his continued attention." Jakobson, *Language in Literature*, 68. According to Bernhard Siegert (following Michel Serres), a reorientation of Jakobson's model such that "not the poetic or the referential function . . . dominates the others, but the *phatic function*, the reference to the channel," is the foundational conceptual move of "media theory—of any media theory." Siegert, *Cultural Techniques*, 21.
7. On the modernist literary magazine as poetic medium, see Brinkman, *Poetic Modernism*.
8. Zukofsky, "Sincerity and Objectification," 274.
9. In the summary judgment of Brent Hayes Edwards, "*lyric* is a literary term that signals a certain musicality or suggests a mode of writing informed by, imbued with, or redolent of the ephemerality and affective force of musical performance." Edwards, *Epistrophies*, 58.
10. Budelmann, "Introducing Greek Lyric," 2-3.
11. Hayles, *My Mother Was a Computer*, 31. In his work on vocal music and early modern poetic culture, Scott A. Trudell appeals to a concept of "intermedia" largely

cognate with my own, though with less emphasis on the intermedial potential even of poetry on the page. Trudell, *Unwritten Poetry*. Cara L. Lewis's excellent study of modernist intermediality wields the term to describe literature's engagements with sculpture, painting, film, and photography, and variations of the concept also play significant roles in Continental media theory. Lewis, *Dynamic Form*; Jensen, "Intermediality." For Jessica E. Teague, the term aptly identifies the "text-recording hybrids" foregrounded in her work on sound technologies and twentieth-century American literature. Teague, *Sound Recording Technology*, 20. What is perhaps the word's most prominent acceptation derives from the postwar art world—and yet there, too, poetry precedes us. When the Fluxus artist Dick Higgins introduced the word "intermedia" in the mid-1960s to describe cross-disciplinary art practices like happenings, action music, and concrete poetry, he credited the term's earliest appearance to Samuel Taylor Coleridge, for whom it serves to characterize the cast of Spenserian allegory: "Narrative allegory is distinguished from mythology as reality from symbol; it is, in short, the proper intermedium between person and personification." Higgins, "Intermedia," 52; Coleridge, "Spenser," 38.

12. "Looking back over the combined poetic-alphabetic tradition," from Homer on, "it seems hard to imagine any point (including its point of origin) at which letters seemed the building blocks of language alone"—"one way or another, writing recorded not only the singer's sense, but also no small part of his sensuous sound." Butler, *The Ancient Phonograph*, 17, 16. By emphasizing the intermedial character of alphabetic writing, I do not mean to minimize the role of sound in the poetry of nonalphabetic languages. For a sample of contemporary approaches to sound in Chinese poetry, for instance, see Cai, "Introduction."

13. Such, I take it, is one central claim of *Reading Voices*, Garrett Stewart's important discussion of silent reading's subvocalized "phonotext," wherein he distinguishes between the "articulatory stream" of "phonemic *values*"—those "linguistically functional differences between sounds" (28)—and the speech noises that are *not* linguistically marked in a differential system (the contingent sounds of language's embodied production in performance). My sense of lyric intermediality encompasses both the phonemic and extra-phonemic orders of sonorous speech. For approaches to poetic prosodies that attend to sounds "processed both acoustically and cognitively," see Blasing, *Lyric Poetry*, 28; Abrams, *The Fourth Dimension*. Instructively, Abrams lavishes his attention equally on what he calls the poem's second dimension, "the sounds of the words" read aloud or silently, and on the oftener neglected fourth dimension, "the mobile and tactile" activity of "enunciating the great variety of speech-sounds that constitute the words of a poem" (2). To complete this scrupulous regard for the oral production of the poem, we need only emphasize, too, the aural experience of *listening* to a performance, either live or recorded, for non-semantic, often somatically processed features of the poem. The field-defining effort on this latter front is Charles Bernstein's collection *Close*

Listening: Poetry and the Performed Word. See also Perloff and Dworkin, *The Sound of Poetry*.

14. Pound, *ABC of Reading*, 14.

15. The ablest synoptic treatment of literary-musical relations in the Western tradition remains James Anderson Winn's *Unsuspected Eloquence: A History of the Relations between Poetry and Music*.

16. I gesture here to the pervasive habit of titling poems as "songs," as in Plath's "Morning Song." Plath, *Ariel*, 5. For Sextus Propertius, of course, the Latin *carmen* meant both "song" and "poem." Propertius, *The Complete Elegies*, xxviii.

17. Genette, *The Architext*, 64–65.

18. See Jackson, *Dickinson's Misery*; Jackson and Prins, *The Lyric Theory Reader*; de Man, *The Rhetoric of Romanticism*, 239–62. For a useful review of the new lyric studies, see Burt, "What Is This Thing Called Lyric?," 422–40.

19. Culler, *Theory of the Lyric*, 119; Horace, *Satires*, 479.

20. Jackson and Prins, *The Lyric Theory Reader*, 452.

21. I urge a sharp distinction between this book's descriptive minimalism, which is designed to characterize a wide gamut of modern poetic practices, and the suspect desire for universalizing accounts that cannot help but affirm essentialist arguments about poetry as such. Of the many reasons for urging this distinction, one is particularly urgent: the need to avert the audist belief that audible sound is somehow the necessary or quintessential "elixir of poetry," to borrow a phrase from the poet and scholar John Lee Clark ("Melodies Unheard," 7). Though this book does focus on audible sound, and its definition of intermediality foregrounds the relation between sonorous events and silent writing, I insist on two qualifying points. First, sonic phenomena are "necessarily multisensory," "intermaterial," and "multimodal," as sound theorists engaging closely with disability studies continue to substantiate by contesting "ear-centric" models of listening. Eidsheim, *Sensing Sound*, 3; Ceraso, *Sounding Composition*, 6. Second, poets can realize the intermedial dimension of poems via material practices that have little if anything to do with audible sound, and any media-theoretical account of modern poetry must acknowledge the medium of the performing body. For instance, in his essay debunking the "sound theory of poetry," whose unsoundness is exposed by the range of deaf poetries across written and sign languages both, Clark quotes the Welsh poet Dorothy Miles, who "tried to blend words with sign-language as closely as lyrics and tunes are blended in song" ("Melodies Unheard," 9). In this case, intermedial accompaniment is managed not by sound, narrowly construed, but by the signing body's dynamic interplay with semantic meaning.

22. On historical poetics, see Prins, "Historical Poetics," 229–34; Prins, "'What Is Historical Poetics?,'" 13–40.

23. For a study of Anglophone modernism's metrical cultures that rigorously avoids these pitfalls, see Glaser, *Modernism's Metronome*.

24. Trotter, *Literature in the First Media Age*, 2.

25. Goble, *Beautiful Circuits*, 13. The present study limits its analysis to phonography and radio because lyric writing chiefly reckons with the media problem of sound, though it would be an error indeed to discount the manifold routes of medial influence running between poetic production and the period's new visual technologies. Happily, poetry's relation with mass print, film, and other communicational and representational media continue to draw the resourceful attention of excellent scholars of modernism. For instance, in addition to Goble's *Beautiful Circuits*, see Lewis, *Dynamic Form*; Edward Allen, *Modernist Invention*; Murphet, *Multimedia Modernism*; Brinkman, *Poetic Modernism*; McCabe, *Cinematic Modernism*; Biers, *Virtual Modernism*; Goody, *Modernist Poetry*; Perlow, *The Poem Electric*; Chasar, *Poetry Unbound*.

26. Auden, *The Complete Works*, 105. See Perloff, *The Dance of the Intellect*, 228; McGann, *Black Riders*. Notoriously, Kittler argues that the phonograph c. 1900 renders poetry, an old technology for storing human voices, functionally obsolete. Kittler, *Gramophone*, 78–83. Charles Bernstein describes this functional obsolescence as a shift in writing's social function from a memorial or "transcriptive" emphasis to one that is medium specific and so distinctly "textual." He locates the condition of possibility for this transformation at a significantly earlier date in history. Though supercharged indeed in our "photo/phono electric, postliterate age," this cultural process began with the widespread adoption of alphabetic writing itself. Bernstein refers to writing's particularly textual vocation as the "art of immemorability": no longer oriented toward the "non-oral, non-speech-based forms of writing," "textual" writing "takes on the work of memory rather than being an aid to memory, and this function is not compromised by writing that is difficult or impossible to memorize." Even if the sounding of poetry is no longer linked to that "memory function essential for poetry in cultures where oral art is the primary technology for language storage and retrieval," Bernstein acknowledges that the "significance of speech for textual poetry [remains] fundamental." Bernstein, *Attack of the Difficult Poems*, 94, 104.

27. Kittler, *Gramophone*, 79.

28. Raymond Williams, *Marxism and Literature*, 122. I employ "residual" with reference to Williams's threefold paradigm for "authentic historical analysis," in which "it is necessary at every point to recognize the complex interrelations between movements and tendencies," and therefore to supplement an account of "dominant" or hegemonic social tendencies with those that appear "emergent," and those that are identifiably "residual" (121). "The residual," writes Williams, "has been effectively formed in the past, but is still active in the cultural process, not only and often not at all as an element of the past, but as an effective element of the present," thus maintaining a potentially "oppositional relation to the dominant culture" (122). See also Acland, *Residual Media*, 2007.

29. I rely throughout this book on a definition of technical media drawn from cultural-studies-inflected accounts of technology's social history. With Lisa Gitelman, Jonathan Sterne, and others who carry forward the media-analytic tradition of Raymond Williams, I understand media as "socially realized structures of communication, where structures include both technological forms and their associated protocols." Gitelman, *Always Already New*, 7. See also Sterne, *The Audible Past*, 182; Raymond Williams, *Television*.

30. Sterling and Kittross, *Stay Tuned*, 156, 183; Loviglio, *Radio's Intimate Public*.

31. Douglas, *Listening In*, 174-75.

32. Niedecker, *Collected Works*, 110, 5.

33. Niedecker, "Letters to *Poetry* Magazine," 188.

34. DuPlessis, "Lorine Niedecker," 134. Niedecker may have been inspired to mine the critical potential of Mother Goose by exposés like Katherine Elwes Thomas's *The Real Personages of Mother Goose* (1930).

35. Quoted in Margot Peters, *Lorine Niedecker*, 76.

36. Niedecker, "Mother Geese," n.p.

37. Niedecker, Letter to Zukofsky, in Penberthy, *Lorine Niedecker*, 147.

38. I refer to the poem "Poet's work." Niedecker, *Collected Works*, 194. See DuPlessis, "Lorine Niedecker's 'Paean to Place,'" 164; Quartermain, "Reading Niedecker," 226.

39. Khlebnikov, *The King of Time*, 155; Marinetti and Masnata, "La Radia," 267; Kittler, "Observations on Public Reception," 75.

40. Cantril and Allport, *The Psychology of Radio*, 259.

41. Berman, *All That Is Solid*, 13, 16.

42. Lacey, *Listening Publics*, 126. See also Fickers, "Visibly Audible," 411-39.

43. Brecht, "The Radio," 52.

44. Benjamin, *Radio Benjamin*, 363. See also Mowitt, *Radio*, 48-76.

45. Adorno, *Current of Music*, 112, 95, 96, 94.

46. Adorno, *Current of Music*, 112, 113.

47. Sartre, *Critique of Dialectical Reason*, 271, 273, 272.

48. Kittler, "Observations on Public Reception," 75; Lacey, *Listening Publics*, 125.

49. Niedecker, *Collected Works*, 195.

50. Niedecker wrote for Madison station WIBA in 1941 and 1942 as a member of the Radio Scripts division of the Wisconsin Writers' Project, and she composed two experimental radio dramas in the early 1950s. Peters, *Lorine Niedecker*, 67. Scholars like Brook Houglum have used this biographical evidence to evolve persuasive arguments for radio as a "sustained subtending component of Niedecker's composition practices" and therefore a crucial context for understanding her finely calibrated innovations in "aural collage, speech reportage, and voice experiment." Houglum, "'Speech without Practical Locale,'" 222. See also Robertson, "In Phonographic Deep Song," 83-90; Goody, *Modernist Poetry*, 279-88.

51. Penberthy, *Lorine Niedecker*, 191.

52. "Entextualization" is a meta-discursive cultural operation by which one "extract[s] a portion of ongoing social action—discourse or some nondiscursive but nevertheless semiotic action—from its infinitely rich, exquisitely detailed context, and draw[s] a boundary around it, inquiring into its structure and meaning." Silverstein and Urban, "The Natural History of Discourse," 1.

53. Nicholls, "Lorine Niedecker," 208.

54. Niedecker, *Collected Works*, 96.

55. On "Mary Had a Little Lamb" as an "exhibit standard" at early demonstrations of the machine, see Rubery, "Thomas Edison's Poetry Machine."

56. See, for instance, Gruenberg, "Radio and the Child," 123-28. In his writings for the Princeton Radio Research Project, Adorno suggests that the young child's mystified attitude toward the radio as a surrogate parent is actually an index of the infantilizing automatization of the radio voice. Adorno, *Current of Music*, 371. See also Mowitt, *Radio*, 34-36.

57. Nachman, *Raised on Radio*, 450.

58. Pound, *Letters*, 442.

59. McLuhan, *Counterblast*, 50.

60. In his treatment of Pound's career-long investment in a fetishistic and death-obsessed conception of the Image, Tiffany argues that "long before radio became a significant element of his cultural and political practice, Pound displayed in his theorization of the Image an affinity for figures that are essential to the history of phonographic and telephonic media (inscription, transmission, radiation, and re-animation)." Tiffany, *Radio Corpse*, 235.

61. Latouche, "The Muse and the Mike," 124.

62. Mathiesen, *Apollo's Lyre*, 235.

63. Sappho, *If Not, Winter*, 241.

64. Hermogenes, *On Types of Style*, 77.

65. Culler, *Theory of the Lyric*, 213. See also Culler, *The Pursuit of Signs*, 35-54. On prosopopoeia and anthropomorphism, see Paul de Man's essays "Autobiography as De-facement" and "Anthropomorphism and Trope in the Lyric," in *The Rhetoric of Romanticism*, 67-82, 239-62.

66. Culler, *The Pursuit of Signs*, 139, 142.

67. Gumbrecht and Pfeiffer, *Materialities of Communication*.

68. *Oxford English Dictionary*, 3rd ed. (2007), s.v. "prosopopoeia," https://www.oed.com/view/Entry/153015.

69. Mathiesen, *Apollo's Lyre*, 159-286.

70. "Although one should refrain from assuming that performances of lyric poetry ceased to exist completely in the Hellenistic era, the Alexandrians' understanding of sixth- and fifth-century lyric poetry was in all probability not based on experiencing for themselves a performance of a sixth- and fifth-century lyric song.

It was rather based presumably on the correlation between the names of the poets who composed this kind of poetry and their poetic compositions which the Alexandrian scholars possessed as material texts in the Library." Hadjimichael, *The Emergence of the Lyric Canon*, 6. See also Barbantani, "Lyric in the Hellenistic Period," 297-318. Barbantani summarizes the lack of scholarly consensus regarding the possession of ancient lyric sound: "Some scholars believe that ancient music and metrical knowledge was handed down to the Alexandrian editors, while others argue that the advent of the 'New Music' stopped the preservation of ancient scores" (301).

71. Budelmann, "Introducing Greek Lyric," 2-3.
72. Mackey, *Blue Fasa*, xi.
73. Horace, *Odes with Carmen Saeculare*, 53.
74. Paz, *El arco y la lira*.
75. Hollander, *The Untuning of the Sky*, 204.
76. Campion, *The Works*, 28.
77. Campion, *The Works*, 12.
78. Shelley, *Queen Mab*, 5.
79. Shelley, "A Defence of Poetry," 511.
80. Morton, *Realist Magic*, 209; Morton, "An Object-Oriented Defense," 205-24.
81. Shelley, "A Defence of Poetry," 511.
82. Kittler, "Number and Numeral," 52.
83. Kittler, "Number and Numeral," 56.
84. John Durham Peters, "Assessing Kittler's *Musik und Mathematik*," 33.
85. Rheinberger, *Toward a History*, 28.
86. O'Donnell, "Material Differences," 15; Bede, *Ecclesiastical History*.
87. Bede, *Ecclesiastical History*, 417.
88. "Caedmon's Hymn," 1.
89. Grossman, *The Long Schoolroom*, 4, 3, 5.
90. Bede, *Ecclesiastical History*, 419.
91. O'Keeffe, "Orality and Literacy," 123.
92. Niles, "The Myth," 44. See also O'Keefe, "Orality and Literacy," 126.
93. Grossman, *The Long Schoolroom*, 16; Niles, "The Myth," 16.
94. The theme of writing's fixity and permanence has a long pedigree in intellectual history, reaching back at least to Plato's *Phaedrus*. For a splendid media-theoretical treatment of the matter, see John Durham Peters, *The Marvelous Clouds*, 261-314.
95. Niles, "The Myth," 16.
96. Niles, "The Myth," 14.
97. Bede, *Ecclesiastical History*, 419 (italics mine).
98. Kittler, *Discourse Networks*, 78.

99. See David Nowell Smith, *On Voice in Poetry*; Wheeler, *Voicing American Poetry*.

100. On the "messiness" of media history, see Alan Liu on "new media encounters" in *Friending the Past*, 44.

101. Hollander, *The Untuning of the Sky*, 45.

102. McLane, *Balladeering*, 6.

103. McGill, "What Is a Ballad?," 161.

104. Walter Scott, *Minstrelsy of the Scottish Border*, 156-59.

105. Burns, *Works*, 370-71.

106. Walter Scott, *Minstrelsy of the Scottish Border*, 156. See also Gibson, "'The Flowers of the Forest Are a' Wede Away,'" 104; Crawford, *Society and the Lyric*, 176-77.

107. "The old air was locked away in the Skene MS. (*c.* 1630) and not printed until 1838, but it survived in oral transmission and became transformed into an eighteenth-century tune." Crawford, *Society and the Lyric*, 176. At military commemorations around the British Commonwealth the tune can still be heard today. Gibson, "'The Flowers of the Forest Are a' Wede Away,'" 105.

108. Gibson, "'The Flowers of the Forest Are a' Wede Away,'" 105, 106.

109. McLane, *Balladeering*, 110.

110. "The Flowers of the Forest."

111. Gibson, "'The Flowers of the Forest Are a' Wede Away,'" 106.

112. In his discussion of the blues lyric, Edward's account of an "unresolved tension" between transcription and score eloquently testifies to the dynamic I call lyric intermediality. Edwards, *Epistrophies*, 83, 57-85. Bernstein, *Attack of the Difficult Poems*, 134, 131-57.

113. Grossman and Halliday, *The Sighted Singer*, 210.

114. Penberthy, *Niedecker*, 43.

115. Niedecker, "Local Letters," 94.

116. Eliot, *On Poetry and Poets*, 32; William Carlos Williams, *I Wanted to Write a Poem*, 65.

117. Niedecker, *Collected Works*, 101.

118. Middleton, "Lorine Niedecker's 'Folk Base,'" 177-78.

119. Niedecker, *Collected Works*, 41.

120. On the cultural history of phonography, see, for instance, Sterne, *The Audible Past*; Gitelman, *Always Already New*; Gitelman, *Scripts*; Weheliye, *Phonographies*; Brady, *A Spiral Way*; Eisenberg, *The Recording Angel*.

121. Camlot, *Phonopoetics*, 5; Furr, *Recorded Poetry*; Nardone, "Our Format," 101-24.

122. Parry, "The Inaudibility of 'Good' Sound Editing," 27; Rubery, *The Untold Story*, 192. On Caedmon, see also Furr, *Recorded Poetry*, 40-52; Jacob Smith, *Spoken Word*, 49-78; Roach, "The Two Women of Caedmon," 21-24.

123. See Parry, "The Inaudibility of 'Good' Sound Editing," 28.
124. Qtd. in Christie, *Dylan Thomas*, 164.
125. See Bernstein, *Close Listening*, 3-26.
126. Liu, *Friending the Past*, 44.
127. Maud, introduction to *On the Air*, v.
128. Dylan Thomas, "Poetic Manifesto," 46.
129. Dylan Thomas, *Collected Poems*, 135. See Ivan Phillips, "I Sing the Bard Electric," 14-15.
130. McLuhan, *Counterblast*, 88.
131. McLuhan, *Counterblast*, 71. See also Goodby, *The Poetry of Dylan Thomas*, 412-13. Quoting this passage by McLuhan, Goodby notes that Thomas's work is "suggestive of areas in which print and electronic media merge" (412).
132. See Guillory, "The Genesis of the Media Concept," 321-62; Grusin, "Radical Mediation," 124-48; Siegert, *Cultural Techniques*; Chun, *Updating to Remain the Same*.
133. Adorno, *Notes to Literature*, 67.
134. Adorno, "The Form," 59.

Chapter 1. A Speech-Musical Modernism

1. Partch, *Bitter Music*, 12, hereafter cited parenthetically as *BM*.
2. Partch, *Genesis of a Music*, 5, hereafter cited parenthetically as *G*.
3. Partch, *Exposition*, 50. Originally called the Monophone, the Adapted Viola— a viola body, an elongated neck with tacked-in bradheads, and cello strings, to be played between the knees—inaugurates Partch's astonishing career as a builder of instruments (*G* 198-202).
4. Partch, promotional pamphlet. For detailed introductions to the Monophonic system, see *G* 67-105; Gilmore, "Harry Partch"; Granade, *Harry Partch*, 28-32; and Kassel, "*Barstow* as History," xiii-lxxix.
5. Partch, *Exposition*, 44.
6. Yeats, "Music for the Plays," 757. Despite the theoretical and historical research he would soon undertake on behalf of Monophony, in his autobiographical writings Partch emphasizes the degree to which it was pure "intuition" that spurred his commitment to speech-music in just intonation: when "I abandoned the traditional scale, instruments, and forms in toto . . . This was the positive result of self-examination—call it intuitive, for it was not the result of any intellectual desire to pick up any lost or obscure historical threads. For better or for worse, it was an emotional decision" (*G* 5).
7. Partch found this phrase, his speech-musical touchstone, in the 1907 edition of Yeats's *Dramatical Poems*, volume 2 of *The Poetical Works of William B. Yeats*, 522. In his 1949 theoretical treatise *Genesis of a Music*, Partch traces the fortunes of this

ancient interdiction throughout musical history, "from Emperor Chun and Plato to the Vacant Lot" of a modern listening culture lamentably enthralled to "sciolist and academic Europeanisms," to an exhausted classical repertory and the endless, fruitless technical refinements of massive orchestras and star virtuosos (3, 52). "The origin of music in speech intonation among the early peoples to whom we ascribe civilizations—the Greeks and the Chinese particularly—seems pretty well established," Partch surmises. From this root has grown all those varied musics the composer prizes with the designation "Corporeal": "essentially vocal or verbal music of the individual," "vital to a time and place, a here and now," and "physically allied with poetry or the dance" (8). *Genesis* recounts how, from the second century to the present, this "Corporeal" disposition has been forced to endure a welter of hostile cultural developments that Partch charges with music's progressive abstraction: "Abstract music grows from the root of non-verbal 'form'" to choke out the Corporeal (16). The Abstract villains of Partch's music history include polyphony, transcendent form, equal temperament, hypertrophied technique, the concert hall, the star system, and of course, "word distortion" (21). Resolutely "mental or spiritual," Abstract music is "always 'instrumental,' even when it involves the singing of words," for in such cases "the vitally rendered words" are sacrificed—stretched, clipped, or otherwise shucked of sense—to "non-verbal form" (8). The first chapters of *Genesis* celebrate the champions of the Corporeal attitude who resisted these developments by "consciously trying to bring to music a quality of musical speech," "a music vitally connected to the human body"—that is, by obeying the speech-music interdiction (45). In addition to the ancient Greeks and Chinese, these include performers of Japanese Kabuki, "the troubadours, trouvères, minnesingers, and meistersingers of the eleventh to sixteenth centuries" (20), composers of early Florentine opera (20-22), Claudio Monteverdi (22-23), Jean-Baptiste Lully (25-26), Christoph Willibald Gluck (27), André Grétry (28), Hector Berlioz (28-30), Richard Wagner (30-31), Modest Mussorgsky (31-33), Leoš Janáček (33), Hugo Wolf (34), the Claude Debussy of *Pelléas et Mélisande* (36-37), Yeats (38-40), the Arnold Schoenberg of *Pierrot Lunaire* (40-41), Virgil Thomson (42), and Marc Blitzstein (42). Partch also includes folk and popular musics in his inventory of contemporary Corporeal activity; insofar as the present day "is a fair historic duplicate of the eleventh," all abstract musical fare having failed "to satisfy an earthly this-time-and-this-place musical hunger," "hillbilly, cowboy, and popular music" can be said to reprise the cultural role of the troubadours (52).

8. Yeats, "Speaking to the Psaltery," 16.

9. For a sampling of recent work on twentieth-century poetry informed by sound and media studies, see Camlot, *Phonopoetics*; Furr, *Recorded Poetry*; Teague, *Sound Recording Technology*; Anthony Reed, *Soundworks*; Graham, *The Great American Songbooks*; Edward Allen, *Modernist Invention*; Bloom, *The Wireless Past*; Meta DuEwa Jones, *The Muse Is Music*; Halliday, *Sonic Modernity*; Furlonge, *Race Sounds*;

Murphet, *Multimedia Modernism*; Golston, *Rhythm and Race in Modernist Poetry and Science*; Shaw, *Narrowcast*; Groth, Hone, and Murphet, *Sounding Modernism*; Cohen, Coyle, and Lewty, *Broadcasting Modernism*; Perloff and Dworkin, *The Sound of Poetry*. These latter collections succeed the pathbreaking volumes assembled by Charles Bernstein and Adalaide Morris. Bernstein, *Close Listening*; Morris, *Sound States*.

10. Yeats to Partch, undated.

11. Schuchard, *The Last Minstrels*, 39. See also Genet, *"Words for Music Perhaps"*; and William Brooks, "Historical Precedents for Artistic Research in Music," 185-96. In "All Soul's Night," Yeats memorializes Farr ("Florence Emery") and her 1912 decision to leave London for Ceylon, where she administered the Buddhist College for Girls until her death in 1917. On Yeats's relationship with Farr, see Hassett, *W. B. Yeats and the Muses*, 37-64.

12. Yeats, "Speaking to the Psaltery," 13.

13. Schuchard, *The Last Minstrels*, xxiii. Speech-music was not without its prominent detractors. Famously, George Bernard Shaw heard only a "nerve-destroying crooning like the maunderings of an idiot banshee." Qtd. in Holroyd, *Bernard Shaw*, 308. For a collection of excerpted reviews of the "new art," see Farr, *The Music of Speech*. "We have tried our art, since we first tried it in a theatre, upon many kinds of audiences, and have found that ordinary men and women take pleasure in it and sometimes tell one that they never understood poetry before. It is, however, more difficult to move those, fortunately for our purpose but a few, whose ears are accustomed to the abstract emotion and elaboration of notes in modern music." Yeats, "Literature and the Living Voice," 13.

14. Yeats, "Literature and the Living Voice," 6. See Said, "Yeats and Decolonization," 220-38. In his wide-ranging treatment of representations of sound in modern culture, Sam Halliday observes that Yeats's oral/aural convictions, like those of T. S. Eliot and D. H. Lawrence, evince a "'depth' approach to sound, where 'depth' designates a temporality *'beneath'* or extending back from one's own present." Halliday, *Sonic Modernity*, 31. He adduces the late poem "Hound Voice," in which "those who have 'hound' voices speak to and thereby recognize each other as the bearers of a primordial and (the poem prophesies) ultimately renascent communal legacy."

15. Yeats, "Literature and the Living Voice," 5, 8, 6. See Schuchard, *The Last Minstrels*, 191-218.

16. See McLuhan, *The Gutenberg Galaxy*; Ong, *Orality and Literacy*, 133.

17. Bloom, *Wireless Past*, 34. On Yeats's ambivalence toward devices for sound transmission, see Morin, "'I Beg Your Pardon?,'" 191-219.

18. Yeats, "Literature and the Living Voice," 9, 8.

19. If for McLuhan "the 'content' of any medium is always another medium," I propose that, in a curiously reflexive doubling, the content of any new-old medium is its own self-awareness *as* a historical medium. McLuhan, *Understanding Media*, 23.

20. Farr, *The Music of Speech*, 16.

21. Yeats, "Speaking to the Psaltery," 14.

22. Farr, *The Music of Speech*, 16-17. See William Brooks for an intriguing discussion of the "new art" as "an early prototype of artistic research," and of a recent attempt to reproduce speech on the psaltery ("Historical Precedents for Artistic Research in Music,"191).

23. Farr, *The Music of Speech*, 10.

24. Yeats, "Speaking to the Psaltery,"16.

25. Yeats's "asceticism of the imagination" and his interest in the artistic exploitation of monotony are emblematic of what Benjamin Steege, with an eye toward developments in experimental physiology and acoustic science, terms the period's "heightened attentiveness to sensory marginalia," the widespread "investment in minute sensation" that had come "to typify an ethical and aesthetic culture Walter Pater had characterized as devoted to 'getting as many pulsations as possible into the given time.'" Steege, *Helmholtz and the Modern Listener*, 2. Yeats explains, "If Mr. Symons will borrow one of my psalteries and speak one of his own poems to a notation of his own, he will find . . . all kinds of beautiful or dramatic modulations which would never have occurred to him had not the cruder effects been fixed by the notation . . . Everything in any art that can be recorded and taught should be recorded and taught, for by doing so we take a burden from the imagination, which climbs higher in light armour than heavy." Yeats, "Speaking to Musical Notes," 591.

26. Farr, *The Music of Speech*, 19.

27. Farr, *The Music of Speech*, 17.

28. "Clifford's Inn-Hall," qtd. in Schuchard, *The Last Minstrels*, 127.

29. Here I invoke Siegert's reading of Michel Serres's media-theoretical concept of the parasite. Serres reveals how the traditional communicative scenario of producer and recipient depends upon the constitutive exclusion of a certain third party, the parasite/noise: "In Serres's model of communication the fundamental relationship is not between sender and receiver, but between communication and noise." Siegert, *Cultural Techniques*, 21; Serres, *The Parasite*.

30. See Schuchard, *The Last Minstrels*, 284-321. Though Yeats's "new art" languished after 1912, the campaign did exert a formative influence on the development of Imagism and English vers libre. Schuchard has persuasively identified Ezra Pound's Imagist third principle——"to compose in the sequence of the musical phrase, not in the sequence of a metronome"—as a direct bequest of Yeats and Farr, a "graft of the Yeatsian bell branch on the Imagist tree" (261). It was Farr's energetic involvement in Imagism's earliest incarnations, Yeats's own prestigious intervention at a decisive moment in its development, and Pound's melopoeiac sympathies with both that effected the move away from T. E. Hulme's exclusively visually oriented poetics (256-83). See also Morrisson, *The Public Face of Modernism*, 54-83; Preston, *Modernism's Mythic Pose*, 100-143.

31. Schuchard, *The Last Minstrels*, 350.

32. On Partch's early systems of notation, see Gilmore, "Harry Partch," 91-120.

33. Yeats to Edmund Dulac, November 21, 1934, quoted in Schuchard, *The Last Minstrels*, 352. To other correspondents, Yeats's enthusiasm was more qualified. In a letter to Margot Ruddock, after admiringly describing Partch's method, he advises, "We cannot however use him in our work at present, he is on his way to Spain to perfect his discovery; it is still, I think, immature. He is very young, and very simple." Yeats, *Ah, Sweet Dancer*, 28.

34. Partch to Henry Allen Moe, January 27, 1934.

35. Partch to Moe, January 29, 1934.

36. Partch to Moe, December 24, 1934.

37. Quoted in Kassel, "Harry Partch," 8. See also Koegel, "Preserving the Sounds of the 'Old' Southwest." Charles Fletcher Lummis, who had established the Southwest Museum in 1907, had made these recordings of several southwestern Indigenous communities, including the Isleta Pueblo, in the first decade of the century. Contrary to Hague's expectations, Partch eventually concluded—with an ethnocentric bias typical of proto-ethnomusicologists at this time—that what initially appeared to be microtonal intervals were in fact the result of "a constant striving for very simple intervals. By very simple I mean those possible in our major and minor diatonic scales" (Southwest transcriptions).

38. Kassel, "*Barstow* as History," xx.

39. Lord, *The Singer of Tales*, 3. On two trips to Yugoslavia in 1933 and 1934, Lord and Parry assembled a massive archive of South Slavic oral poetry and song on some 3,500 double-sided aluminum discs. After Parry's untimely death, "research on the music of the collection was for a long time hampered by the unavailability of a real connoisseur of the folk music of eastern Europe." Bartók and Lord, *Serbo-Croatian Folk Songs*, ix. Enter Bartók, modern music's exemplary folklorist, who in the early 1940s was hired to transcribe into conventional musical notation roughly 200 of these discs, putting oral "literature" innovatively collected on recording devices into the alien realm of printed musical notation. Parry and Lord are known chiefly for their work on the "men's," or heroic, songs, as performed by *guslari*, or epic singers, but Bartók focused his efforts on the collections' "women's songs" (248). In his transcriptions, Bartók relies on the diatonic scale and deals with the frequent "deviations" from these given pitches by means of "arrow signs above the notes" (4).

40. Ernst, *Sonic Time Machines*, 77, 72. In his attempts to theorize a temporal "sonicity" distinct from audible sound, Ernst raises the fascinating media-archeological question of how the nature of *guslari* songs *essentially* changes when Lord returns to Yugoslavia in 1950 with a magnetic wire recorder in tow: "it was not simply a continuation of the same recording method with different technological means. Electromagnetic recording subliminally changes the nature of sound memory, from symbolic notation (writing) or mechanical engraving (phonography) to sonic latency, a field very different from writing" (79).

41. Bartók and Lord, *Serbo-Croation Folk Songs*, 3. For Bartók, the value of conventional notation is nevertheless its "simplicity": "The human mind . . . must have as visual impressions conventional symbols of drastic simplicity in order to be able to study and to categorize sound phenomena."

42. On the "social genesis" of these values, see Sterne, *The Audible Past*, 215–334.

43. Gilmore, "Harry Partch," 156.

44. Adorno, "The Form," 59.

45. "Under this new impetus, doubts and ideas achieved some small resolution, and I began to take wing" (*G* vii).

46. Helmholtz, *On the Sensations of Tone*, 1, 4, 3. For a discussion of Helmholtz's "key role in the externalization and instrumentalization of senses, which forms a crucial but largely forgotten backdrop for modern media," see John Durham Peters, "Helmholtz, Edison, and Sound History," 179. For a cultural history of Helmholtz's role in fashioning a "new style of listening," see Steege, *Helmholtz and the Modern Listener*. See also Sterne, *The Audible Past*, 62–67.

47. To grasp the construction of Partch's microtonal system, it is important to register that each note represents an interval, a relation between frequencies: "A tone, in music, is not a hermit, divorced from the society of its fellows. It is always a relation to another tone, heard or implied" (*G* 86). Hence the utility of notating tones as ratios. For instance, Partch designates what in equal temperament would be known as a perfect fifth, which names the interval between the second and third harmonics, by the fraction 3/2, because its two component frequencies vibrate at a ratio of 3:2. One can derive new notes—Partch settled on a scale of forty-three—by observing new vibrational ratios, remarking new intervals.

48. On the elastic "fabric" of Partch's microtonal scales, see Gilmore, *Harry Partch*, 64.

49. Sterne, *The Audible Past*, 64.

50. See Peters, "Helmholtz, Edison, and Sound History," 183.

51. Among the acousticians discussed in *Genesis* is Dayton Clarence Miller, whose *The Science of Musical Sounds* (1916) Partch quotes approvingly on the Pythagorean ideal of musical interdisciplinarity: "When the artist, the artisan, and the scientist shall all work together in unity of purposes and resources, then unsuspected developments and perfections will be realized." Miller, *The Science of Musical Sounds*, 263–64, qtd. in *G* 96. Miller made his contribution to the history of sound with the *phonodeik*, an early device for sound visualization that employed a mirror system to photograph waveforms. Partch reproduces several phonodeik images to illustrate consonance as a function of waveform and to confirm visually what is patently clear to the ear: the "superior consonance" of just intonation as compared with equal temperament (*G* 148). Perhaps unbeknownst to the composer, the phonodeik had in fact intriguingly preceded Partchian speech-music

on something like its own turf. In 1918, the pioneering ethnomusicologist Frances Densmore contracted Miller and his machine to help verify her own musical transcriptions of Northern Ute songs, and the result was a kind of musical-media showdown—a microcosmic picture of that same media situation from which Partchian speech-music would soon arise. Densmore brought to Miller's Cleveland laboratory the wax cylinders on which the Ute songs had been recorded as well as her own transcriptions, which had been rendered "by ear, the only instruments of measurement being the piano . . . and a standardized metronome." Densmore, *Northern Ute Music*, 206. When the phonodeik "filmed" the wax cylinders, it vindicated Densmore's abilities as a transcriber: for her purposes, it turns out "there is no essential difference" between conventional symbolic and waveform representations. Dayton C. Miller, "Analytical Study of Photographs Taken with the Phonodeik," 209. While we know that such material differences are indeed materially "essential," we ought to take note of the fact that Densmore asked the imaging machine to intercede between symbolic notation and the uncoded signal of the phonograph. In other words, the phonodeik got these two essentially different orders talking to one another in a manner that is itself essentially different, but functionally analogous to, the role of Partch's Adapted Viola at the Southwest Museum and that of Monophonic compositions more generally as they mediate between the prerogatives of musical scores and sonic inscriptions.

52. Partch mentions Fletcher's apparently enthusiastic recommendation, now lost, in a 1933 letter to Moe. Partch to Moe, December 28, 1933.

53. Fletcher, *Speech and Hearing*, xiii.

54. Helmholtz, *On the Sensations of Tone*, 5.

55. See Johnston, "The Corporealism of Harry Partch." For excellent treatments of Partch's major theatrical works, see Sheppard, *Revealing Masks*, 180-230.

56. On Lawrence and Spengler, see *G* 15-16; Albright, *Putting Modernism Together*, 211-32.

57. On the "distinct impression of bilocation" that characterizes Partch's lifelong modes of self-fashioning, see Jake Johnson, "'Unstuck in Time,'" 164.

58. Partch, *Exposition*, 44.

59. Partch, interview with Vivian Perlis, 1974.

60. Partch to Moe, December 24, 1934. But for a few foundered plans, Partch might have enjoyed even further involvement with Anglophone literary modernism. In Rapallo, Partch met with Ezra Pound, assumedly to enlist his support at Yeats's suggestion, but found the poet "a most difficult man": "He and I didn't get along together at all, not at all . . . [H]e was such an egoist." Partch to Betty Freeman, July 25, 1968; interview with Vivian Perlis. Moreover, in his letter to Moe, Partch enthuses, "Yeats and T. S. Eliot are writing experimental plays to be produced at the Mercury Theatre, London, next spring. There was talk of my writing music for one of them, nothing definite. Yeats had many suggestions as to material

for me to work with. I seem to have my work planned for me for a good long time." Partch to Moe, December 24, 1934.

61. Partch, "Author's Preface," v.
62. Partch, interview with Vivian Perlis, 1974.
63. Granade's *Harry Partch: Hobo Composer* furnishes the indispensable cultural context for *Bitter Music* and compelling interpretations of his major "Americana" compositions.
64. "You have a narrative gift and a remarkable power of explaining yourself." Yeats to Harry Partch, undated. On the fluky publication history of *Bitter Music*, see *G* 323; Gilmore, *Harry Partch*, 113-35; Granade, *Harry Partch*, 88-109; Raulerson, "'A Fountainhead of Pure Musical Americana.'"
65. Partch advises his readers, "There is a strange alliteration in the three principal desires of bums. Food and flops are the most important, but they get the least attention in conversations" (*BM* 42).
66. Popularized by Pierre Schaeffer and associated with musique concrète, the acousmatic concept "marks the perceptive reality of sound as such." Schaeffer, *Treatise*, 64. On the origins of the acousmatic concept, see Battier, "What the GRM Brought to Music."
67. "My creations . . . are more intimate in their inherent nature than any music I know. I have never held hope for them in a concert hall and have consistently discouraged any suggestion that they be given before any but small groups." Partch to Elizabeth S. Coolidge, January 25, 1932.
68. Harkness, *Songs of Seoul*, 16.
69. Jakobson, *Language in Literature*, 68.
70. Bakhtin, *Dialogic Imagination*, 293. See also Harkness, *Songs of Seoul*, 19.
71. Silverstein, "Shifters, Linguistic Categories and Cultural Description," 11, 24. The concept of linguistic indexicality derives from the tripartite semiotic schema of Charles Sanders Peirce, wherein *symbols* maintain an arbitrary reference to their objects (as in the Saussurean sign), *icons* forge an association of likeness (as a portrait refers to a face), and *indices* assert a "dynamical (including spatial) connection both with the individual object, on the one hand, and with the senses or memory of the person for whom it serves as a sign, on the other hand." Peirce, *Philosophy*, 107. In Silverstein's terminology, the index signifies by virtue of "understood spatio-temporal contiguity" between sign and its signaled object ("Shifters, Linguistic Categories and Cultural Description," 27). We are accustomed to the indexical operations of those indices that leave a material trace—footprints, for instance, or photographs—but according to Silverstein, certain aspects of speech behavior function indexically as well. These "pure" or "nonreferential" indices work to signal the "contextual 'existence' of an entity" or the "structure of the speech context" but "do not contribute to the referential speech event" itself (27-30). The classic

examples of nonreferential indices are "deference" indices, which signal "inequalities of status, rank, age, sex, and the like," though the phenomenon is much broader (31). For examples of how linguistic anthropological methods can be articulated to formalist literary criticism and the sociology of literature, see Lucey and McEnaney, "Language-in-Use and the Literary Artifact," 1-173.

72. Feld et al., "Vocal Anthropology," 340.

73. *Oxford English Dictionary*, 3rd ed. (2007), s.v. "emphasis," https://www.oed.com/view/Entry/61310. "The word phatic is obviously derived from the Greek verbal stem φα- which is known from the verb φημι 'say, affirm, speak' which is also the source of words like *aphasia* and *emphatic*." Haberland, "Communion or Communication?," 165.

74. I refer to Reich's *It's Gonna Rain* (1965) and *Come Out* (1966), which take "exact segments of recorded speech" as their "musical material," such that "speech-melody and meaning are presented as they naturally occur." Reich, *Writings on Music*, 19. As an artifact of the media-historical conditions subtending modern poetry's relationship to sonorous speech, we should note that Reich was purportedly impelled toward these experiments with a speech-music on tape by his failure to set the speech-based poetry of William Carlos Williams, Charles Olson, and Robert Creeley. On Chopin's audiopoems, see Melillo, "The Politics of Noise." For other, even more contemporary examples of electronically and digitally mediated speech-music, see Lane, "Voices from the Past," 3-11; and Lane, *Playing with Words*. Of course, hip-hop constitutes by far the largest cultural arena for contemporary speech-music. For a sonic exploration of how "our ways of listening to American speech music across genres have changed since hip-hop brought speech music to the mainstream," see *Crosstalk* (2008), a remarkable collection of "American Speech Music" assembled by Mendi and Keith Obadike. Obadike and Obadike, "Crosstalk."

75. Agha, "Voice, Footing, Enregisterment," 38.

76. According to Partch, his "own personal Great Depression" came to an end in April 1943. News of a much-desired Guggenheim award reached the composer's temporary residence in Ithaca, New York, just as he was applying the finishing touches to the first draft of *U.S. Highball: A Musical Account of Slim's Transcontinental Hobo Trip*, what he would eventually term "the most creative piece of music I ever wrote" (*BM* 211). Partch would retrospectively bundle *U.S. Highball* together with other settings of "Americana texts" (*Barstow: Eight Hitchhiker Inscriptions from a Highway Railing at Barstow, California* [1941], *San Francisco: A Setting of the Cries of Two Newsboys on a Street Corner* [1943], and *Letter from Hobo Pablo* [1943]) and rechristen them *The Wayward*, a "collection of musical compositions based on the spoken and written words of hobos and other characters—the result of my wanderings in the Western part of the United States from 1935 to 1941." Qtd. in McGeary, *The Music of Harry Partch*, 120. *U.S. Highball* and *The Wayward* express the neat

sublimation of the previous ten years' work: in these compositions Partch submits overheard or found language of the sort that *Bitter Music* pressed for its social disclosures to the full theoretical and instrumental apparatus of his Monophonic practice. For a thorough and perceptive analysis of *U.S. Highball*'s textual, instrumental, and musical development, see Granade, *Harry Partch*, 144-97.

77. William Carlos Williams, "Two Letters to Robert Lawrence Beum," 198.
78. Moore to Celia Zukofsky, undated, 1952.
79. Louis Zukofsky to Niedecker, undated, 1952.
80. Malanga and Van Doren, jacket copy; Malanga, "Some Thoughts on *Bottom* and *After I's*," 64. Begun in 1943, the *Pericles* setting is far from Celia Zukofsky's only musical contribution to the Zukofsky *oeuvre*. In addition to numerous interpretations of individual poems, collected in *Autobiography* (1970), Celia also produced the culminating twenty-fourth section of Zukofsky's life-poem *"A," "A-24,"* or *L.Z. Masque*, which orchestrates a collage of selections from Zukofsky's poetry and prose in a polyvocal arrangement using the music of G. F. Händel's *Harpsicord Suites*. Unlike *Pericles*, *L.Z. Masque* has been staged and sounded on several occasions, including a gala performance at the San Francisco Art Institute in 1978. See Twitchell-Waas, *Z-Site*. Of the thousands of musical treatments of Shakespeare, *Pericles* is among the plays least engaged by musicians and composers, and Celia Zukofsky's score appears to be the first published operatic treatment. Gooch and Thatcher, *A Shakespeare Music Catalogue*, 1289-90.
81. Zukofsky and Zukofsky, *Bottom*, vol. 1, 33, 37.
82. Corman, *Essays*, 168.
83. Louis Zukofsky to Niedecker, undated, 1952.
84. William Carlos Williams to Zukofsky, March 21, 1943. Celia Zukofsky's setting of "Choral: The Pink Church" was published in *Briarcliff Quarterly*, 165-68. See also Williams, *Collected Poems*, 2:478. Celia Zukofsky would also set Williams's "Turkey in the Straw."
85. Zukofsky and Zukofsky, *Bottom*, vol. 1, 333.
86. Louis Zukofsky, "Bottom, a Weaver," 167.
87. Louis Zukofsky, "Sincerity and Objectification," 274.
88. Louis Zukofsky, *"A,"* 24.
89. See Twitchell-Waas, "What Were the 'Objectivist' Poets?"; DuPlessis and Quartermain, introduction to *The Objectivist Nexus*, 1-24.
90. Zukofsky and Zukofsky, *Bottom*, vol. 1, 33. See Pound, *Literary Essays*, 25.
91. Louis Zukofsky, *Prepositions +*, 169. See Adams, *The Degradation of Democratic Dogma*, 267-311; Scroggins, *The Poem of a Life*, 42-45.
92. Louis Zukofsky, *Prepositions +*, 170. Elsewhere in the Zukofsky corpus these ages are represented, respectively, as "sense," "essence," and "non-sense" (note the pejorative emphasis of the lattermost), or by "glyph," "syllabary," and "letters"; they can also be traced in the etymological fortunes of single words, as in the case of

ruthmos: "first it meant *shape,*" Zukofsky notes, "then *rhythm,* now *proportion* or *style.*" Dembo and Zukofsky, "Louis Zukofsky," 215; Louis Zukofsky, *"A,"* 126; Louis Zukofsky, *Prepositions +*, 55. See also Ahearn, *Zukofsky's "A,"* 228.

93. Zukofsky and Zukofsky, *Bottom*, vol. 1, 81.
94. Zukofsky and Zukofsky, *Bottom*, vol. 1, 88.
95. Emerson, *Representative Man*, 130.
96. See Eisenstein, *The Printing Press*; McLuhan, *The Gutenberg Galaxy*.
97. Kenner, *The Stoic Comedians*, xix, xviii.
98. Louis Zukofsky to Kenner, March 29, 1963; Kenner, *The Stoic Comedians*, 44.
99. Joyce, *Finnegan's Wake*, 341.
100. Kenner, *The Stoic Comedians*, 44.
101. Zukofsky and Zukofsky, *Bottom*, vol. 1, 14; Shakespeare, *The Sonnets*, 99.
102. Zukofsky and Zukofsky, *Bottom*, vol. 1, 18.
103. Zukofsky and Zukofsky, *Bottom*, vol. 1, 19.
104. Zukofsky and Zukofsky, *Bottom*, vol. 1, 36.
105. Zukofsky and Zukofsky, *Bottom*, vol. 1, 423.
106. Shakespeare, *Troilus and Cressida*, 4.5.64-66.
107. Zukofsky and Zukofsky, *Bottom*, vol. 1, 181.
108. Louis Zukofsky, *Prepositions +*, 159.
109. Zukofsky and Zukofsky, *Bottom*, vol. 1, 16, 15, 16.
110. Louis Zukofsky, *Prepositions +*, 159.
111. See Scroggins, *Louis Zukofsky*; Comens, *Apocalypse and After*, 158-74; Bernstein, "Words and Pictures," 9-34; Melnick, "The 'Ought' of Seeing," 55-65; Twitchell-Waas, "Keep Your Eyes on the Page," 209-47.
112. Zukofsky and Zukofsky, *Bottom*, vol. 1, 15.
113. Zukofsky and Zukofsky, *Bottom*, vol. 1, 17.
114. Louis Zukofsky, *"A,"* 151; Zukofsky and Zukofsky, *Bottom*, vol. 1, 17.
115. Jonson, "Ode to Himself," 283. See, for instance, Skeele, *"Pericles* in Criticism and Production," 1-33. A bona fide hit in the seventeenth century, the play fell into obscurity for two hundred years, emerging only in the past century as a work more ostensibly unified and more characteristically Shakespearean than the "miserable archaic fragment" critics had often held it to be. Strachey, *Books and Characters*, 60. But in the first half of the twentieth century especially, the play drew the fervid attention of modernist writers like Zukofsky, James Joyce, and T. S. Eliot. *Pericles* was rescued from Victorian disdain by scholarly attempts to see the play not as a work marred by dubious authorship and "structural discombobulation" but as a complex dramatic whole. Skeele, *Thwarting the Wayward Seas*, 23. Joyce's *Ulysses* offers a snapshot of this critical turning point when Stephen Daedalus, gabbing in the National Library, adduces recent scholarship on *Pericles* as grist for his own Shakespeare theories. The play's totemic significance for Stephen stems from its act 5 recognition scene, which Eliot will later call "the finest of all [Shakespeare's]

'recognition scenes'" and re-canonize for English literature in the 1930 poem "Marina." Eliot, "The Development of Shakespeare's Verse," 555; Eliot, *Collected Poems*, 105-6. Zukofsky included "Marina" in *An "Objectivist" Anthology*, 160-61.

116. Zukofsky and Zukofsky, *Bottom*, vol. 1, 327.
117. Zukofsky and Zukofsky, *Bottom*, vol. 1, 327.
118. Zukofsky and Zukofsky, *Bottom*, vol. 1, 18.
119. Zukofsky and Zukofsky, *Bottom*, vol. 2, 6.
120. Zukofsky and Zukofsky, *Bottom*, vol. 1, 333.
121. Louis Zukofsky, *"A,"* 138.
122. Zukofsky and Zukofsky, *Bottom*, vol. 1, 412.
123. Saussure, *Course in General Linguistics*, 15.
124. See, for instance, Woods, *The Poetics of the Limit*.
125. Zukofsky and Zukofsky, *Bottom*, vol. 1, 38.
126. Marianne Moore, jacket copy for *Bottom*.
127. Shakespeare, *Pericles*, 4.8; 5.1.45-46, 47-48.
128. Shakespeare, *Pericles*, 5.1.75, 81-82, 90.
129. Shakespeare, *Pericles*, 5.1.224-36.
130. See, for example, Ortiz, *Broken Harmony*, 160-64.
131. Zukofsky and Zukofsky, *Bottom*, vol. 2, 216-17.
132. Zukofsky and Zukofsky, *Bottom*, vol. 2, 333.
133. Shakespeare, *Pericles*, 2.5.30.
134. The heteronormative strains in Zukofsky's late work have been submitted to powerful critique by Libbie Rifkin, who has shown how *Bottom* "poses marriage as a vessel for containing excesses of subjectivity, desire, and mutability within the bounds of artistic mastery." Rifkin, *Career Moves*, 98. Nick Salvato has brilliantly queered this interpretation, disclosing in what sense Zukofsky is "the bottom to Celia's top"—his wife "the master not only of music but also of him." Salvato, *Uncloseting Drama*, 68. See also Cole, "'(Wo)Man (Critic) Is but an Ass?'"
135. Silliman, "Louis Zukofsky," 405.
136. Benjamin, *The Work of Art*, 23. See Eisenberg, *The Recording Angel*, 41; Bryan Wagner, *Disturbing the Peace*, 194.
137. Louis Zukofsky, *Anew*, 157.

Chapter 2. Latent Remediation

1. Latouche, "The Muse and the Mike," 124.
2. The "first established instance of poetry written specifically for the radio," as Milton Kaplan reports, was D. G. Bridson's verse-drama "The March of the '45," broadcast in 1936 on the BBC. Kaplan, *Radio and Poetry*, 6.
3. Barnouw, *Handbook of Radio Writing*, 157.
4. Directed by the *Workshop*'s creator, Irving Reis, and featuring twenty-two-

year-old Orson Welles, *The Fall of the City* was transmitted live from the massive Seventh Regiment Armory on New York's Upper East Side. Verma, *Theater of the Mind*, 18.

5. Wheelwright, "Toward the Recovery of Speech," 164.
6. MacLeish, *The Fall of the City*, ix.
7. MacLeish, *The Fall of the City*, xii.
8. MacLeish, "A Letter," 1. See also, for instance, Meltzer, "Poetry on the Air," 29.
9. MacLeish, *The Fall of the City*, ix-x.
10. See Ong, *Orality and Literacy*, 133.
11. Monroe, "The Radio and the Poets," 35. The influential publisher James Laughlin concurs in 1939: "Cheap print for several generations has weakened our faculty of aural appreciation. The mind cannot grasp in one sudden hearing anything very complex." Laughlin, *New Directions in Prose & Poetry*, xix. See also Lechlitner, "Verse-Drama for Radio,"110-15.
12. Bolter and Grusin, *Remediation*, 45.
13. *Oxford English Dictionary*, 3rd ed. (2007), s.v. "remedy, n," https://www.oed.com/view/Entry/162129.
14. On "remediation as reform," see Bolter and Grusin, *Remediation*, 59-62.
15. A sample of important contributions might include Keane, *Ireland and the Problem of Information*; Hollenbach, *Poetry FM*; Bloom, *The Wireless Past*; Cohen, Coyle, and Lewty, *Broadcasting Modernism*; Morris, *Sound States*; Tiffany, *Radio Corpse*; Chasar, *Poetry Unbound*; Edward Allen, *Modernist Invention*; Avery, *Radio Modernism*; Campbell, *Wireless Writing in the Age of Marconi*; Chasar, *Everyday Reading*; Dinsman, *Modernism at the Microphone*; Feldman, Mead, and Tonning, *Broadcasting in the Modernist Era*; Kahn, *Noise, Water, Meat*; Selch, "Engineering Democracy."
16. Barnouw, *Handbook of Radio Writing*, 157.
17. Rockefeller Foundation, *Annual Report*, 326. See Hilmes, *Network Nations*, 115-20; Cramer, "The Rockefeller Foundation," 77-99.
18. Lyons, "After Leap of 300 Miles"; "Quicker Fox," *Time*, 47.
19. On W1XAL as a vehicle for the Rockefeller-funded "Pan American Broadcasting Project," see Cramer, "The Rockefeller Foundation," 77-99.
20. Lyons, "After Leap of 300 Miles."
21. Lyons, "After Leap of 300 Miles."
22. Winfield Townley Scott, "John Wheelwright," 189.
23. See Wald, *The Revolutionary Imagination*, 42. Wald's book remains the indispensable guide to the poet's life and art. In 2015 Wald returned to the poet's biography and to the particular matter of his sexuality in "Wheelwright and His Kind," 194-208.
24. Edmund March Wheelwright (1854-1912), a celebrated architect, died in a Connecticut sanitarium where he had been institutionalized two years for "extreme

depression." Wald, *The Revolutionary Imagination*, 44. His notable constructions include the Harvard Lampoon Building and the Longfellow Bridge, the setting of his son's exquisite poem "Father," which ends: "Come home. Wire a wire of warning without words./Come home and talk to me again, my first friend. Father,/come home, dead man, who made your mind my home." Wheelwright, *Collected Poems*, 78, hereafter cited parenthetically.

25. Wheelwright published two collections prior to *Political Self-Portrait*: 1933's *Rock and Shell* and *Mirrors of Venus*, a novel in sonnets from 1938. All three books were brought out in small runs by the Boston publisher Bruce Humphries. James Laughlin at *New Directions* published a slim posthumous *Selected Poems* in 1941, but it was not until 1972 that the full *Collected Poems* appeared, complete with Wheelwright's final volume, *Dusk to Dusk*, as well as additional uncollected poems.

26. As Alan Wald details, the widespread radicalization of intellectuals in the early 1930s effectively split the poet's Ivy League cohort: R. P. Blackmur and Wheelwright's brother-in-law S. Foster Damon, for instance, eschewed express commitment, while Wheelwright himself followed the likes of Malcolm Cowley, Kenneth Burke, and Horace Gregory decidedly to the left, even as his own sympathies lay from the beginning more particularly with the anti-Stalinist currents of east coast radicalism later associated with the *Partisan Review*. Wald, *The Revolutionary Imagination*, 89-90.

27. Quoted in Damon and Rosenfeld, "John Wheelwright," 321.

28. Raymond Williams, *Marxism and Literature*, 205, 204.

29. Raymond Williams, *Marxism and Literature*, 204, 205.

30. Damon and Rosenfeld, "John Wheelwright," 316.

31. In his explanatory gloss of the poem, Wheelwright cites Engel's *Anti-Dühring*, though in fact the latter quotes Goethe in his introduction to *Socialism: Utopian and Scientific*. Engels, *Socialism*, 19.

32. Winfield Townley Scott, "John Wheelwright," 179; Wheelwright, radio script for Patchen broadcast; Wald, *The Revolutionary Imagination*, 38, 99.

33. Josephson, "Improper Bostonian," 535.

34. Wheelwright to unknown recipient, January 9, 1933.

35. See Gramsci, "Intellectuals and Education," 300-11.

36. Wheelwright to unknown recipient, January 9, 1933. For an exhibit of Wheelwright's association with modernist writers both in New York and abroad, see Rosenfeld's treatment of the poet's tumultuous editorial stint at the expatriate magazine *Secession*: "John Wheelwright,"13-40.

37. Winfield Townley Scott, "John Wheelwright," 191. Scott quotes from the notes to *Political Self-Portrait* (154).

38. Wheelwright to unknown recipient, January 4, 1933.

39. Wheelwright, "U.S. 1," 55, 54. For a judicious assessment of Wheelwright's review, see Dayton, *Muriel Rukeyser's* The Book of the Dead, 122-30.

40. Wheelwright, "U.S. 1," 56. "A limited philosophy limits poetry," Wheelwright contends, and "political and aesthetic failings have one root" (55).

41. Wheelwright, "Verse + Radio = Poetry," reprinted in Damon and Rosenfeld, "John Wheelwright," 324.

42. Andrews to Wheelwright, September 19, 1938.

43. It is unclear to which "broadcast school" of poetry—the conventional sentimental fare of Malone and Won or the populist stylings of Corwin—Wheelwright is referring when he criticizes those "whose genius is guided by the spirit of the kindergarten" and who "aim to win the public to poetry by spooning them out unpoetic verses." Wheelwright, "Talking Machines."

44. Eberhart to Wheelwright.

45. Loring to Wheelwright, January 6, 1939. The first letter alerting Wheelwright to his termination refers to Lemmon's "dissatisfaction with the unevenness of quality of the Poetry Corner." Loring to Wheelwright, December 15, 1938.

46. While the evidence is limited to just a few dozen pages of script material and a handful of letters from Loring to Wheelwright, Damon and Rosenfeld agree that his "dismissal from W1XAL seems to have been at least partly a matter of his political views, both presumed and actual." Damon and Rosenfeld, "John Wheelwright," 317.

47. Damon and Rosenfeld, "John Wheelwright," 322.

48. Wheelwright, "'Poetry Noon Hour.'"

49. Wheelwright, "'Poetry Noon Hour.'"

50. Sarton, *Collected Poems*, 45.

51. Hilmes, *Network Nations*, 118. See also Van Loon, "WRUL," 23.

52. Cramer, "The Rockefeller Foundation," 78; Clements, "Foreign Language Broadcasting of 'Radio Boston,'" 175.

53. Hilmes, *Network Nations*, 117. See also Boyd, "The Secret Persuaders."

54. Stephenson, *British Security Coordination*, 60, qtd. in Hilmes, *Network Nations*, 117.

55. "Propaganda from the U.S.," 43-46; Hilmes, *Network Nations*, 118.

56. Wheelwright, radio script; Wheelwright, "Verse + Radio = Poetry," 1.

57. Wheelwright, "Verse + Radio = Poetry," 4.

58. Wheelwright, "Talking Machines," 3.

59. Given his emphasis on the integrative and associative capacities of verse, Wheelwright's example largely substantiates Mark Wollaeger's thesis that "modernism and propaganda, as incipient languages of the new information age, are related yet ultimately divergent mechanisms for processing information within modernity's new regimes of rationalization," though Wheelwright's poetics represent a politically interested attempt to refuse that ultimate divergence. Wollaeger, *Modernism, Media, and Propaganda*, 10.

60. See Adorno, *Current of Music*, 41-143; Adorno, *Notes to Literature*, 59-73.

61. Brecht, "The Radio as an Apparatus of Communication," 51–53; Benjamin, *Radio Benjamin*. Echoing Brecht's call for "radio as an apparatus of communication," Wheelwright laments that as of 1939, "all serious-minded culture lovers desire to reduce the radio (which is world conversation) to a world lecture platform (horrid thought)." Wheelwright, "Talking Machines," 2.

62. Wheelwright, "Verse + Radio = Poetry," 2.

63. Wheelwright, "Talking Machines," 1, 3, 2.

64. Raymond Williams, *Marxism and Literature*, 204.

65. Wheelwright, "Verse + Radio = Poetry," 2. On the "moral meaning of communication" as "the establishment and extension of God's kingdom on earth," see Carey, *Communication as Culture*, 13. See also Sturken et al., *Technological Visions*.

66. Wheelwright, "Toward the Recovery of Speech," 164.

67. Ashbery, *Other Traditions*, 92; Howard, "Four Originals," 351. To the extent that Wheelwright has enjoyed an afterlife in postwar US poetry, it's largely thanks to poets of the New York School. Ashbery devoted one of his 1989–90 Charles Eliot Norton lectures to the poet, and Frank O'Hara a wreath of five poems. See Shand-Tucci, *The Crimson Letter*, 216; O'Hara, "A Wreath for John Wheelwright," 75–77. Susan Howe staged a late twentieth-century reunion between Wheelwright and radio when she devoted an episode of her WBAI-Pacifica show *Poetry* to the poet in 1981. Howe, "The Bostonian."

68. Charles Foster Smith, "Stephen Phillips," 385.

69. Stephen Phillips, *Poems*, 21.

70. Stephen Phillips, *Poems*, 3, 15, 44, 8, 51.

71. Stephen Phillips, *Poems*, 84, 88.

72. *OED*, s.v. "resume, v."

73. Marx, *Reader*, 277.

74. Marx, *Reader*, 279.

75. Marx, *Der achtzehnte Brumaire des Louis Bonaparte*, 4.

76. Stephen Phillips, *Poems*, 53.

77. Trotsky, *Writings*, 320.

78. See also Wald, *The Revolutionary Imagination*, 103.

79. Wheelwright, "Toward the Recovery of Speech," 167.

80. *The Columbia Program Book*, 30. For fine accounts of Corwin's early career and *Words without Music*, see Cummings, "Media Primer," 151–70; and Selch, "Engineering Democracy."

81. Qtd. in Cummings, "Media Primer," 160.

82. Van Doren, preface to Corwin, *Thirteen by Corwin*, vii; Kaplan, *Radio and Poetry*, 5–6.

83. Cummings, "Media Primer," 160; Barnouw, qtd. in Kaplan, *Radio and Poetry*, 9. Corwin's choral techniques were indebted to the contemporary vogue for pedagogical verse choirs in schools around the country, a claim borne out by the

letters Corwin received from schoolteachers and verse choir leaders asking for copies of his radio scripts. Cummings, "Media Primer," 161; Corwin, Correspondence (119) 1938-1946.

84. Morrison, "Words without Music," 42.

85. Selch, "Engineering Democracy," 172-240. See also Chasar, *Everyday Reading*, 84.

86. *Words without Music* premiered on December 4, 1938, though this same pilot episode was broadcast on CBS a month prior, on November 3, under the title of Corwin's previous poetry program for Long Island's WQXR, *Poetic License*. Cummings, "Media Primer," 160, 170, 155.

87. Lindsay's conception of a "poem game" involves the participation of a dancer, a designated chanter, and sometimes the audience itself. See Lindsay, *The Chinese Nightingale*, 93-97. Also published under the title "Daniel," "The Daniel Jazz" first appeared in *Golden Whales*, 91-94.

88. Lindsay's instruction regarding "negro dialect" can be found at the beginning of "A Negro Sermon:—Simon Legree," a poem also arranged by Corwin for the December 11, 1938, broadcast of *Words without Music*. Lindsay, *The Chinese Nightingale*, 104. On Lindsay's own recording career, see Mustazza, "Vachel Lindsay."

89. Hoffman, *American Poetry in Performance*, 63. For the phrase "higher vaudeville," see Lindsay, *The Congo*, vii. On Lindsay and race, see Gubar, *Racechanges*, 139-48; DuPlessis, "'HOO, HOO, HOO.'"

90. Goble, *Beautiful Circuits*, 22.

91. For a discussion of the poem's musical sources, see Taylor, "The Folk Imagination," 146-49.

92. Savage, *Broadcasting Freedom*, 7.

93. Lindsay, *Golden Whales*, 92.

94. I have transcribed this passage in order to convey a rough sense of Corwin's adaptation, but I refer readers to the digitized recording of Corwin's pilot broadcast (one of the few extant transcription recordings of *Words without Music*) available online at *The Old Time Radio Catalog* under the title *Poetic License*: www.otrcat.com/p/norman-corwin.

95. Lindsay, *Golden Whales*, 91.

96. Stoever, *The Sonic Color Line*, 230, 11.

97. Stoever, *The Sonic Color Line*, 235.

98. Savage, *Broadcasting Freedom*, 12, 10.

99. Lott, *Black Mirror*. This quip is Sterling Brown's: "Negro Characters," in Sanders, *A Son's Return*, 163.

100. Ellis, "How Radio Discriminates against Negroes," 65. See also Lowe, "More Negroes in Radio," 2. For more on the extent to which Corwin's Popular Front politics extended to issues of race, see Judith E. Smith, "Literary Radicals in

Radio's Public Sphere," 320-22; and for a critical discussion of the way Corwin's politics of inclusion via "a technological color blindness" nevertheless "relied upon the sonic color line's distinction," see Stoever, *The Sonic Color Line*, 239-41.

101. Qtd. in Lowe, "More Negroes in Radio," 2.

102. "Poems Become a Radio Show: New Literary Possibilities in Words and Music," *New York Times*, March 26, 1939, 144.

103. Corwin, script for *Words without Music*, January 15, 1939. Corwin dedicated a portion of the broadcast to Johnson—in Corwin's estimation one of the nation's "ablest and most popular poets"—who had died in a car accident the previous June (1).

104. "Did you catch *Go Down Death* on the CBS *Words without Music* show last week? The narrator fooled me completely. I remarked to Norm Corwin, the writer and director of the show, that the narrator was swell, and how impossible it was for any white man to catch the true spirit and feeling of the Negro dialect. Norm seemed amazed and asked, 'You're not kidding me are you? That was ME.'" Lesser, "Radio Talent," 9. The script in the Norman Corwin collection names, in addition to the Deep River Boys, Eric Burroughs, Don Costello, Canada Lee, Leonard De Paur, Service Bell, Paul Johnson, Maurice Ellis, and Anne Freeman.

105. Stoever, *The Sonic Color Line*, 164, 165. White-authored dialect by writers like Russell represented, in effect, the literary front of the coordinated campaigns to dismantle Reconstruction, "taking readers imaginatively back in time as the South was being taken politically back in time." North, *The Dialect of Modernism*, 22. For an introduction to Russell, see Jean Wagner, *Black Poets of the United States*, 51-59.

106. Corwin, script for *Words without Music*, 14.

107. Ellis, "How Radio Discriminates against Negroes," 65. See Wittke, *Tambo and Bones*, 187-88, qtd. in Wagner, *Black Poets of the United States*, 56. Despite Corwin's keen good intentions, the specific conditions of the episode's audio production caused the endeavor promptly to backfire: apparently some listeners were convinced they were listening to a minstrel show. Following the broadcast, for instance, one chairman of his lodge's masonic entertainment committee wrote Corwin from Danville, Illinois, to ask for a copy of the script: "We the committee are going to put on a negro minstrel show to be given free." Letter to Corwin, Box 24, Norman Corwin Papers. See also Selch, "Engineering Democracy," 191; Meltzer, "Poetry on the Air," 29.

108. Joel Chandler Harris, introduction to Russell, *Poems*, x, xi.

109. Sterling Brown, *Negro Poetry and Drama*, 88-89.

110. Sterling Brown, "Negro Character," 156, 174.

111. Sterling Brown, "Our Literary Audience," 145.

112. Lamothe, *Inventing the New Negro*, 97.

113. Gates, *The Signifying Monkey*, 192.

114. Gates, *The Signifying Monkey*, 177. See also Sanders, *Afro-Modernist Aesthetics*, 3.

115. James Weldon Johnson, "Preface to the First Edition," in *The Book of American Negro Poetry*, 41.

116. James Weldon Johnson, "Preface to the Revised Edition," in Johnson, *The Book of American Negro Poetry*, 3.

117. James Weldon Johnson, "Preface to the Revised Edition," in Johnson, *The Book of American Negro Poetry*, 4.

118. Brown, *Negro Poetry and Drama*, 42-43.

119. James Weldon Johnson, *God's Trombones*, 11, 9. On the "audio legacy" of these verse sermons, see Furr, *Recorded Poetry*, 117. See also Mustazza, "James Weldon Johnson."

120. Sanders, *Afro-Modernist Aesthetics*, 31.

121. Sterling Brown, *Negro Poetry and Drama*, 68; Sterling Brown, "The New Negro in Literature," 190.

122. In addition to Sanders's *Afro-Modernist Aesthetics*, see, for instance, Kindley, *Poet-Critics*, 85-108; Carmody, "Sterling Brown," 820-40; Skinner, "Sterling Brown," 186-92; Anderson, "Sterling Brown's Southern Strategy," 193-208.

123. Brown's *Outline* has lately elicited several illuminating arguments from critics engaged with issues of lyric reading and reception. See Posmentier, "Blueprints for Negro Reading," 119-35; Glaser, "Folk Iambics," 417-34.

124. Posmentier, "Blueprints for Negro Reading," 124.

125. Sterling Brown, *Outline for the Study*, 18. On the popularity and influence of commercially recorded folk sermons, see Martin, *Preaching on Wax*.

126. Connelly, *The Green Pastures*, xv.

127. Sterling Brown, "Folk Literature," 229; Sterling Brown, *Negro Poetry and Drama*, 89.

128. Sterling Brown, *Negro Poetry and Drama*, 119.

129. Sterling Brown, *Outline for the Study*, 18.

130. Biers, *Virtual Modernism*, 112.

131. Moten, *In the Break*, 185. See also Weheliye, *Phonographies*, 34.

132. Biers, *Virtual Modernism*, 112.

133. Weheliye, *Phonographies*, 12.

134. Bryan Wagner, *Disturbing the Peace*, 192, 194.

135. See also Eisenberg, *The Recording Angel*, 41.

136. Carmody, "Sterling Brown," 821, 832.

137. Bryan Wagner, *Disturbing the Peace*, 193.

138. Sterling Brown, "Folk Literature," 211.

139. Gates, *Figures in Black*, 187.

140. Gavin Jones, *Strange Talk*, 184.

141. Sterling Brown, *Collected Poems*, 52-53.

142. Odum and Johnson, *The Negro and His Songs*; Newman Ivey White, *American Negro Folk-Songs*.
143. Karl Hagstrom Miller, *Segregating Sound*, 58.
144. Corwin, script for *Words without Music*, 7.
145. See Gabbin, *Sterling A. Brown*, 151-52.
146. Henderson, "The Heavy Blues," 49.
147. Kelley, "Notes on Deconstructing 'The Folk,'" 1402.
148. Handy, *Memphis Blues*, 1913. Steven C. Tracy points to Esther Bigeou's 1921 recording for OKeh Records as the first and most plausible reference point for Brown and his readers. Tracy, "Vernacular Recordings," 452.
149. Benston, "Listen Br'er Sterling," 96.
150. Sterling Brown, *Collected Poems*, 60-61.
151. Posmentier, *Cultivation and Catastrophe*, 138.
152. In his *Outline for the Study of the Poetry of American Negroes*, Brown asks students to consider "What spiritual is suggested by this poem?" (36). Sanders suggests "What You Gonna Do?" as collected in Howard Odum and Guy Johnson's 1926 *Negro Workaday Songs*. Sanders, *Afro-Modernist Aesthetics*, 64. We might also mention "What Yo' Gwine to do when Yo' Lamp Burn Down?," which Johnson includes in his first *Book of American Negro Spirituals*. Johnson and Johnson, *The Books of American Negro Spirituals*, 170-71. Listening beyond the spirituals, Henderson reminds us that "this particular call is one of the most deeply engrained in the oral tradition" more generally construed, adducing Big Bill Broonzy ("Watcha gonna do when the pond goes dry?"), Blind Lemon Jefferson ("Watcha gonna do when they send your man to war?"), and Furry Lewis ("Watcha gonna do, your troubles get like mine?"). Henderson, "The Heavy Blues," 51-52.
153. Ellison, "Richard Wright's Blues," 199.
154. James Weldon Johnson, *God's Trombones*, 55.
155. Sanders, *Afro-Modernist Aesthetics*, 64.
156. Henderson, "The Heavy Blues," 53-54.
157. Sterling Brown, "Negro Character," 150, 183. See also Sanders, foreword to *A Son's Return*, xviii-xix.
158. Sanders, *Afro-Modernist Aesthetics*, 64-65.
159. *OED*, s.v. "skin, n."
160. Du Bois, *The Souls of Black Folk*, 254; Johnson and Johnson, *The Books of American Negro Spirituals*, 11-12.
161. Johnson and Johnson, *The Books of American Negro Spirituals*, 17.
162. Eliot, *Collected Poems*, 60. Brown's "Cabaret" addresses itself explicitly to Eliot's *The Waste Land*. Sterling Brown, *Collected Poems*, 111-13. See Callahan, "'A Brown Study,'" 244-47.
163. Saidiya V. Hartman, "The Time of Slavery," 758.

Chapter 3. Langston Hughes's Songwork

1. Hughes, "My Adventures as a Social Poet," 205.
2. Hughes, qtd. in Rampersad, *The Life*, vol. 2, 120. Hughes, *Fields of Wonder*.
3. Rampersad, *The Life*, vol. 2, 133.
4. Hughes, *Fields of Wonder*, 26.
5. Hughes, *Fields of Wonder*, 16.
6. Hughes, *Collected Poems*, 50.
7. Adorno employs this figure to arrest the dialectical relation between culture industrial artifacts and autonomous works of art, perhaps nowhere more memorably than in his epistolary critique of Walter Benjamin's "The Work of Art in the Age of its Technological Reproducibility": "Both bear the stigmata of capitalism, both contain elements of change . . . Both are torn halves of an integral freedom, to which however they do not add up." Adorno et al., *Aesthetics and Politics*, 123.
8. Greenberg, *Art and Culture*, 3.
9. Huyssen, *After the Great Divide*; Mao and Walkowitz, "The New Modernist Studies," 744. On *The Waste Land* and popular musical culture, see Graham, *The Great American Songbooks*, 57-71; Chinitz, *T. S. Eliot*, 19-52.
10. For Pete Astor, it was Richard Goldstein's classic 1969 *The Poetry of Rock* anthology that signaled an epochal "difference in the engagement with lyrics . . . [I]t was the beginning of the time when the rock and pop fan could situate their tastes alongside more 'high' cultural values." Astor, "The Poetry of Rock," 144. An even more illustrative example, according to Adam Bradley, may be David R. Pichaske's textbook anthology *Beowulf to Beatles* (1972), where Sly Stone sits recto/verso with W. B. Yeats. Bradley, *The Poetry of Pop*, 21. Among US literary critics, Bradley is perhaps the most visible scholar working in this formalist vein. See also Bradley, *Book of Rhymes*; Bradley and DuBois, *The Anthology of Rap*; Robbins, *Equipment for Living*; See also Frith, *Performing Rites*; Frith, "Why Do Songs Have Words?," 77-106.

Often pop songs solicit the attention of formalist literary critics because comparison with song holds out the pedagogical promise of a broader, more receptive audience for poetry on the page, or else because pop songs furnish an attractive archive to which scholars and students can apply the critical tools developed for that poetry. For example, Charlotte Pence, editor of "the first collection of academic essays that treats songs as literature," argues that "songs, not poetry itself, might be the key to increasing poetry's readership." Pence, introduction to *The Poetics of American Song Lyrics*, xvi. Under the pressure of these investments, song can yield up subtly contrary aspects, offering a magic mirror of sorts for critics' most closely held convictions about poetry or literary studies more generally. Literary scholars working on pop songs are thus likely to claim for lyrics at least some degree of "unexpected complexity," in Bradley's phrase, whether of a formal or thematic nature, sufficient to sustain a hermeneutic gaze trained on printed matter. Bradley, *The Poetry of Pop*, 41. Bradley disavows any interest in "dignify[ing] or defend[ing] pop lyrics," but he's quite up front about the extent to which his book—with its chapters on rhythm, rhyme, figurative language, voice, style, and story—relies on "readily acquired skills of literary analysis": "Accepting the poetics of pop songs simply admits that lyrics are made of much the same stuff and respond to the same kinds of attention generally reserved for poems on the page" (6, 16). As a result, the implicit or explicit claims made for song as a cultural artifact deserving of serious literary-critical study prove inextricable from claims made on behalf of close reading itself. Meanwhile, other critics like Jonathan Culler locate in "the historical connection" between lyric poetry and song "a salutary corrective" to the hyper-interpretive engines of reigning critical habits. "With song we allow ourselves to be seduced without much guilt by sensuous form and to dwell in the realm of sonorous patterning without an insistent quest for meaning," observes Culler; our relationship with popular song may therefore model a method of study where the object is not to produce new and ingenious "readings" of poems but rather, following the template of our relationship with song, to "develop considerable expertise" as discerning readers and gain "confidence in our ability to appreciate what secure[s] our attention" to the poem in the first place. Culler, *Theory of the Lyric*, 352. In short, for scholars of a formalist bent, popular song may promise a break from literary-critical business as usual, an incitement to different readerly affects and attitudes, and a means for enlarging poetry's audience, but in most cases these rewards will be determined by any critic's particular motivations and by the specific ideas about song to which they appeal.

11. Pence, introduction to *The Poetics of American Song Lyrics*, xiv.
12. Christgau, "Rock Lyrics Are Poetry (Maybe)," 562.
13. Frith, *Performing Rites*, 181.
14. Valéry, *Collected Works*, 214.
15. See, for instance, Ong, *The Presence of the Word*; McLuhan, *The Gutenberg Galaxy*.
16. Booth, *The Experience of Songs*, 9, 7.
17. Ramazani, *Poetry and Its Others*, 191-92.
18. Karlin, *The Figure of the Singer*, xvii.
19. Albright, *Panaesthetics*, 212. Albright borrows the term from Adorno, though not the negative connotations the latter attaches to it. See Adorno, "On Some Relationships," 66-79. To explain why so many printed poems self-identify as "songs," one would need to countenance a historical variety too immense even to sketch in paraphrase. Demonstrating how "it is possible to write poems shaped by the idea of song even in the absence of a musical setting," Elizabeth Helsinger attempts a partial list for the English nineteenth century, adducing as motivations

the "heightened attention to a poem's potential for verbal music" that comparisons with song provide, a "scientific or philosophical interest in the poem as a kind of thinking analogous to music," and aspirations toward the communal or ritual experiences of specific song genres and musical cultures. Helsinger, "Poem into Song," 687. See also Helsinger, *Poetry and the Thought of Song*. Ultimately, it's unmanageable to describe at any level of generality the cultural force exerted by hypostasized notions of poetry's ancient union with music. As John Hollander noted several decades ago, while "it is extremely difficult for a modern reader to grasp the significance" of the unity of song and poem in Greek *mousike*, "all of Western literary history has conspired to make [them] accept a metaphorical identification of the two." Hollander, *The Untuning of the Sky*, 13. *Songs of Innocence and of Experience*, "The Love Song of J. Alfred Prufrock," and *Song of the Andoumboulou* are works by William Blake, T. S. Eliot, and Nathaniel Mackey, respectively.

20. *Eidolon* is Albright's term "for the phantoms generated by the transposition of a work in one artistic medium into an alien one." Albright, *Panaesthetics*, 213.

21. Kittler, *Gramophone*, 80, 22, 80–83; Kittler, "Benn's Poetry," 7.

22. In a discussion of the poet Gottfried Benn, Kittler helps us account for those twentieth-century poems that counterfactually deny their media conditions and self-identify as "songs." With reference to poems like "Chanson" and "Distant Songs," Kittler demonstrates how Benn's high modernist verse makes recourse to the mass media, and especially radio, as a poetic "data source": "Benn's poems . . . are the consummation and consumption of all the media sources they use and sketch in a freely associative manner until even the cold medium of type and print can finally compete with hits on the charts." They cannot *actually* compete, of course, on the level of cultural influence. Rather, Kittler credits printed poetry with the negative agency secured by its "Zeigarnik-effect" vis-à-vis technical systems, referring to the hypothesis in experimental psychology that interrupted tasks are easier to recall than completed ones. "Lyrical practice . . . equals an interruption in computer systems." Like Adornian aesthetic negativity, then, even when poetry forgoes the pseudomorphic strategy of intermedial mimicry and marks its distance from the media, it participates in the totalizing system by all the more dramatically throwing it into relief. Kittler, "Benn's Poetry," 13, 10, 14.

23. Carolyn R. Miller, "Genre as Social Action," 151–67.

24. Gitelman, *Paper Knowledge*, 2.

25. Gitelman, *Always Already New*, 7.

26. For example, "comic books, mysteries, westerns, science fiction, pornography, greeting card verse, newspaper reports, academic criticism, advertising texts, movie and TV scripts, [and] popular song lyrics." Delaney, *Shorter Views*, 210. See McGill, "What Is a Ballad?," 161.

27. Delaney, *Shorter Views*, 211, 205.

28. See McGill, "What Is a Ballad?," 160, 161. See also McLane, *Balladeering*.

29. Jackson and Prins, *The Lyric Theory Reader*, 452.

30. Hughes, *Fine Clothes to the Jew*, xiii, 18. "It is questionable whether any book of American poetry, other than *Leaves of Grass*, had ever been greeted so contemptuously." Rampersad, "Langston Hughes's *Fine Clothes*," 151.

31. The requirement of thematic or narrative coherence in the blues song developed alongside the phonographic recording of classic blues on "race records," as did the length of the typical song. Whereas the duration of much "folk" blues relied primarily on the singer's capacity for continuous improvisation, professional recording dictated new constraints, as jazz critic Martin Williams explains: "At their own right tempo each of these singers could get in about four blues stanzas on a ten-inch recording. Many singers . . . responded to the limitations of time on records simply by stringing together four stanzas on (more or less) the same subject; others . . . attempted some kind of narrative continuity." Qtd. in Baraka, *Blues People*, 103.

32. See Hollenbach, "Phonography," 301-16; Tracy, *Langston Hughes*.

33. Susan Stewart, *Crimes of Writing*, 67-101.

34. McLane, *Balladeering*, 6.

35. See Lamothe, *Inventing the New Negro*.

36. McLane, *Balladeering*, 234.

37. Wordsworth, *The Major Works*, 85-88, 319-20.

38. Hughes, *Collected Poems*, 50; Rampersad, *The Life*, vol. 1, 65.

39. Chinitz, *Which Sin to Bear?*, 44.

40. Hughes, *Fine Clothes to the Jew*, xiii.

41. See Newman, "Moderation in the *Lyrical Ballads*," 185-210.

42. Jeff Titon, qtd. in Barlow, *"Looking Up at Down,"* 131. On Smith's "Crazy Blues," see Gussow, "'Shoot Myself a Cop,'" 8-44.

43. Locke, "Common Clay and Poetry," 55; Chinitz, *Which Sin to Bear?*, 5. See also Edwards, *Epistrophies*, 79-85; Borshuk, *Swinging the Vernacular*, 21-60; Emanuel, *Langston Hughes*, 137-43; Michlin, "Langston Hughes's Blues," 236-53; Tracy, *Langston Hughes*; Ponce, "Langston Hughes's Queer Blues," 505-38; Tidwell, "'Greatest Is the Song,'" 55-66.

44. Hughes and Kingsley, "Hard Luck Blues."

45. See Stoever, *The Sonic Color Line*, 11.

46. A digital recording of the full broadcast can be heard online courtesy of the New York Public Library. WNYC, "Blues Symposium." Though Hughes's poems were frequently set to music throughout his career, the vast majority of these collaborations, as W. S. Tkweme notes, resulted in art songs or jazz numbers. Blues settings are rarer. One major exception is Big Miller's *Did You Ever Hear the Blues?* (1959), an entire album of Hughes's blues. Tkweme, "Blues in Stereo," 504.

47. Edwards, *Epistrophies*, 57, 79, 83.

48. WNYC, "Blues Symposium."

49. Composer Elie Siegmeister, qtd. in Rampersad, *The Life*, vol. 2, 172.

50. This is not to mention Hughes's work as an archivist, critic, and advocate of popular music, as well as his multi-medial efforts in the last decade and a half of his life reading poems to jazz accompaniment—matters I bracket here but to which scholars are now paying increasing attention. See, for instance, Tkweme, "Blues in Stereo"; Meta DuEwa Jones, *The Muse Is Music*, 33-84; Anthony Reed, *Soundworks*, 61-102.

51. Qtd. in Walker, "Langston Hughes." In a section of his late article "Ten Ways to Use Poetry in Teaching" entitled "The Poetry of Popular Song," Hughes affirms his view that pop songs *can* be poems, provided they rise above the status of mere "jingle": "Questions for the teacher to raise: when is a song lyric a poem and when is it just a jingle? What gives *Bali Ha'i* a true lyrical quality as contrasted to a song like *Music, Music, Music*, which is mere rhyming? Why do the words of *The Tennessee Waltz* have a poetic quality? What makes W. C. Handy's *St. Louis Blues* a fine poem even without music?" Hughes, "Ten Ways," 276.

52. Hughes collected "Ballad of the Girl Whose Name Is Mud" in 1942's *Shakespeare in Harlem*. In 1949, Hughes added several new verses to the poem, now set to music by Albert Hague and renamed "Dorothy's Name Is Mud," for the Dwight Deere Wiman Revue, and then revised the lyric again for Burl Ives shortly thereafter. Hughes, "Dorothy's Name Is Mud." For examples of "poem versions" that appear in the *Collected Poems*, see "Refugee Road," "America's Young Black Joe," and "Just an Ordinary Guy." Hughes, *Collected Poems*, 256, 565-66, 579-80.

53. Surveying Hughes scholarship, David Chioni Moore has proposed four overlapping frames through which critics can approach the poet's massive oeuvre: "the bardic-demotic, left-internationalist/Afro-planetary, professional, and sublimated-closeted frameworks." Moore, "Langston Hughes," 702. Studying Hughes's "professional" songwork demands that one cut obliquely across these modes of methodological address. The songs make good on Hughes's most "bardic-demotic" claims, for instance, since "words to music reach so many more people than mere poetry on a printed page." Qtd. in Rampersad, *The Life*, vol. 2, 49. Likewise, one can trace ostensible shifts in Hughes's various political commitments by analyzing his use of song forms as vehicles for protest and propaganda, from his "Park Bench" workers' round (1933) to "Let My People Go" (1944), his campaign song for Adam Clayton Powell, to his 1966 collaboration with Nina Simone on the civil rights anthem "Backlash Blues." Finally, one suspects that Hughes's songwriting—brimming with camp artificiality and pronominal ambiguity—may inform "sublimated-closeted frameworks" and queer readings of Hughes more generally. Moore wonders at the fact that "Hughes seems never to have attempted a full-fledged love poem in a half-century literary career," but that is patently *not* the case if we widen our view to his pop songs ("Langston Hughes," 718). On "Park Bench," see Lee, "The Poetic Voice," 2.

54. Bontemps, "Negro Poets," 360. Karen Jackson Ford notes the irony to

which it's not clear Bontemps is alert: "Paul Laurence Dunbar . . . could find work only as an elevator operator after graduating high school with honors, paid to publish his own first volume of poetry, was subsidized thereafter by white patrons, and died in his early thirties before it could reasonably be claimed he made a *living* at writing." Ford, in "Making Poetry Pay," 275.

55. Bontemps, "Negro Poets," 360.

56. Qtd. in Rampersad, *The Life*, vol. 2, 49. Responding to a survey circulated by Harry Roskolenko to nearly two dozen poets on the subject of "what poets do for a living," Hughes mentions, in addition to the lecture circuit and his weekly *Defender* column, his recent collaborations on the operas *Street Scene* and *Troubled Island*: "Neither commercial successes—and no movie sales. However manage to live from such stints, at least related to poetry. My ASCAP rating helps." Hughes, "What Poets Do for a Living."

57. Throughout his life Hughes would demonstrate a preference for the commercial blues, citing most often in his lists of favorite blues performers the famous Smiths—Mamie, Bessie, Clara, and Trixie. Tracy, *Langston Hughes*, 119.

58. See Karl Hagstrom Miller, *Segregating Sound*, 187–214.

59. Davis, *Blues Legacies and Black Feminism*, 3; Carby, *Cultures in Babylon*, 11.

60. See Hughes, *The Big Sea*, 264; Rampersad, "Langston Hughes's *Fine Clothes*," 150.

61. Hughes and Kingsley, "Hard Luck Blues."

62. Hughes, *Collected Poems*, 215-16.

63. Hughes published three major collections in the decade, all with Knopf: *Shakespeare in Harlem* (1942), the aforementioned *Fields of Wonder* (1947), and *One-Way Ticket* (1949). Though published in 1951, *Montage of a Dream Deferred* was composed in 1948. He begins self-identifying as a "literary sharecropper" in letters of the 1940s and 1950s. The phrase refers to Hughes's contractual obligations and his commitment to others' projects, typically those of an expressly commercial nature. Presumably the virtue of the phrase, which was often employed in moods of comic exasperation, is to conjugate a lack of ownership over the means of his own creative production with a principal institution of Black subjugation. To Ina Steele in 1952: "I retire to the country for another summer of book writing and composers, still having that biography to finish and the show of last season. After which I'd like to retire for GOOD, these past two years having been about the most contracted and committed years of my whole life! Nothing but a literary sharecropper! I do not intend again to work on anybody's books or shows but my black own! If them!" Hughes, *Selected Letters*, 308. "I'm a literary sharecropper tied to a publisher's plantation" (321). See also Berry, *Langston Hughes*, 314.

64. Rampersad, *The Life*, vol. 2, 5. This "public repudiation" was precipitated by the controversy over the resurfaced poem "Goodbye, Christ." I borrow the phrase "poetics of indirection" from John Edgar Tidwell and Cheryl Ragar, "Langston

Hughes," 7-8; and "ethics of compromise" from Chinitz, *Which Sin to Bear?*, 85. Numerous scholars have considerably qualified Rampersad's account. See, for instance, Smethurst, "'Don't Say Goodbye,'" 1225-36; Dolinar, "Langston Hughes," 47-56; Jonathan Scott, *Socialist Joy*; Kernan, *New World Maker*.

65. Rampersad, *The Life*, vol. 2, 23; Hughes to Maxim Lieber, December 28, 1940, qtd. in Rampersad, *The Life*, vol. 1, 392; Hughes "America's Young Black Joe."

66. Kathleen E. R. Smith, *God Bless America*, ix. See also Hajduk, "Tin Pan Alley," 497-512.

67. Hughes, "Go-and-Get-the-Enemy Blues"; Hughes, "Rights of Democracy"; Hughes, "Honolulu Yaka-Hula Dixie. "I hope we can make a good one out of this," writes Hughes to W. C. Handy, his collaborator on "Go and Get the Enemy Blues," "something that the soldiers, white and colored, and the public, too, can sing." Hughes to W. C. Handy, September 6, 1942. Though Hughes was indeed the first Black writer to attend Yaddo, it was the composer Nathaniel Dett, in residence for several weeks by the time Hughes arrived, who first broke the colony's color barrier. Dett collaborated with Hughes on the song "Salute from Old Broadway." See Ben Alexander, "'This Really Is a Delightful Place,'" 81-99.

68. For a brilliant treatment of the Double V campaign and its relation to aesthetic production, see Griffin, *Harlem Nocturne*.

69. Hughes, "Democracy," in 210-11. See also Hughes, "My America," 232-39.

70. Hughes, "Negro Writers," 216, 217.

71. Hughes, "Salute from Old Broadway."

72. Hughes, "Enemy."

73. Hughes, "Dixie Negro to Uncle Sam"; Hughes, "Message to the President."

74. Hughes, *Collected Poems*, 290.

75. Hughes, *Collected Poems*, 297-98, 299-300; Hughes, "Salute to Soviet Armies."

76. Rampersad, *The Life*, vol. 2, 48. To Bontemps from Yaddo in August 1942: "It looks like I am becoming a song writer at last! I shall have four, possibly six numbers published this fall . . . Let's hope some one of them makes some MONEY$$$$$$$. I need it as I am just as overcoatless as I can be and winter is on the way." Hughes and Bontemps, *Letters*, 106. Hughes's songwork at Yaddo was by no means limited to war numbers. He was also penning sentimental ballads like "Just a Little Spot of Sunlight" ("Just a little spot of sunlight/When you came into my heart./Love opened like a flower/For you gave me a new start"), novelty songs like "Pop-Corn Balls" ("It wasn't at a soda fountain,/Certainly wasn't at a bar./It was before the days of motors,/So it wasn't in a car—/'Twas//Sitting behind the kitchen stove,/Eating pop corn balls/I became acquainted with the one I love/Eating pop corn balls"), and revue numbers for French singer and actress Marianne Oswald. Hughes, "Just a Little Spot of Sunlight"; Hughes, "Pop-Corn Balls."

77. Qtd. in Rampersad, *The Life*, vol. 2, 79.

78. Hughes, "Here to Yonder," 12. In a letter to one Private Otis Hudley, July 15,

1945, Hughes stresses the point: "The problem is to get bands like Cab Calloway's or singers like Lena Horne to use [the songs], and that is just about like trying to get hold of a solid gold wrist watch set with diamonds, if you are not able to afford one. In fact, it is much easier to get the solid gold wrist watch than to get your songs in the limelight." Hughes, "Fan Mail."

79. Hughes exclaims in a letter, with reference to "America's Young Black Joe": "If somebody doesn't publish it soon, I will use up all my income on photo-copies of it." Hughes, "Fan Mail."

80. On network radio's enlistment in the war effort, see Horten, *Radio Goes to War*.

81. I make recourse to Adorno, that "hyperbolic hater" of pop music, in much the same spirit as Agnès Gayraud, whose *Dialectic of Pop* uncovers how "at the very moment when Adorno identifies the fundamental contradictions of popular music, his objective hostility creates a subjective alliance with pop music's longstanding critical relation to itself." In fact, "Adorno's hostile criticism" is "the constitutive negative that no aesthetics of pop can do without." Gayraud, *Dialectic of Pop*, 24, 26, 27.

82. Adorno, "On Popular Music," 444-45.

83. On pop music automation, see Cope, *Computer Models*.

84. Rampersad, *The Life*, vol. 2, 36, 59.

85. "Total War" was published, by way of the Associated Negro Press, in the Baltimore *Afro-American* in February 1943, though a copy is also preserved in the "Song Lyrics" section of the Hughes papers. Hughes, *Collected Poems*, 577n693. Hughes, "Total War."

86. For a discussion of the insufficiency of applying evaluative standards of poetic prosody to song lyrics, see Furia, *The Poets*, 39-40.

87. Hughes's relation to kitsch exceeds the matter of his songwork. Most criticisms of the poet, after the scandalized middle-class reaction to his blues poems of the 1920s and the widespread dismissal of his radical verse in the 1930s, boil down to accusations of kitsch. Karen Jackson Ford writes of the "chronic critical scorn" exercised on Hughes's apparent "simplicity," reprising "a litany of faults" that all land somewhere in kitsch's semantic ballpark: "The poems are superficial, infantile, silly, small, unpoetic, common, jejune, iterative, and, of course, simple." Ford, "Do Right to Write Right," 437. James Baldwin leveled perhaps the most notorious of these allegations in his review of the 1959 *Selected Poems*. "This book contains a great deal which a more disciplined poet would have thrown into the waste-basket," writes Baldwin, gesturing particularly to the book's last section, "Words like Freedom," wherein poems like "Freedom Train" and "Freedom's Plow" exploit the demotic plain speech also characteristic of Hughes's 1940s songwork. Add to Baldwin's "waste-basket" his further complaint that many of the poems "take refuge, finally, in a *fake simplicity* in order to avoid the very difficult simplicity

of the experience" (my italics) and we can discern the clear profile of kitsch via two of its most prominent definitions: trash art, following the word's disputed etymological tie to the German *kitschen*, "to pick up street trash," and false art, kitsch as counterfeit or imitation. Baldwin, "Sermons and Blues," 6; Călinescu, *Five Faces of Modernity*, 234. Clement Greenberg famously defined kitsch as "vicarious experience and faked sensations." Greenberg, *Art and Culture*, 10. For a survey of the literature on kitsch, see Gillo Dorfles's *Kitsch: The World of Bad Taste*, which reprints Greenberg's essay in addition to equally significant comments by Hermann Broch, "Notes on the Problem of Kitsch," 49-76.

88. Tiffany, *My Silver Planet*, 8.
89. Tiffany, *My Silver Planet*, 13, 11.
90. Tiffany, *My Silver Planet*, 12.
91. Tiffany, *My Silver Planet*, 11, 69, 32.
92. Tiffany, *My Silver Planet*, 15.
93. Tiffany, *My Silver Planet*, 21.
94. Tiffany, *My Silver Planet*, 75-76.
95. In the specific case of Hughes's wartime songs, the kitsch quality is amplified by their nationalistic content—an example of what Saul Friedlander identifies as "uplifting kitsch," kitsch with a "clear mobilizing function." "Uplifting kitsch" expresses what is "easily understood and accessible to the great majority of people," "calls for an unreflective emotional response," and "handles the core values of a political regime or ideological system as a closed, harmonious entity which has to be endowed with 'beauty' to be made more effective." Friedlander, "Preface to a Symposium," 203.
96. Greenberg, *Art and Culture*, 15; Broch, "Notes," in Dorfles, *Kitsch*, 73; Lindberg, "Popular Modernism?," 283; Adorno, "On Popular Music," 443.
97. Tiffany, *My Silver Planet*, 64-65.
98. See Furia, *The Poets*, 5. At least once—in the case of "Freedom Land," which he composed in full for his 1964 gospel musical *Jerico-Jim Crow*—Hughes dispensed with a collaborating composer. With some puzzlement as to how a poet who was unable to "write music, play an instrument, or carry a tune past his lips" managed to compose the music for such a successful song, Rampersad recounts how during writing sessions Hughes's assistants "would listen intently to his moaning, then pick out a tune on the piano. 'Is this it, Mr. Hughes?' 'That's it!' Langston would assert, impounding the melody." Rampersad, *The Life*, vol. 2, 372.
99. Hughes, note to Elie Siegmeister.
100. Hughes, "My Heart Is a Lone Ranger."
101. Hughes to W. C. Handy, September 6, 1942; Handy to Hughes, September 11, 1942.
102. "Freedom Road" was broadcast on *The March of Time* in July 1942 and recorded by Josh White in 1944 for *Songs of Citizen C.I.O.*, an Asch Records album

sponsored by the National CIO War Relief Committee. Hughes, "Freedom Road"; Various Artists, *That's Why We're Marching*.

103. For one introduction to the complicated issue of the musical artwork's ontological status, see Davies, *Musical Works*. Critics like Peter Middleton and Charles Bernstein, who insist on any poem's "fundamentally plural existence," may dispute my characterization of poetry's ontological stability vis-à-vis song. Bernstein, introduction to *Close Listening*, 9; Middleton, *Distant Reading*.

104. Wimsatt, *The Verbal Icon*, 81. The phrase is Archibald MacLeish's, from the poem "Ars Poetica." MacLeish, *Collected Poems*, 107.

105. Furia, *Ira Gershwin*, 36; Miles, *Paul McCartney*, 202.

106. Hughes, Notes on songwriting with sample lyrics.

107. Hughes, Notes on songwriting with sample lyrics.

108. Tiffany, *My Silver Planet*, 72.

Chapter 4. "In Lieu of the Lyre"

1. Howe, "Every Mark on Paper Is an Acoustic Mark."

2. Raymond Williams, *Marxism and Literature*, 160.

3. That process of spiritualization is itself a "protest," Williams notes with a stirring turn of the dialectic, against the degradation of labor under industrial capitalism. Raymond Williams, *Marxism and Literature*, 160-62.

4. Raymond Williams, *Marxism and Literature*, 162.

5. Ponge, *Le parti pris des choses*. On modernism and the object world, see especially Mao, *Solid Objects*. On "the matter of modernism," see also Bill Brown, *Other Things*, 45-151.

6. Pound, "A Retrospect," 3.

7. Miller, *Poets of Reality*, 8.

8. Cleanth Brooks, *The Well Wrought Urn*; Wimsatt, *The Verbal Icon*. See also Mao, "The New Critics," 227-54.

9. Altieri, "What Is Living," 775; Fried, *Art and Objecthood*; Belgrad, *The Culture of Spontaneity*.

10. Kunitz, "Process and Thing," 103. See also Rasula, *The American Poetry Wax Museum*, 238-39.

11. To the extent this shift has been confirmed and compounded by theoretical developments in literary and cultural studies, one has to agree with Stephanie Burt as to the "inescapable" sense, among poets and critics, that today "poems ask to be understood as ongoing processes, not as complete objects." Burt, "Is American Poetry Still a Thing?," 279. See Donald Allen, *The New American Poetry*.

12. Kahn, *Noise, Water, Meat*, 123-56; Connor, *Beckett*, 84.

13. Middleton, "Poetry Reading," 1068-70. On poetry and performance in the postwar period, see also Wheeler, *Voicing American Poetry*; Meta DuEwa Jones, *The*

Muse Is Music; Hoffman, *American Poetry in Performance*; Perloff and Dworkin, *The Sound of Poetry*; Middleton, *Distant Reading*; Grobe, *The Art of Confession*; Allison, *Bodies on the Line*; Blount, "The Preacherly Text," 582-93. MacArthur, "Monotony," 38-63.

14. See Camlot, *Phonopoetics*; Picker, *Victorian Soundscapes*.
15. Pound, "A Retrospect," 4; Burt, "Is American Poetry Still a Thing?," 282.
16. McGann, "The Text," 274.
17. Jauss, "Poiesis," 594.
18. See de Man, *The Rhetoric of Romanticism*, 239-62; Jackson, *Dickinson's Misery*; Jackson and Prins, *The Lyric Theory Reader*.
19. Marianne Moore, *New Collected Poems*, 27-28.
20. Lowell, "Acceptance Speech." Though Lowell's distinction between the raw and the cooked rhymes with Claude Lévi-Strauss's binary structuration in 1964's *Mythologiques: Le cru et le cuit*, he may have been recalling Virginia Woolf's complaint against James Joyce's *Ulysses*: "When one can have the cooked flesh, why have the raw?" Woolf, *A Writer's Diary*, 47. See Rasula, *The American Poetry Wax Museum*, 233-47.
21. Marianne Moore, "Poetry Reading."
22. My gratitude to Stuart Davis for confirming Moore's version of events.
23. I refer to Moore's celebration of the salamander, whose "shield was his humility" in "his/unconquerable country of unpompous gusto" ("His Shield"), and to Moore's declaration, in the poem "Tell Me, Tell Me," that "deference may be my defense." Marianne Moore, *New Collected Poems*, 179, 241. Nearly twenty-five years prior, in December 1941, Moore had stood before an earlier generation of Harvard students and announced "humility, concentration, and gusto" as the "three foremost aids to persuasion" in poetry. A published version of these remarks was later collected in Moore's *Complete Prose* (420). Leavell, *Holding On Upside Down*, 311.
24. Sargeant, "Humiliation," 38. As a result of her first-reader's agreement with the magazine's poetry editor Howard Moss, the *New Yorker* published twenty-three new poems by Moore from 1953 to 1970. On Moore and the *New Yorker*, see Green, "Marianne Moore," 58-78; Gregory, "'Combat Cultural,'" 208-21.
25. Marianne Moore, *Reader*, 258.
26. For another probing reading of this poem and its occasion, see Gregory, *Apparition of Splendor*, 117-23.
27. On Moore and phonography, see especially Edward Allen, "'Spenser's Ireland,'" 451-74; Edward Allen, *Modernist Invention*, 135-205; Neal, "Marianne Moore's Tone Technologies."
28. Howarth, "Marianne Moore's Performances," 557.
29. Howarth, "Marianne Moore's Performances," 575; Grobe, "The Breath of the Poem," 217.

30. Marianne Moore, *Complete Prose*, 606-7.

31. The first several examples are drawn from Moore's reading of "O to Be a Dragon" and "The Arctic Ox (or Goat)" at the Library of Congress in 1963. Marianne Moore, "Marianne Moore Reading," October 21, 1963. The final example comes from her reading of "Armor's Undermining Modesty" in October 1957, just weeks after the Little Rock Nine desegregated Little Rock Central High School. Marianne Moore, "Marianne Moore," October 18, 1957.

32. Gunn's comments are reported and paraphrased by Philip Levine, *My Lost Poets*, 107. See also Donald Hall's comment on a Harvard reading in 1948: "When she finished one poem and introduced another, it was difficult to tell when poem ended and commentary began. Listeners always had trouble distinguishing her poems from her talk—attesting not so much to the prosiness of her poems as to the finish of her speech—but in this reading the distinction was narrower than ever." Hall, *Their Ancient Glittering Eyes*, 154.

33. Marianne Moore, "Marianne Moore," October 10, 1957; Moore, *New Collected Poems*, 54.

34. Moore was catapulted into public view in the early 1950s, when her *Collected Poems* (1951) earned the National Book Award and the Pulitzer and Bollingen Prizes. Linda Leavell's biography captures the improbability of Moore's trajectory from "insufferable high brow" modernist of the 1920s, in Mark Van Doren's phrase, to American cultural icon, fit for appearances in *Vogue*, *Harper's Bazaar*, *Ladies' Home Journal*, *McCall's*, on *Today* and *The Tonight Show*, and on the mound at Yankee Stadium. Leavell, *Holding On Upside Down*, xii. Taking up the consensus view, Leavell observes that "public life took a toll on [Moore's] poetry," as the latter wrote more "quickly and prolifically" than ever before. "From a poetry of understatement and inconclusive encounter with the larger literary tradition," writes Jeanne Heuving, "Moore begins to compose a poetry of overstatement and positive assertion of values." Heuving, *Omissions Are Not Accidents*, 141. Compared to previous work, the late poetry's "slightness is conceded universally," according to John M. Slatin. Slatin, *The Savage's Romance*, 13. Happily, this consensus has been queried and qualified in recent years by critics eager to heed Elizabeth Gregory's suggestion that "rather than involving a decline in her poetic capacities ... Moore's movement into the popular arena represents a complex handling of her paradoxical position as a poet who both aspired to authoritative status and questioned the hierarchical term on which such authority circulated." Gregory, "Stamps," 235. See also Vincent, *Queer Lyrics*, 89-120; Gregory, "Is Andy Warhol Marianne Moore?," 237-51; Finch, "'Passion for the Particular,'" 221-35.

35. Leavell, *Holding On Upside Down*, xii; Edward Allen, "'Spenser's Ireland'"; Neal, "Marianne Moore's Tone Technologies."

36. Sargeant, "Humiliation," 38.

37. Sargeant, "Humiliation," 38, 48.

38. This phrase appears in Bishop's poem "Invitation to Marianne Moore." Bishop, *Poems*, 81. Sargeant, "Humiliation," 48-49.

39. Sargeant, "Humiliation," 49.

40. The medieval notion that Joseph stored "wheat in the pyramids" has its roots in Genesis 41, where Joseph directs the Pharaoh to gather crops that "shall be for store to the land against the seven years of famine." Gen. 41:36 KJV.

41. While the version in *Tell Me, Tell Me* (1966) reprints the form in which the poem appeared in the *Harvard Advocate* 100 (Fall 1965), Moore revised "In Lieu of the Lyre" slightly for *The Complete Poems*. Heather Cass White's indispensable *New Collected Poems*, my own copy text, reverts to the earlier version. Marianne Moore, *New Collected Poems*, 252.

42. Cristanne Miller, *Marianne Moore*, 3.

43. Levin, "A Note on Her French Aspect," 40-43. See Leavell, *Holding On Upside Down*, 326, 337-39; Stapleton, *Marianne Moore*, 158-83.

44. The poem appeared under the title "I've Been Thinking . . ." in the *New York Review of Books*, October 31, 1963, 19, and as "Avec Ardeur" in the *Complete Poems* (1967) and subsequent editions. Moore, *New Collected Poems*, 418.

45. Fang, "Rhymeprose on Literature," 527-66. Moore also includes the phrase "sentir avec ardeur" in "Tom Fool at Jamaica." In her notes to the latter poem, she includes Fang's footnote, quoted below, as well as the full text of Mme de Boufflers's poem. Marianne Moore, *Complete Poems*, 286, 393-94.

46. Horace, *Odes*, 153. On Bewick and Moore, see "'In Lieu of the Lyre,'" 5-7.

47. Miller, *Marianne Moore*, 269.

48. See Trachtenberg, *Brooklyn Bridge*, 68-71.

49. Marianne Moore, "In Lieu of the Lyre."

50. On Moore, collage, and quotation, see especially Brinkman, *Poetic Modernism*, 105-40; Gregory, *Quotation*, 129-85.

51. Fang, "Rhymeprose on Literature," 529.

52. Pound, *New Selected Poems*, 281. I quote from Ezra Pound's English, for his translation of Mme de Boufflers's poem is surely the relevant intertext. In the *Complete Poems* (1967), when Moore adopts the new title "Avec Ardeur" for "Occasionem Cognosce," mentioned above (a witty, rhyming, didactic effort that much resembles Mme de Boufflers's), she appends the dedication: *"Dear Ezra, who knows what cadence is."* Marianne Moore, *Complete Poems*, 237-39. For Moore's note to "In Lieu of the Lyre," see p. 295.

53. Bernstein, introduction to *Close Listening*, 12.

54. Heather Cass White, introduction to *New Collected Poems*, xv. On Moore and revision, see, for instance, Kappel, "Complete with Omissions," in 125-56; Schulze, "Textual Darwinism," 270-305.

55. Marianne Moore, *Complete Poems*, 36.

56. Schulze, "*Others*," 474.

57. Marianne Moore, *New Collected Poems*, 27-28.
58. Bernstein, introduction to *Close Listening*, 9. See also Middleton, *Distant Reading*, 1-24.
59. Marianne Moore, "The World in Books," 14.
60. Marianne Moore, "The World in Books," 15-16. When Atkins published versions of this poem in *Objects* (1961) and *Heretofore* (1968), he called it "Trainyard at Night." Atkins, *Russell Atkins*, 62-64; Atkins, *World'd Too Much*, 176-78.
61. Marianne Moore, "The World in Books," n.p.
62. See Nielsen, "Russell Atkins," 119.
63. Rosemont and Kelley, *Black, Brown and Beige*, 218; Hughes and Bontemps, *The Poetry of the Negro*, 208.
64. Hughes, "Some Practical Observations," 309. Hughes offered Atkins advice about publishing ("I should think that NEW DIRECTIONS would be among the publishers most likely to be interested in your type of poetry"), aesthetics ("If I were you, I would not worry about being a social poet . . . Only if you think and feel socially should you try to write in that way"), and networking (the addresses of Lloyd Addison, Leroi Jones [Amiri Baraka], and Gloria Oden). Hughes to Atkins. In a 1947 letter to Bontemps, Hughes writes, "In Cleveland I got some good poetry from Helen Johnson. Also Russell Atkins . . . Looks like we are in for a set of Negroes who will out-do the most *avant-garde* whites! Which delights me!" Hughes and Bontemps, *Letters*, 226. See also Atkins's elegy for Hughes, "May Twenty-Second, Nineteen Sixty-Seven," which includes the acknowledgment: "Few fugitives had been sheltered by him/as had I." Atkins, *World'd Too Much*, 115.
65. Atkins, *Phenomena*, 79. On Atkins and the Black Arts Movement, see Smethurst, *The Black Arts Movement*, 219-20; Nielsen, "Objects: For Russell Atkins." Atkins has been subject to criticism of the kind leveled by this writer for the *Negro Digest*: "If Brother Atkins's picture were not on the cover [of his book], one would have a difficult time knowing he is a black poet," for even the poems that appear "black-related" are "written as if by a Victorian era observer and not a blkman dealing with his history as he should be about doing (blkartists are responsible to the blkcommunity)." Amini, "Heretofore," 95. On the Muntu Poets of Cleveland, a workshop organized in the late 1960s by Atkins and Black Arts–affiliated poet Norman Jordan, see the Muntu Poets of Cleveland, *The Muntu Poets*.
66. Nielsen, *Black Chant*, 58. See also Nielsen, *Integral Music*, 8.
67. In 2013, Kevin Prufer and Michael Dumanis released *Russell Atkins: On the Life & Work of an American Master* in Pleiades Press's Unsung Masters Series. In 2019, Cleveland State University Press released the more substantial collection *World'd Too Much: The Selected Poetry of Russell Atkins*, edited by Prufer and Robert E. McDonough. See also Dworkin, *Radium of the Word*, 101-42.
68. Conrad Kent Rivers, quoted in Redmond, *Drumvoices*, 328. On *Free Lance*, see Arthur and Johnson, *Propaganda and Aesthetics*, 155-58.

69. Prior to the recent selected volumes—*Russell Atkins* and *World'd Too Much*—Atkins published just one full collection, *Here in The* (1967) with Cleveland State University Press.
70. Atkins, "Russell Atkins," 10, 3.
71. Atkins, "Russell Atkins," 3, 9.
72. Atkins, "Russell Atkins," 13.
73. Atkins, "Russell Atkins," 10.
74. Atkins, "Russell Atkins," 8.
75. Atkins, "The Invalidity," 32. See also Nielsen, *Integral Music*, 30-34; Nielsen, "Black Deconstruction," 86-103.
76. Atkins, "The Invalidity," 31, 25, 24.
77. Atkins, "The Invalidity," 31, 25, 24.
78. Atkins, "The Invalidity," 22.
79. Atkins, "Russell Atkins," 13.
80. Atkins, "A Psychovisual Perspective," in Prufer and Dumanis, *Russell Atkins*, 90. Originally published in *Free Lance* 3, no. 2 (1955).
81. Atkins, "A Psychovisual Perspective," 89-91.
82. Atkins, "A Psychovisual Perspective," 96.
83. Atkins, "Russell Atkins," 14.
84. Atkins, "Russell Atkins," 15. On the wider vogue for Gestalt therapy among postwar avant-gardists, see Belgrad, *The Culture of Spontaneity*, 142-56.
85. Writing in the early 1990s, Atkins felt the need to clarify his relationship to Derridean poststructuralism: "The word 'deconstruction' was not in current use during 1954 to 1958 . . . I resurrected the word and redefined it for my theory." Atkins, "Russell Atkins," 14-15. On Derrida and Atkins, see Nielsen, "Black Deconstruction."
86. Atkins, "A Psychovisual Perspective," 91.
87. Atkins, "A Psychovisual Perspective," 90.
88. Atkins himself looked to Gestalt psychologists like G. W. Hartman and their experimental pursuit of a fundamental "unity of the senses." See, for instance, Hartman, "Changes in Visual Acuity," 393-407; Hartman, *Gestalt Psychology*, 141-51.
89. See Fernandez-Prieto et al., "Does Language Influence," 1-11.
90. Atkins, "A Psychovisual Perspective," 92.
91. The poet Billera ventriloquizes Atkins: "My eyes do not close to sound/I read my voice upon cleff/It promises a new meaning [*sic*]." Billera, "Russell Atkins's Object Forms," 9.
92. Pryer, "Graphic Notation"; Cage, *Notations*.
93. Atkins, "A Psychovisual Perspective," 40.
94. Atkins's only published musical scores are the four short *Objects for Piano* from 1969.

95. For exceptions to Atkins's otherwise silent musicological reception, see Stuckenschmidt, "Contemporary Techniques in Music," 11; Stuckenschmidt, *Twentieth Century Music*, 229.

96. Kahn, *Noise, Water, Meat*, 18, 363-64.

97. See Williams, "'Really, Music.'"

98. Schaeffer, *Treatise*, 66, 64.

99. Kane, *Sound Unseen*, 94.

100. Chion, "Reflections on the Sound Object," 23.

101. Schaeffer, *Treatise*, 67; Dack, "Pierre Schaeffer," 34.

102. Schaeffer, *Treatise*, 69.

103. Kane, "The Fluctuating Sound Object," 63.

104. Steintrager and Chow, "Sound Objects," 7.

105. For an attempt to reinsert "history and mediation in Schaeffer's ontology," see Kane, "The Fluctuating Sound Object," 68.

106. Atkins, "A Psychovisual Perspective," 14.

107. Atkins, "A Psychovisual Perspective," 13.

108. Atkins, "A Psychovisual Perspective," 13-14.

109. Atkins, "A Psychovisual Perspective," 17, 23.

110. Chion, "Reflections on the Sound Object," 30.

111. Atkins, "A Psychovisual Perspective," 25.

112. Jordan and Stefanski, "1955-1960," 19.

113. McLuhan, *Understanding Media*, 33.

114. Jordan, afterword to *Phenomena*, 77. See also Nielsen, *Black Chant*, 32.

115. As if to insist on the theoretical continuity of his musical and poetic compositions, Atkins's 1969 score *Objects for Piano* constitutes the third part in a series whose first two installments, *Objects* and *Objects 2*, are poetry chapbooks.

116. See William J. Harris, *The Poetry and Poetics*, 104; interview with Steve McCaffery by Abigail Lang, "The Poem as Score," 17-33; Carruthers, *Notational Experiments*.

117. Marianne Moore to Russell Atkins, 11 April 1953.

118. Ashbery, *Selected Prose*, 112.

119. Atkins, *Phenomena*, 79.

120. Atkins, "Russell Atkins," 14.

121. Atkins, "Russell Atkins," 10.

122. Atkins, *Phenomena*, 79.

123. Atkins, *Juxtapositions*, n.p.

124. Staten, "The Origin of the Work," 57. See also Sterne, "Communication as Technē," 91-98.

125. Atkins, *Juxtapositions*, n.p.

126. Atkins, *Juxtapositions*, n.p.

127. Atkins, *Heretofore*, 30-31.

128. Reagan, *When Abortion Was a Crime*, 216-20.
129. Atkins, "The Abortionist," 77-87; Myrsiades, *Splitting the Baby*, 110-38.
130. Myrsiades, *Splitting the Baby*, 130. On Atkins and the natural world, see Shockley, "Death Is Only Natural," 108-15.
131. Gwendolyn Brooks, *Blacks*, 21-22.
132. Marianne Moore, *New Collected Poems*, 40.
133. Bornstein, *Material Modernism*, 93, 103.
134. Leavell, *Holding On Upside Down*, 157; Marianne Moore, *New Collected Poems*, 52.
135. Barbara Johnson, "Apostrophe," 34.

Coda

1. Jackson, *Dickinson's Misery*, 1. See also Burt, "What Is This Thing Called Lyric?," 422-40; Culler, *Theory of the Lyric*.
2. Jackson, *Dickinson's Misery*, 1.
3. Jackson, *Dickinson's Misery*, 53, 7.
4. Jackson, *Dickinson's Misery*, 90.
5. Adam appeared in von Praunheim's *Death Magazine, or How to Be a Flowerpot* (*Das Todesmagazin oder: Wie werde ich ein Blumentopf?*, 1979) and *Our Corpses Still Live* (*Unsere Leichen leben noch*, 1981), as well as Marianne Enzenberger's *The Bite* (*Der Biß*, 1984).
6. Adam, *Reader*, 117-18.
7. Blake, *Selected Poems*, 48, 58.
8. Blake, *Selected Poems*, 59.
9. "Side by side (because our fate/Damned us ere our birth)/We stole out of Limbo Gate/Looking for the Earth." Kipling, *Rudyard Kipling's Verse*, 678. Identifying Kipling as a source, Kristin Prevallet opines that the "play on her own name (the Helen in 'Helen All Alone' sliding into the Adam in 'Limbo Gate') is almost certainly intentional (although in an interview with Susan Howe in 1977, she denies being aware of any symbolic significance to her own name)." Adam, *Reader*, 469.
10. Adam, "Helen Adam & Robert Duncan Reading."
11. Grundberg, introduction to *Songs with Music*, 12.
12. Grundberg, introduction to *Songs with Music*, 12.
13. Adam, *Reader*, 337. "Anyone who has heard Helen perform realizes how much she communicates in the moment of the song; how much is lost when the bare words are rendered on the printed page." Grundberg, introduction to *Songs with Music*, 12.
14. See McLane, "On the Use and Abuse," 136; McLane, *Balladeering*.
15. Susan Stewart, *Crimes of Writing*, 67.
16. Susan Stewart, *Crimes of Writing*, 74, 68.

17. Susan Stewart, *Crimes of Writing*, 74, 67.
18. Coleridge, *Biographia Literaria*, 169.
19. Owen, "Helen Adam, 1909-1993," 382.
20. Quoted in Prevallet, introduction to Adam, *Reader*, 10.
21. Davidson, *The San Francisco Renaissance*, 180.
22. Duncan, *Fictive Certainties*, 30.
23. Duncan to Helen Adam, July 17, 1955, 304.
24. Donald Allen, *The New American Poetry*, 435.
25. Adam, *Reader*, 65-67, 147-48.
26. Finkelstein, "Helen Adam," 131. See also Finkelstein, *The Utopian Moment*, 83-90.
27. Finkelstein, "Helen Adam," 136.
28. See McGann, *The Romantic Ideology*.
29. Hirsch, *How to Read a Poem*, 4; Vendler, *Poems, Poets, Poetry*, xi. See Poovey, "The Model System," 408-38; Jackson and Prins, *The Lyric Theory Reader*.
30. On "the hermeneutic circle of lyricized and racialized reading," see Anthony Reed, *Freedom Time*, 97; Javadizadeh, "The Atlantic Ocean," 475-90.
31. I owe a debt to Claire Seiler's superb discussion of suspension's "varied and sometimes contradictory meanings" in *Midcentury Suspension*, 28.
32. Benjamin, *The Arcades Project*, 475, 462.
33. Davidson, *The San Francisco Renaissance*, 179.
34. Davidson, *The San Francisco Renaissance*, 181.
35. Marvell, *Complete Poems*, 101.
36. Ps. 137:2, KJV; Sannazaro, *Arcadia and Piscatorial Eclogues*, 102-5; Parker, *Literary Fat Ladies*, 54-66.
37. Spenser, *The Faerie Queene*, 2.7.80.
38. Parker, *Literary Fat Ladies*, 57.
39. Parker, *Literary Fat Ladies*, 59.
40. Parker, *Literary Fat Ladies*, 61.
41. Ramazani, "Traveling Poetry," 283; Auden, *Complete Works*, 105.
42. Gumbrecht and Pfeiffer, *Materialities of Communication*.
43. Susan Stewart, *Poetry*, 104.
44. Susan Stewart, *Poetry*, 68, 105.
45. Adam, "A Few Notes," 356.
46. Adam, *Reader*, 94-97.
47. See Thompson, "Transformation Flight," 77.
48. Finkelstein, *The Utopian Moment*, 87.

Bibliography

Abrams, M. H. *The Fourth Dimension of a Poem: And Other Essays*. New York: Norton, 2012.
Acland, Charles R., ed. *Residual Media*. Minneapolis: U of Minnesota P, 2007.
Adam, Helen. "A Few Notes on the Uncanny in Narrative Verse." In *A Helen Adam Reader*, edited by Kristin Prevallet, 351-77. Orono, ME: National Poetry Foundation, 2007.
Adam, Helen. "Helen Adam & Robert Duncan Reading." June 9, 1976, Naropa University, MP3/cassette tape, 1:27, Naropa University Audio Archive. http://archives.naropa.edu/digital/collection/p16621coll1/id/8/.
Adam, Helen. *A Helen Adam Reader*. Edited by Kristin Prevallet. Orono, ME: National Poetry Foundation, 2007.
Adam, Helen, and Carl Grundberg. *Songs with Music*. Edited and transcribed by Carl Grundberg. San Francisco: Aleph Press, 1982.
Adams, Henry. *The Degradation of Democratic Dogma*. New York: MacMillan, 1919.
Adorno, Theodor W. *Current of Music*. Edited by Robert Hullot-Kentor. Cambridge: Polity Press, 2009.
Adorno, Theodor W. "The Form of the Phonograph Record." Translated by Thomas Y. Levin. *October* 55 (1990): 56-61.
Adorno, Theodor W. *Notes to Literature*. Edited by Rolf Tiedemann. Translated by Shierry Weber Nicholsen. New York: Columbia UP, 2019.
Adorno, Theodor W. "On Popular Music." In *Essays on Music*, edited by Richard Leppert, 437-69. Berkeley: U of California P, 2002.
Adorno, Theodor W. "On Some Relationships between Music and Painting." Translated by Susan Gillespie. *Musical Quarterly* 79, no. 1 (1995): 66-79.
Adorno, Theodor, Walter Benjamin, Ernst Bloch, Bertolt Brecht, and Georg Lukács. *Aesthetics and Politics*. London: Verso, 2007.
Agha, Asif. "Voice, Footing, Enregisterment." *Journal of Linguistic Anthropology* 15, no. 1 (2005): 38-59.
Ahearn, Barry. *Zukofsky's "A": An Introduction*. Berkeley: U of California P, 1982.

Albright, Daniel. *Panaesthetics: On the Unity and Diversity of the Arts*. New Haven, CT: Yale UP, 2014.

Albright, Daniel. *Putting Modernism Together: Literature, Music, and Painting, 1872-1927*. Baltimore: Johns Hopkins UP, 2015.

Alexander, Ben. "'This Really Is a Delightful Place to Work': Yaddo, Langston Hughes, and a Revaluation of the Black Radical Left." In *Black Writers and the Left*, edited by Kristen Moriah, 81-99. Newcastle upon Tyne: Cambridge Scholars, 2013.

Allen, Donald. *The New American Poetry*. New York: Grove Press, 1960.

Allen, Edward. *Modernist Invention: Media Technology and American Poetry*. Cambridge: Cambridge UP, 2020.

Allen, Edward. "'Spenser's Ireland': December 1941: Scripting a Response." *Twentieth-Century Literature* 63, no. 4 (2017): 451-74.

Allison, Raphael. *Bodies on the Line: Performance and the Sixties Poetry Reading*. Iowa City: U of Iowa P, 2014.

Altieri, Charles. "What Is Living and What Is Dead in American Postmodernism: Establishing the Contemporaneity of Some American Poetry." *Critical Inquiry* 22, no. 4 (1996): 764-89.

Amini, Johari. "Heretofore." *Negro Digest*, September 1969, 95.

Anderson, David R. "Sterling Brown's Southern Strategy: Poetry as Cultural Evolution in *Southern Road*." In *After Winter: The Art and Life of Sterling Brown*, edited by John Edgar Tidwell and Steven C. Tracy, 193-208. Oxford: Oxford UP, 2009.

Andrews, Loring. Letter to John Wheelwright, September 19, 1938. Box 11, Folder 1, John Wheelwright Papers 1920-1940. John Hay Library, Brown University, Providence, RI.

Andrews, Loring. Letter to John Wheelwright, December 15, 1938. Box 11, Folder 6, John Wheelwright Papers 1920-1940. John Hay Library, Brown University, Providence, RI.

Andrews, Loring. Letter to John Wheelwright, January 6, 1939. Box 11, Folder 9, John Wheelwright Papers 1920-1940. John Hay Library, Brown University, Providence, RI.

Arthur, Abby, and Ronald Maberry Johnson. *Propaganda and Aesthetics: The Literary Politics of African-American Magazines in the Twentieth Century*. Amherst: U of Massachusetts P, 1979.

Ashbery, John. *Other Traditions*. Cambridge, MA: Harvard UP, 2000.

Ashbery, John. *Selected Prose*. Edited by Eugene Richie. Ann Arbor: U of Michigan P, 2004.

Astor, Pete. "The Poetry of Rock: Song Lyrics Are Not Poems but the Words Still Matter; Another Look at Richard Goldstein's Collection of Rock Lyrics." *Popular Music* 29, no. 1 (2010): 143-48.

Atkins, Russell. "The Abortionist: A Poetic Drama to Be Set to Music." In *Russell*

Atkins: On the Life and Work of an American Master, edited by Kevin Prufer and Michael Dumanis, 77-87. Warrensburg, MO: Pleiades Press, 2013.
Atkins, Russell. *Heretofore*. Heritage: London, 1968.
Atkins, Russell. "The Invalidity of Dominant-Group 'Education' Forms for 'Progress' for Non-dominant Ethnic Groups as Americans." *Free Lance* 7, no. 2 (1963): 19-32.
Atkins, Russell. *Juxtapositions*. Cleveland, OH: self-published, 1991.
Atkins, Russell. *Objects for Piano*. Cleveland, OH: Free Lance Press, 1969.
Atkins, Russell. *Phenomena*. Wilberforce, OH: Wilberforce UP, 1961.
Atkins, Russell. "A Psychovisual Perspective for 'Musical' Composition." *Free Lance* 3, no. 2 (1955): 7-47.
Atkins, Russell. "A Psychovisual Perspective for 'Musical' Composition." In *Russell Atkins: On the Life and Work of an American Master*, edited by Kevin Prufer and Michael Dumanis, 88-107. Warrensburg, MO: Pleiades Press, 2013.
Atkins, Russell. "Russell Atkins." In *Contemporary Authors Autobiography Series*, vol. 16, edited by Joyce Nakamura, 1-19. Detroit: Gale Research, 1992.
Atkins, Russell. *Russell Atkins: On the Life and Work of an American Master*, edited by Kevin Prufer and Michael Dumanis. Warrensburg, MO: Pleiades Press, 2013.
Atkins, Russell. *World'd Too Much: The Selected Poetry of Russell Atkins*. Edited by Kevin Prufer and Robert E. McDonough. Cleveland, OH: Cleveland State U Poetry Center, 2019.
Auden, W. H. *The Complete Works of W. H. Auden*. Vol. 1, *Prose and Travel Books in Prose and Verse, 1926-1938*. Edited by Edward Mendelson. Princeton, NJ: Princeton UP, 1997.
Avery, Todd. *Radio Modernism: Literature, Ethics, and the BBC, 1922-1938*. Aldershot: Ashgate, 2006.
Bakhtin, Mikhail. *Dialogic Imagination: Four Essays*. Edited by Michael Holquist. Translated by Caryl Emerson and Michael Holquist. Austin: U of Texas P, 1981.
Baldwin, James. "Sermons and Blues." *New York Times Book Review*, March 29, 1959, 6.
Baraka, Amiri. *Blues People: Negro Music in White America*. New York: William Morrow, 1963.
Barbantani, Silvia. "Lyric in the Hellenistic Period and Beyond." In *The Cambridge Companion to Greek Lyric*, edited by Felix Budelmann, 297-318. Cambridge: Cambridge UP 2009.
Barlow, William. *"Looking Up at Down": The Emergence of Blues Culture*. Philadelphia: Temple UP, 1989.
Barnouw, Erik. *Handbook of Radio Writing: An Outline of Techniques and Markets in Radio Writing in the United States*. Boston: Little, Brown, 1947.
Bartók, Béla, and Albert B. Lord. *Serbo-Croatian Folk Songs: Texts and Transcriptions*

of Seventy-Five Folk Songs from the Milman Parry Collection and a Morphology of Serbo-Croatian Folk Melodies. New York: Columbia UP, 1951.

Battier, Marc. "What the GRM Brought to Music: From Musique Concrète to Acousmatic Music." *Organised Sound* 12, no. 3 (2007): 189-202.

Bede. *Bede's Ecclesiastical History of the English People*. Edited by Bertram Colgrave and R. A. B. Mynors. Oxford: Clarendon Press, 1969.

Belgrad, Daniel. *The Culture of Spontaneity: Improvisation and the Arts in Postwar America*. Chicago: U of Chicago P, 1998.

Benjamin, Walter. *The Arcades Project*. Translated by Howard Eiland and Kevin McLaughlin. Cambridge, MA: Belknap, 1999.

Benjamin, Walter. *Radio Benjamin*. Edited by Lecia Rosenthal. London: Verso, 2014.

Benjamin, Walter. *The Work of Art in the Age of Its Technological Reproducibility: And Other Writings on Media*. Edited by Michael W. Jennings, Brigid Doherty, and Thomas Y. Levin. Cambridge, MA: Harvard UP, 2008.

Benston, Kimberly W. "Listen Br'er Sterling: The Critic as Liar [a Pre(r)amble to Essays on Sterling Brown]." In *After Winter: The Art and Life of Sterling Brown*, edited by John Edgar Tidwell and Steven C. Tracy, 95-104. Oxford: Oxford UP, 2009.

Berman, Marshall. *All That Is Solid Melts into Air: The Experience of Modernity*. New York: Verso, 1983.

Bernstein, Charles, ed. *Close Listening: Poetry and the Performed Word*. New York: Oxford UP, 1998.

Bernstein, Charles. *Attack of the Difficult Poems: Essays and Inventions*. Chicago: U of Chicago P, 2011.

Bernstein, Charles. Introduction to *Close Listening: Poetry and the Performed Word*, edited by Charles Bernstein, 3-26. New York: Oxford UP, 1998.

Bernstein, Charles. "Words and Pictures." *Sagetrieb* 2, no. 1 (1983): 9-34.

Berry, Faith. *Langston Hughes: Before and Beyond Harlem*. New York: Citadel, 1992.

Biers, Katherine. *Virtual Modernism: Writing and Technology in the Progressive Era*. Minneapolis: U of Minnesota P, 2013.

Billera, Jau. "Russell Atkins's Object Forms," *Kaleidoscope*, March 1-14, 1968, 9.

Bishop, Elizabeth. *Poems*. New York: Farrar, Straus and Giroux, 2011.

Blake, William. *Selected Poems*. Edited by G. E. Bentley Jr. New York: Penguin, 2005.

Blasing, Mutlu Konuk. *Lyric Poetry: The Pain and the Pleasure of Words*. Princeton, NJ: Princeton UP, 2007.

Bloom, Emily C. *The Wireless Past: Anglo-Irish Writers and the BBC, 1931-1968*. Oxford: Oxford UP, 2016.

Blount, Marcellus. "The Preacherly Text: African American Poetry and Vernacular Performance." *PMLA* 107, no. 3 (1992): 582-93.

Bök, Christian. *The Xenotext: Book 1*. Toronto: Coach House Books, 2015.

Bolter, Jay David, and Richard Grusin. *Remediation: Understanding New Media*. Cambridge, MA: MIT Press, 1999.
Bontemps, Arna. "Negro Poets, Then and Now." *Phylon* 11, no. 4 (1950): 355-60.
Booth, Mark. *The Experience of Songs*. New Haven, CT: Yale UP, 1981.
Bornstein, George. *Material Modernism: The Politics of the Page*. Cambridge: Cambridge UP, 2001.
Borshuk, Michael. *Swinging the Vernacular: Jazz and African American Modernist Literature*. New York: Routledge, 2006.
Boyd, William. "The Secret Persuaders." *Guardian*, August 19, 2006.
Bradley, Adam. *Book of Rhymes: The Poetics of Hip Hop*. New York: Basic Civitas, 2009.
Bradley, Adam. *The Poetry of Pop*. New Haven, CT: Yale UP, 2017.
Bradley, Adam, and Andrew DuBois. *The Anthology of Rap*. New Haven, CT: Yale UP, 2010.
Brady, Erika. *A Spiral Way: How the Phonograph Changed Ethnography*. Jackson: U of Mississippi P, 1999.
Brecht, Bertolt. "The Radio as an Apparatus of Communication." In *Brecht on Theatre: The Development of an Aesthetic*, edited and translated by John Willett, 51-53. New York: Hill and Wang, 1964.
Brinkman, Bartholomew. *Poetic Modernism in the Culture of Mass Print*. Baltimore: Johns Hopkins UP, 2016.
Broch, Hermann. "Notes on the Problem of Kitsch." In *Kitsch: The World of Bad Taste*, edited by Gillo Dorfles, 49-76. New York: University Books, 1969.
Brooks, Cleanth. *The Well Wrought Urn: Studies in the Structure of Poetry*. London: Dobson Books, 1947.
Brooks, Gwendolyn. *Blacks*. Chicago: Third World Press, 1967.
Brooks, William. "Historical Precedents for Artistic Research in Music: The Case of William Butler Yeats." In *Artistic Experimentation in Music: An Anthology*, edited by Darla Crispin and Bob Gilmore, 185-96. Leuven: Leuven UP, 2014.
Brown, Bill. *Other Things*. Chicago: U of Chicago P, 2015.
Brown, Sterling. *The Collected Poems of Sterling Brown*. Evanston: TriQuarterly, 1980.
Brown, Sterling. "Folk Literature." In *A Son's Return: Selected Essays of Sterling A. Brown*, edited by Mark A. Sanders, 207-31. Boston: Northeastern UP, 1996.
Brown, Sterling. "Negro Character as Seen by White Authors." In *A Son's Return: Selected Essays of Sterling A. Brown*, edited by Mark A. Sanders, 149-83. Boston: Northeastern UP, 1996.
Brown, Sterling. *Negro Poetry and Drama*. New York: Arno, 1969.
Brown, Sterling. "The New Negro in Literature (1925-1955)." In *A Son's Return: Selected Essays of Sterling A. Brown*, edited by Mark A. Sanders, 184-203. Boston: Northeastern UP, 1996.
Brown, Sterling. "Our Literary Audience." In *A Son's Return: Selected Essays of*

Sterling A. Brown, edited by Mark A. Sanders, 139-48. Boston: Northeastern UP, 1996.
Brown, Sterling. *Outline for the Study of the Poetry of American Negroes*. New York: Harcourt, Brace, 1931.
Budelmann, Felix. "Introducing Greek Lyric." *The Cambridge Companion to Greek Lyric*, edited by Felix Budelmann, 1-18. Cambridge: Cambridge UP, 2009.
Burns, Robert. *The Works of Robert Burns; With Dr. Currie's Memoir of the Poet, and an Essay on His Genius and Character by Professor Wilson*. Vol. 2. Glasgow: Blackie and Son, 1844.
Burt, Stephanie. "Is American Poetry Still a Thing?" *American Literary History* 28, no. 2 (2016): 271-87.
Burt, Stephanie. "What Is This Thing Called Lyric?" *Modern Philology* 113, no. 3 (2016): 422-40.
Butler, Shane. *The Ancient Phonograph*. Brooklyn: Zone Books, 2015.
Caedmon. "Caedmon's Hymn." Translated by John Pope. In *The Norton Anthology of Poetry: Sixth Edition*, edited by Margaret Ferguson, Tim Kendall, and Mary Jo Salter, 1. New York: Norton, 2018.
Cage, John. *Notations*. New York: Something Else Press, 1969.
Cai, Zong-Qi. "Introduction: The Primacy of Sound in Chinese Poetry." *Journal of Chinese Literature and Culture* 2, no. 2 (2015): 251-57.
Călinescu, Matei. *Five Faces of Modernity: Modernism, Avant-Garde, Decadence, Kitsch, Postmodernism*. Durham, NC: Duke UP, 1987.
Callahan, John F. "'A Brown Study': Sterling Brown's Legacy of Compassionate Connections." In *After Winter: The Art and Life of Sterling Brown*, edited by John Edgar Tidwell and Steven C. Tracy, 239-54. Oxford: Oxford UP, 2009.
Camlot, Jason. *Phonopoetics: The Making of Early Literary Recordings*. Stanford, CA: Stanford UP, 2019.
Campbell, Timothy C. *Wireless Writing in the Age of Marconi*. Minneapolis: U of Minnesota P, 2006.
Campion, Thomas. *The Works of Thomas Campion: Complete Songs, Masques, and Treatises with a Selection of the Latin Verse*. Edited by Walter R. Davis. New York: Doubleday, 1967.
Cantril, Hadley, and Gordon W. Allport. *The Psychology of Radio*. New York: Harper & Brothers, 1935.
Carby, Hazel V. *Cultures in Babylon: Black Britain and African America*. London, Verso, 1999.
Carey, James. *Communication as Culture: Essays on Media and Society*. Rev. ed. New York: Routledge, 1992.
Carmody, Todd. "Sterling Brown and the Dialect of New Deal Optimism." *Callaloo* 33, no. 3 (2010): 820-40.

Bibliography

Carruthers, A. J. *Notational Experiments in North American Long Poems, 1961–2011.* Cham: Palgrave Macmillan, 2017.
Ceraso, Steph. *Sounding Composition: Multimodal Pedagogies for Embodied Listening.* Pittsburgh, PA: Pittsburgh UP, 2018.
Chasar, Mike. *Everyday Reading: Poetry and Popular Culture in Modern America.* New York: Columbia UP, 2012.
Chasar, Mike. *Poetry Unbound: Poems and New Media from the Magic Lantern to Instagram.* New York: Columbia UP, 2020.
Chinitz, David E. *T. S. Eliot and the Cultural Divide.* Chicago: U of Chicago Press, 2003.
Chinitz, David E. *Which Sin to Bear? Authenticity and Compromise in Langston Hughes.* Oxford: Oxford UP, 2013.
Chion, Michel. "Reflections on the Sound Object and Reduced Listening." In *Sound Objects*, edited by James A. Steintrager and Rey Chow, 23–32. Durham, NC: Duke UP, 2019.
Christgau, Robert. "Rock Lyrics Are Poetry (Maybe)." In *Rock and Roll Is Here to Stay: An Anthology*, edited by William McKeen, 561–68. New York: Norton, 2000.
Christie, William. *Dylan Thomas: A Literary Life.* Basingstoke, Hampshire: Palgrave Macmillan, 2014.
Chun, Wendy Hui Kyong. *Updating to Remain the Same: Habitual New Media.* Cambridge, MA: MIT Press, 2017.
Clark, John Lee. "Melodies Unheard: Deaf Poets and Their Subversion of the 'Sound' Theory of Poetry." *Sign Language Studies* 7, no. 1 (2006): 4–10.
Clements, Robert J. "Foreign Language Broadcasting of 'Radio Boston.'" *Modern Language Journal* 27, no. 3 (1943): 175–79.
Cohen, Debra Rae, Michael Coyle, and Jane Lewty, eds. *Broadcasting Modernism.* Gainesville: UP of Florida, 2009.
Cole, Barbara. "'(Wo)Man (Critic) Is but an Ass?' The Bottom Line for Gender Criticism on Zukofsky." Paper presented at Re-Reading Louis Zukofsky's *Bottom: On Shakespeare*, University of Buffalo, October 31, 2003.
Coleridge, Samuel Taylor. *Biographia Literaria, or Biographical Sketches of My Literary Life and Opinions.* London: Dent, 1975.
Coleridge, Samuel Taylor. "Spenser." In *Coleridge's Miscellaneous Criticism*, edited by Thomas Middleton Raysor, 32–39. Cambridge, MA: Harvard UP, 1936.
The Columbia Program Book: January 1939. New York: Columbia Broadcasting System, 1939.
Comens, Bruce. *Apocalypse and After: Modern Strategy and Postmodern Tactics in Pound, Williams, and Zukofsky.* Tuscaloosa: U of Alabama P, 1995.
Connelly, Marc. *The Green Pastures.* New York: Farrar & Rinehart, 1929.

Connor, Steven. *Beckett, Modernism and the Material Imagination*. New York: Cambridge UP, 2014.
Cope, David. *Computer Models of Musical Creativity*. Cambridge, MA: MIT Press, 2006.
Corman, Cid. *Essays on the Arts of Language*. Vol. 1, *Word for Word*. Santa Barbara: Black Sparrow, 1977.
Corwin, Norman. Correspondence (119) 1938-1946. Box 24, Norman Corwin Papers. Special Collections Research Center, Syracuse University, Syracuse, NY.
Corwin, Norman. Script for *Words without Music*. January 15, 1939. Norman Corwin Collection, University of California Santa Barbara Library, Santa Barbara, CA.
Cramer, Gisela. "The Rockefeller Foundation and Pan American Radio." In *Patronizing the Public: American Philanthropy's Transformation of Culture, Communication, and the Humanities*, edited by William J. Buxton, 77-99. Lanham, MD: Lexington, 2009.
Crawford, Thomas. *Society and the Lyric: A Study of the Song Culture of Eighteenth-Century Scotland*. Edinburgh: Scottish Academic Press, 1979.
Culler, Jonathan. *The Pursuit of Signs: Semiotics, Literature, and Deconstruction*. Ithaca, NY: Cornell UP, 1981.
Culler, Jonathan. *Theory of the Lyric*. Cambridge, MA: Harvard UP, 2015.
Cummings, Troy. "Media Primer: Norman Corwin's Radio Juvenilia." In *Anatomy of Sound: Norman Corwin and Media Authorship*, edited by Jacob Smith and Neil Verma, 151-70. Oakland: U of California P, 2016.
Dack, John. "Pierre Schaeffer and the (Recorded) Sound Source." In *Sound Objects*, edited by James A. Steintrager and Rey Chow, 33-52. Durham, NC: Duke UP, 2019.
Damon, S. Foster, and Alvin H. Rosenfeld. "John Wheelwright: New England's Colloquy with the World." *Southern Review* 8, no. 2 (1972): 310-28.
Davidson, Michael. *The San Francisco Renaissance: Poetics and Community at Midcentury*. Cambridge: Cambridge UP, 1989.
Davies, Stephen. *Musical Works and Performances: A Philosophical Exploration*. New York: Oxford UP, 2001.
Davis, Angela Y. *Blues Legacies and Black Feminism: Gertrude "Ma" Rainey, Bessie Smith, and Billie Holiday*. New York: Pantheon Books, 1998.
Dayton, Tim. *Muriel Rukeyser's The Book of the Dead*. Columbia: U of Missouri P, 2003.
Delaney, Samuel R. *Shorter Views: Queer Thoughts and the Politics of the Paraliterary*. Middleton, CT: Wesleyan UP, 1999.
de Man, Paul. *The Rhetoric of Romanticism*. New York: Columbia, 1984.
Dembo, L. S., and Louis Zukofsky. "Louis Zukofsky." *Contemporary Literature* 10, no. 2 (1969): 203-19.

Bibliography

Densmore, Frances. *Northern Ute Music*. Washington, DC: Government Printing Office, 1922.
Dinsman, Melissa. *Modernism at the Microphone: Radio, Propaganda, and Literary Aesthetics during World War II*. London: Bloomsbury, 2015.
Dolinar, Brian. "Langston Hughes in the New American Century." *Langston Hughes Review* 20 (2006): 47-56.
Dorfles, Gillo. *Kitsch: The World of Bad Taste*. New York: University Books, 1969.
Douglas, Susan J. *Listening In: Radio and the American Imagination*. Minneapolis: U of Minnesota P, 2004.
Du Bois, W. E. B. *The Souls of Black Folk*. Chicago: A. C. McClurg, 1904.
Duncan, Robert. *Fictive Certainties: Essays*. New York: New Directions, 1985.
Duncan, Robert, to Helen Adam, July 17, 1955. In *A Helen Adam Reader*, edited by Kristin Prevallet, 304-5. Orono, ME: National Poetry Foundation, 2007.
DuPlessis, Rachel Blau. "'HOO, HOO, HOO': Some Episodes in the Construction of Modern Whiteness." *American Literature* 67, no. 4 (1995): 667-700.
DuPlessis, Rachel Blau. "Lorine Niedecker's 'Paean to Place' and Its Reflective Fusions." In *Radical Vernacular: Lorine Niedecker and the Poetics of Place*, edited by Elizabeth Willis, 151-79. Iowa City: U of Iowa P, 2008.
DuPlessis, Rachel Blau. "Lorine Niedecker, The Anonymous: Gender, Class, Genre and Resistances" In *Lorine Niedecker: Woman and Poet*, edited by Jenny Penberthy, 113-37. Orono, ME: National Poetry Foundation, 1996.
DuPlessis, Rachel Blau, and Peter Quartermain. Introduction to *The Objectivist Nexus: Essays in Cultural Poetics*, edited by Rachel Blau DuPlessis and Peter Quartermain, 1-24. Tuscaloosa: U of Alabama P, 1999.
Dworkin, Craig. *Radium of the Word: A Poetics of Materiality*. Chicago: U of Chicago P, 2021.
Eberhart, Richard. Letter to John Wheelwright, n.d. Box 11, Folder 6, John Wheelwright Papers 1920-1940. John Hay Library, Brown University, Providence, RI.
Edwards, Brent Hayes. *Epistrophies: Jazz and the Literary Imagination*. Cambridge, MA: Harvard UP, 2017.
Eidsheim, Nina Sun. *Sensing Sound: Singing and Listening as Vibrational Practice*. Durham, NC: Duke UP, 2015.
Eisenberg, Evan. *The Recording Angel: Music, Records and Culture from Aristotle to Zappa*. 2nd ed. New Haven, CT: Yale UP, 2005.
Eisenstein, Elizabeth. *The Printing Press as an Agent of Change: Communications and Cultural Transformations in Early-Modern Europe*. Cambridge: Cambridge UP, 1979.
Eliot, T. S. *Collected Poems: 1909-1962*. Orlando, FL: Harcourt, 1963.
Eliot, T. S. "The Development of Shakespeare's Verse: Two Lectures." In *The Complete Prose of T. S. Eliot: The Critical Edition: Tradition and Orthodoxy, 1934-1939*,

edited by Iman Javadi, Ronald Schuchard, and Jayme Stayer, 531-61. Baltimore: Johns Hopkins UP, 2017.

Eliot, T. S. *On Poetry and Poets*. New York: Farrar, Straus and Giroux, 1943.

Ellis, Robert. "How Radio Discriminates against Negroes." *Negro Digest*, June 1949, 64-66.

Ellison, Ralph. "Richard Wright's Blues." *Antioch Review* 5, no. 2 (1945): 198-211.

Emanuel, James A. *Langston Hughes*. New York: Twayne, 1967.

Emerson, Ralph Waldo. *Representative Man: Seven Lectures*. London: Routledge, 1850.

Engels, Friedrich. *Socialism: Utopian and Scientific*. Translation by Edward Aveling. Chicago: Charles Kerr, 1910.

Ernst, Wolfgang. *Sonic Time Machines: Explicit Sound, Sirenic Voices, and Implicit Sonicity*. Amsterdam: Amsterdam UP, 2016.

Fang, Achilles. "Rhymeprose on Literature: The Wên-Fu of Lu Chi (A.D. 261-303)." *Harvard Journal of Asiatic Studies* 14, nos. 3/4 (19510): 527-66.

Farr, Florence. *The Music of Speech, Containing the Words of Some Poets, Thinkers and Music-Makers regarding the Practice of the Bardic Arts Together with Fragments of Verse Set to Its Own Melody by Florence Farr*. London: Elkin Matthews, 1909.

Feld, Steven, Aaron A. Fox, Thomas Porcello, and David Samuels. "Vocal Anthropology: From the Music of Language to the Language of Song." In *A Companion to Linguistic Anthropology*, edited by Alessandro Duranti, 321-45. Malden, MA: Blackwell, 2004.

Feldman, Matthew, Henry Mead, and Erik Tonning, eds. *Broadcasting in the Modernist Era*. London: Bloomsbury, 2014.

Fernandez-Prieto, Irune, Charles Spence, Ferran Pons, and Jordi Navarra. "Does Language Influence the Vertical Representation of Auditory Pitch and Loudness?" *i-Perception* 8, no. 3 (2017): 1-11.

Fickers, Andreas. "Visibly Audible: The Radio Dial as Mediating Interface." In *The Oxford Handbook of Sound Studies*, edited by Trevor Pinch and Karin Bijsterveld, 411-39. Oxford: Oxford UP, 2012.

Finch, Zachary. "'Passion for the Particular': Marianne Moore, Henry James, Beatrix Potter, and the Refuge of Close Reading." In *Twenty-First Century Marianne Moore: Essays from a Critical Renaissance*, edited by Elizabeth Gregory and Stacy Carson Hubbard, 221-35. New York: Palgrave Macmillan, 2018.

Finkelstein, Norman. "Helen Adam and Romantic Desire." *Credences* 3, no. 3 (1985): 125-37.

Finkelstein, Norman. *The Utopian Moment in Contemporary American Poetry*. Bucknell, PA: Bucknell UP, 1993.

Fletcher, Harvey. *Speech and Hearing*. New York: D. Van Nostrand, 1929.

"The Flowers of the Forest." In *The Halfpenny Lyre* 13. Leith: R. W. Hume, c. 1840.

Bibliography

Inglis Collection of Printed Music. National Library of Scotland. Edinburgh, Scotland, UK.
Ford, Karen Jackson. "Do Right to Write Right: Langston Hughes's Aesthetics of Simplicity." *Twentieth Century Literature* 38, no. 4 (1992): 436-56.
Ford, Karen Jackson. "Making Poetry Pay: The Commodification of Langston Hughes." In *Marketing Modernisms: Self-Promotion, Canonization, Rereading*, edited by Kevin J. H. Dettmar and Stephen Watt, 275-96. Ann Arbor: U of Michigan P, 1996.
Fried, Michael. *Art and Objecthood: Essays and Reviews*. Chicago: U of Chicago P, 1998.
Friedlander, Saul. "Preface to a Symposium: Kitsch and the Apocalyptic Imagination." *Salmagundi* 85/86 (1990): 201-6.
Frith, Simon. *Performing Rites: On the Value of Popular Music*. Cambridge, MA: Harvard UP, 1996.
Frith, Simon. "Why Do Songs Have Words?" *Sociological Review* 34, no. 1 (1986): 77-106.
Furia, Phillip. *Ira Gershwin: The Art of the Lyricist*. New York: Oxford UP, 1996.
Furia, Phillip. *The Poets of Tin-Pan Alley: A History of America's Great Lyricists*. New York: Oxford UP, 1990.
Furlonge, Nicole Brittingham. *Race Sounds: The Art of Listening in African American Literature*. Iowa City: Iowa UP, 2018.
Furr, Derek. *Recorded Poetry and Poetic Reception from Edna Millay to the Circle of Robert Lowell*. New York: Palgrave Macmillan, 2010.
Gabbin, Joanne V. *Sterling A. Brown: Building the Black Aesthetic Tradition*. Westport, CT: Greenwood Press, 1985.
Gates, Henry Louis, Jr. *Figures in Black: Words, Signs, and the "Racial" Self*. New York: Oxford UP, 1989.
Gates, Henry Louis, Jr. *The Signifying Monkey: A Theory of African-American Literary Criticism*. 25th anniversary ed. Oxford: Oxford UP, 1988.
Gayraud, Agnès. *Dialectic of Pop*. Translated by Robin Mackay, Daniel Miller, and Nina Power. Falmouth, UK: Urbanomic, 2019.
Genet, Jacqueline. *"Words for Music Perhaps": Yeats's "New Art."* Villeneuve d'Ascq: Septentrion, 2010.
Genette, Gérard. *The Architext: An Introduction*. Translated by Jane E. Lewin. Berkeley: U of California P, 1992.
Gibson, Corey. "'The Flowers of the Forest Are a' Wede Away': The Dispersal of a Familiar Refrain." *Scottish Literary Review* 11, no. 1 (2019): 103-24.
Gilmore, Bob. *Harry Partch: A Biography*. New Haven, CT: Yale UP, 1998.
Gilmore, Bob. "Harry Partch: The Early Vocal Works 1930-33." PhD diss., Queen's University of Belfast, 1992.
Gitelman, Lisa. *Always Already New: Media, History, and the Data of Culture*. Cambridge, MA: MIT Press, 2006.

Gitelman, Lisa. *Paper Knowledge: Toward a Media History of Documents*. Durham, NC: Duke UP, 2014.
Gitelman, Lisa. *Scripts, Grooves, and Writing Machines: Representing Technology in the Edison Era*. Stanford, CA: Stanford UP, 2000.
Glaser, Ben. "Folk Iambics: Prosody, Vestiges, and Sterling Brown's *Outline for the Study of the Poetry of American Negroes*." *PMLA* 129, no. 3 (2014): 417-34.
Glaser, Ben. *Modernism's Metronome: Meter and Twentieth-Century Poetics*. Baltimore: Johns Hopkins UP, 2020.
Goble, Mark. *Beautiful Circuits: Modernism and the Mediated Life*. New York: Columbia UP, 2010.
Golston, Michael. *Rhythm and Race in Modernist Poetry and Science*. New York: Columbia UP, 2008.
Gooch, Bryan N. S., and David Thatcher. *A Shakespeare Music Catalogue*. 5 vols. Oxford: Clarendon, 1991.
Goodby, John. *The Poetry of Dylan Thomas: Under the Spelling Wall*. Liverpool: Liverpool UP, 2013.
Goody, Alex. *Modernist Poetry, Gender and Leisure Technologies: Machine Amusements*. New York: Palgrave Macmillan, 2019.
Graham, T. Austin. *The Great American Songbooks: Musical Texts, Modernism, and the Value of Popular Culture*. Oxford: Oxford UP, 2013.
Gramsci, Antonio. "Intellectuals and Education." In *The Gramsci Reader: Selected Writings 1916-1935*, edited by David Forgacs, 300-22. New York: New York UP, 2000.
Granade, S. Andrew. *Harry Partch: Hobo Composer*. Rochester, NY: U of Rochester P, 2014.
Green, Fiona. "Marianne Moore and the Hidden Persuaders." In *Writing for the New Yorker: Critical Essays on an American Periodical*, edited by Fiona Green, 58-78. Edinburgh: Edinburgh UP, 2015.
Greenberg, Clement. *Art and Culture: Critical Essays*. Boston: Beacon, 1961.
Gregory, Elizabeth. *Apparition of Splendor: Marianne Moore Performing Democracy through Celebrity, 1952-1970*. Newark: U of Delaware P, 2001.
Gregory, Elizabeth. "'Combat Cultural': Marianne Moore and the Mixed Brow." In *Critics and Poets on Marianne Moore: "A Right Good Salvo of Barks,"* edited by Linda Leavell, Cristanne Miller, and Robin G. Schulze, 208-21. Lewisburg, PA: Bucknell UP, 2005.
Gregory, Elizabeth. "Is Andy Warhol Marianne Moore? Celebrity, Celibacy and Subversion." In *Twenty-First Century Marianne Moore: Essays from a Critical Renaissance*, edited by Elizabeth Gregory and Stacy Carson Hubbard, 237-51. New York: Palgrave Macmillan, 2018.
Gregory, Elizabeth. *Quotation and Modern American Poetry: "Imaginary Gardens with Real Toads."* Houston, TX: Rice UP, 1996.

Gregory, Elizabeth. "Stamps, Money, Pop Culture and Marianne Moore." In *The Critical Response to Marianne Moore*, edited by Elizabeth Gregory, 234-46. Westport, CT: Praeger, 2003.

Griffin, Farah Jasmine. *Harlem Nocturne: Women Artists & Progressive Politics during World War II*. New York: Basic Civitas, 2013.

Grobe, Christopher. *The Art of Confession: The Performance of Self from Robert Lowell to Reality TV*. New York: New York UP, 2017.

Grobe, Christopher. "The Breath of the Poem: Confessional Print/Performance circa 1959." *PMLA* 127, no. 2 (2012): 215-30.

Grossman, Allen. *The Long Schoolroom: Lessons in the Bitter Logic of the Poetic Principle*. Ann Arbor: U of Michigan P, 1997.

Grossman, Allen, and Mark Halliday. *The Sighted Singer: Two Works on Poetry for Readers and Writers*. Baltimore: Johns Hopkins UP, 1992.

Groth, Helen, Penelope Hone, and Julian Murphet, eds. *Sounding Modernism: Rhythm and Sonic Mediation in Modern Literature and Film*. Edinburgh: Edinburgh UP, 2017.

Gruenberg, Sidonie Matsner. "Radio and the Child." *Annals of the American Academy of Political and Social Science* 177 (1935): 123-28.

Grundberg, Carl. Introduction to *Songs with Music: Helen Adam*, by Helen Adam and Carl Grundberg, 11-12. San Francisco: Aleph Press, 1982.

Grusin, Richard. "Radical Mediation." *Critical Inquiry* 42, no. 1 (2015): 124-48.

Gubar, Susan. *Racechanges: White Skin, Black Face in American Culture*. New York: Oxford, 1997.

Guillory, John. "The Genesis of the Media Concept." *Critical Inquiry* 36, no. 2 (2010): 321-62.

Gumbrecht, Hans Ulrich, and K. Ludwig Pfeiffer. *Materialities of Communication*. Translated by William Whobrey. Stanford, CA: Stanford UP, 1994.

Gussow, Adam. "'Shoot Myself a Cop': Mamie Smith's 'Crazy Blues' as Social Text." *Callaloo* 25, no. 1 (2002): 8-44.

Haberland, Harmut. "Communion or Communication? A Historical Note on One of the 'Founding Fathers' of Pragmatics." In *Theoretical Linguistics and Grammatical Description*, edited by Robin Sackmann and Monika Budde, 163-66. Amsterdam: John Benjamins, 1996.

Hadjimichael, Theodora A. *The Emergence of the Lyric Canon*. Oxford: Oxford UP, 2019.

Hajduk, John C. "Tin Pan Alley on the March: Popular Music, World War II, and the Quest for a Great War Song." *Popular Music and Society* 26, no. 4 (2003): 497-512.

Hall, Donald. *Their Ancient Glittering Eyes: Remembering Poets and More Poets*. New York: Ticknor & Fields, 1992.

Halliday, Sam. *Sonic Modernity: Representing Sound in Literature, Culture and the Arts*. Edinburgh: Edinburgh UP, 2013.

Handy, W. C. Letter to Langston Hughes, September 11, 1942. Box 74, Folder 1420, Langston Hughes Papers, James Weldon Johnson Memorial Collection. Beinecke Library, Yale University. New Haven, CT.

Handy, W. C. *Memphis Blues*. New York: Joe Morris Music, 1913.

Harkness, Nicholas. *Songs of Seoul: An Ethnography of Voice and Voicing in Christian South Korea*. Berkeley: U of California P, 2014.

Harris, Joel Chandler. Introduction to *Poems by Irwin Russell*, by Irwin Russell, ix-xi. New York: Century, 1888.

Harris, William J. *The Poetry and Poetics of Amiri Baraka: The Jazz Aesthetic*. Columbia: U of Missouri Press, 1985.

Hartman, G. W. "Changes in Visual Acuity through Simultaneous Stimulation of Other Sense Organs." *Journal of Experimental Psychology* 16, no. 3 (1933): 393-407.

Hartman, G. W. *Gestalt Psychology: A Survey of Facts and Principles*. New York: Ronald Press Company, 1935.

Hartman, Saidiya V. "The Time of Slavery." *South Atlantic Quarterly* 101, no. 4 (2002): 757-77.

Hassett, Joseph M. *W. B. Yeats and the Muses*. Oxford: Oxford UP, 2010.

Hayles, N. Katherine. *My Mother Was a Computer: Digital Subjects and Literary Texts*. Chicago: U of Chicago P, 2005.

Helmholtz, Hermann von. *On the Sensations of Tone as a Physiological Basis for the Theory of Music*. Translated by Alexander J. Ellis. London: Longmans, Green, 1885.

Helsinger, Elizabeth. "Poem Into Song." *New Literary History* 46, no. 4 (2015): 669-90.

Helsinger, Elizabeth. *Poetry and the Thought of Song in Nineteenth-Century Britain*. Charlottesville: U of Virginia P, 2015.

Henderson, Stephen E. "The Heavy Blues of Sterling Brown: A Study of Craft and Tradition." In *After Winter: The Art and Life of Sterling Brown*, edited by John Edgar Tidwell and Steven C. Tracy, 31-57. Oxford: Oxford UP, 2009.

Hermogenes. *On Types of Style*. Translated by Cecil W. Wooten. U of North Carolina P, 1987.

Heuving, Jeanne. *Omissions Are Not Accidents: Gender in the Art of Marianne Moore*. Detroit, MI: Wayne State UP, 1992.

Higgins, Dick "Intermedia." *Leonardo* 34, no. 1 (2001): 49-54.

Hilmes, Michele. *Network Nations: A Transnational History of British and American Broadcasting*. New York: Routledge, 2012.

Hirsch, Edward. *How to Read a Poem: And Fall in Love with Poetry*. New York: Harvest, 1999.

Hoffman, Tyler. *American Poetry in Performance: From Walt Whitman to Hip Hop*. Ann Arbor: U of Michigan P, 2011.

Hollander, John. *The Untuning of the Sky: Ideas of Music in English Poetry, 1500-1700*. Princeton, NJ: Princeton UP, 1961.
Hollenbach, Lisa. "Phonography, Race Records, and the Blues Poetry of Langston Hughes." In *A Companion to the Harlem Renaissance*, edited by Cherene Sherrard-Johnson, 301-16. Malden, MA: John Wiley & Sons, 2015.
Hollenbach, Lisa. *Poetry FM: American Poetry and Radio Counterculture*. Iowa City: U of Iowa P, 2023.
Holroyd, Michael. *Bernard Shaw*. Vol. 1, *1856-1898: The Search for Love*. London: Chatto and Windus, 1988.
Horace. *Odes with Carmen Saeculare*. Translated by Stanley Lombardo. Indianapolis, IN: Hackett, 2018.
Horace. *Satires, Epistles, and Ars Poetica*. Translated by H. Rushton Fairclough. Cambridge, MA: Harvard UP, 1929.
Horten, Gerd. *Radio Goes to War: The Cultural Politics of Propaganda during World War II*. Berkeley: U of California P, 2002.
Houglum, Brook. "'Speech without Practical Locale': Radio and Lorine Niedecker's Aurality." In *Broadcasting Modernism*, edited by Debra Rae Cohen, Michael Coyle, and Jane Lewty, 221-37. Gainesville: UP of Florida, 2009.
Howard, Richard. "Four Originals." *Poetry* 122, no. 6 (1973): 351-57.
Howarth, Peter. "Marianne Moore's Performances." *ELH* 87, no. 2 (2020): 553-79.
Howe, Susan. "The Bostonian." Poetry. New York: WBAI-Pacifica, November 23, 1981. PennSound. https://writing.upenn.edu/pennsound/x/Howe-Pacifica.php.
Howe, Susan. "Every Mark on Paper Is an Acoustic Mark: Susan Howe on Emily Dickinson's Envelope Poems, at McNally Jackson Picture Room, 2016." *New Directions*. www.ndbooks.com/article/every-mark-on-paper-is-an-acoustic-mark-by-susan-howe/.
Hughes, Langston. "America's Young Black Joe," 1940. Box 390, Folder 7225, Langston Hughes Papers, James Weldon Johnson Memorial Collection. Beinecke Library, Yale University. New Haven, CT.
Hughes, Langston. *The Big Sea*. New York: Hill and Wang, 1993.
Hughes, Langston. *Collected Poems of Langston Hughes*. Edited by Arnold Rampersad and David Roessel. New York: Vintage, 1994.
Hughes, Langston. "Democracy, Negroes, and Writers." In *The Collected Works of Langston Hughes*. Vol. 9, *Essays on Art, Race, Politics and World Affairs*, edited by Christopher C. De Santis, 210-12. Columbia: U of Missouri P, 2002.
Hughes, Langston. "Dixie Negro to Uncle Sam." N.d. Box 392, Folder 7383, Langston Hughes Papers, James Weldon Johnson Memorial Collection. Beinecke Library, Yale University. New Haven, CT.
Hughes, Langston. "Dorothy's Name is Mud." 1949. Box 392, Folder 7397, Langston Hughes Papers, James Weldon Johnson Memorial Collection. Beinecke Library, Yale University. New Haven, CT.

Hughes, Langston. "Enemy." October 9, 1942. Box 393, Folder 7424, Langston Hughes Papers, James Weldon Johnson Memorial Collection. Beinecke Library, Yale University. New Haven, CT.

Hughes, Langston. "Fan Mail." Box 209, Folder 3543, Langston Hughes Papers, James Weldon Johnson Memorial Collection. Beinecke Library, Yale University. New Haven, CT.

Hughes, Langston. *Fields of Wonder: A Book of Lyric Poems*. New York: Knopf, 1947.

Hughes, Langston. *Fine Clothes to the Jew*. New York: Knopf, 1927.

Hughes, Langston. "Freedom Road." 1942. Box 393, Folder 7446, LHP, Langston Hughes Papers, James Weldon Johnson Memorial Collection. Beinecke Library, Yale University. New Haven, CT.

Hughes, Langston. "Go-and-Get-the-Enemy Blues." 1942-43. Box 393, Folder 7462, Langston Hughes Papers, James Weldon Johnson Memorial Collection. Beinecke Library, Yale University. New Haven, CT.

Hughes, Langston. "Here to Yonder: The Song Writing Game." *Chicago Defender*, April 21, 1945, 12.

Hughes, Langston. "Honolulu Yaka-Hula Dixie." August 6, 1942. Box 394, Folder 7527, Langston Hughes Papers, James Weldon Johnson Memorial Collection. Beinecke Library, Yale University. New Haven, CT.

Hughes, Langston. "Just a Little Spot of Sunlight." August 18, 1942. Box 395, Folder 7610, Langston Hughes Papers, James Weldon Johnson Memorial Collection. Beinecke Library, Yale University. New Haven, CT.

Hughes, Langston. Letters to Russell Atkins, 1946-1964. Author's copy, courtesy of Kevin Prufer.

Hughes, Langston. Letter to W. C. Handy, September 6, 1942. Box 74, Folder 1420, Langston Hughes Papers, James Weldon Johnson Memorial Collection. Beinecke Library, Yale University. New Haven, CT.

Hughes, Langston. "Message to the President," March 28, 1941. Box 397, Folder 7730, Langston Hughes Papers, James Weldon Johnson Memorial Collection. Beinecke Library, Yale University. New Haven, CT.

Hughes, Langston. "My Adventures as a Social Poet." *Phylon*, 8, no. 3 (1947): 205-12.

Hughes, Langston. "My America." In *The Collected Works of Langston Hughes*. Vol. 9, *Essays on Art, Race, Politics and World Affairs*, edited by Christopher C. De Santis, 232-39. Columbia: U of Missouri P, 2002.

Hughes, Langston. "My Heart Is a Lone Ranger." August 20, 1942. Box 397, Folder 7765, Langston Hughes Papers, James Weldon Johnson Memorial Collection. Beinecke Library, Yale University. New Haven, CT.

Hughes, Langston. "Negro Writers and the War." In *The Collected Works of Langston Hughes*. Vol. 9, *Essays on Art, Race, Politics and World Affairs*, edited by Christopher C. De Santis, 215-19. Columbia: U of Missouri P, 2002.

Hughes, Langston. Notes on songwriting with sample lyrics. N.d. Box 511, Folder 12680, Langston Hughes Papers, James Weldon Johnson Memorial Collection. Beinecke Library, Yale University. New Haven, CT.
Hughes, Langston. Note to Elie Siegmeister on "When a Soldier Writes a Letter Home." June 22, 1942. Box 401, Folder 8055, Langston Hughes Papers, James Weldon Johnson Memorial Collection. Beinecke Library, Yale University. New Haven, CT.
Hughes, Langston. "Pop-Corn Balls." October 11, 1942. Box 398, Folder 7844, Langston Hughes Papers, James Weldon Johnson Memorial Collection. Beinecke Library, Yale University. New Haven, CT.
Hughes, Langston. "Rights of Democracy." September 5, 1942. Box 398, Folder 7875, Langston Hughes Papers, James Weldon Johnson Memorial Collection. Beinecke Library, Yale University. New Haven, CT.
Hughes, Langston. "Salute from Old Broadway." August 7, 1942. Box 398, Folder 7883, Langston Hughes Papers, James Weldon Johnson Memorial Collection. Beinecke Library, Yale University. New Haven, CT.
Hughes, Langston. "Salute to Soviet Armies." February 15, 1944. Box 398, Folder 7884, Langston Hughes Papers, James Weldon Johnson Memorial Collection. Beinecke Library, Yale University. New Haven, CT.
Hughes, Langston. *Selected Letters of Langston Hughes*. Edited by Arnold Rampersad and David Roessel. New York: Knopf, 2015.
Hughes, Langston. "Some Practical Observations: A Colloquy." *Phylon* 11, no. 4 (1950): 307-311.
Hughes, Langston. "Ten Ways to Use Poetry in Teaching." *CLA Journal* 11, no. 4 (1968): 273-79.
Hughes, Langston. "Total War." January 30, 1943. Box 400, Folder 8022, Langston Hughes Papers, James Weldon Johnson Memorial Collection. Beinecke Library, Yale University. New Haven, CT.
Hughes, Langston. "What Poets Do for a Living." N.d. Box 1, Folder 87, Harry Roskolenko Collection. Beinecke Library, Yale University. New Haven, CT.
Hughes, Langston, and Arna Bontemps. *Arna Bontemps-Langston Hughes Letters, 1925-1967*. Edited by Charles Harold Nichols. New York: Dodd, Mead, 1980.
Hughes, Langston, and Arna Bontemps, eds. *The Poetry of the Negro: 1746-1949*. Garden City, NY: Doubleday, 1951.
Hughes, Langston, and Herbert Kingsley. "Hard Luck Blues." N.d. Box 394, Folder 7496, Langston Hughes Papers, James Weldon Johnson Memorial Collection. Beinecke Library, Yale University. New Haven, CT.
Huyssen, Andreas. *After the Great Divide: Modernism, Mass Culture, Postmodernism*. Bloomington: Indiana UP, 1986.
"'In Lieu of the Lyre': Two Engravings." *Marianne Moore Newsletter* 5, no. 2 (1981): 5-7.

Jackson, Virginia. *Dickinson's Misery: A Theory of Lyric Reading*. Princeton, NJ: Princeton UP, 2005.
Jackson, Virginia, and Yopie Prins, eds. *The Lyric Theory Reader*. Baltimore: Johns Hopkins UP, 2014.
Jakobson, Roman. *Language in Literature*. Edited by Krystyna Pomorska and Stephen Rudy. Cambridge, MA: Belknap, 1987.
Jauss, Hans Robert. "Poiesis." Translated by Michael Shaw. *Critical Inquiry* 8, no. 3 (1982): 591-608.
Javadizadeh, Kamran. "The Atlantic Ocean Breaking on Our Heads: Claudia Rankine, Robert Lowell, and the Whiteness of the Lyric Subject." *PMLA* 134, no. 3 (2019): 475-90.
Jensen, Klaus Bruhn. "Intermediality." In *The International Encyclopedia of Communication Theory and Philosophy*. Wiley Online Library, 2006.
Johnson, Barbara. "Apostrophe, Animation, and Abortion." *Diacritics* 16, no. 1 (1986): 28-47.
Johnson, Jake. "'Unstuck in Time': Harry Partch's Bilocated Life." *Journal of the Society for American Music* 9, no. 2 (May 2015): 163-77.
Johnson, James Weldon. *God's Trombones: Seven Negro Sermons in Verse*. New York: Viking Press, 1927.
Johnson, James Weldon. "Preface to the First Edition." In *The Book of American Negro Poetry*, edited by James Weldon Johnson, 9-48. San Diego, CA: Harvest/Harcourt Brace Jovanovich, 1969.
Johnson, James Weldon. "Preface to the Revised Edition." In *The Book of American Negro Poetry*, edited by James Weldon Johnson, 3-8. San Diego, CA: Harvest/Harcourt Brace Jovanovich, 1969.
Johnson, James Weldon, and J. Rosamond Johnson, eds. *The Books of American Negro Spirituals*. New York: Da Capo, 1969.
Johnston, Ben. "The Corporealism of Harry Partch." *Perspectives of New Music* 13, no. 2 (1975): 85-97.
Jones, Gavin. *Strange Talk: The Politics of Dialect Literature in Gilded Age America*. Berkeley: U of California P, 1999.
Jones, Meta DuEwa. *The Muse Is Music: Jazz Poetry from the Harlem Renaissance to Spoken Word*. Champaign: U of Illinois P, 2013.
Jonson, Ben. "Ode to Himself." In *Ben Jonson: The Complete Poems*, edited George Parfitt, 282-84. New York: Penguin, 1975.
Jordan, Casper L. Afterword to *Phenomena*, by Russell Atkins, 76-78. Wilberforce, OH: Wilberforce UP, 1961.
Jordan, Casper L., and J. Stefanski, eds. "1955-1960: Theories." *Free Lance* 14, no. 2 (1970): 18-22.
Josephson, Matthew. "Improper Bostonian: John Wheelwright and His Poetry." *Southern Review* 7, no. 2 (1971): 509-40.

Joyce, James. *Finnegan's Wake*. New York: Penguin, 1999.
Kahn, Douglas. *Noise, Water, Meat: A History of Sound in the Arts*. Cambridge, MA: MIT Press, 1999.
Kane, Brian. "The Fluctuating Sound Object." In *Sound Objects*, edited by James A. Steintrager and Rey Chow, 53-70. Durham, NC: Duke UP, 2019.
Kane, Brian. *Sound Unseen: Acousmatic Sound in Theory and Practice*. New York: Oxford UP, 2014.
Kaplan, Milton. *Radio and Poetry*. New York: Columbia UP, 1949.
Kappel, Andrew J. "Complete with Omissions: The Text of Marianne Moore's *Complete Poems*." In *Representing Modernist Texts: Editing as Interpretation*, edited by George Bornstein, 125-56. Ann Arbor: U of Michigan P, 1991.
Karlin, Daniel. *The Figure of the Singer*. Oxford: Oxford UP, 2013.
Kassel, Richard. "*Barstow* as History: An Introduction to the Sound World of Harry Partch." In *Barstow: Eight Hitchhiker Inscriptions from a Highway Railing at Barstow, California (1968 Version)*, edited by Richard Kassel, xiii-lxxix. Madison, WI: A-R Editions, 2000.
Kassel, Richard. "Harry Partch: In the Field." *Musicworks*, no. 51 (Autumn 1991): 6-15.
Keane, Damien. *Ireland and the Problem of Information: Irish Writing, Radio, Late Modernist Communication*. University Park: Penn State UP, 2014.
Kelley, Robin D. G. "Notes on Deconstructing 'The Folk.'" *American Historical Review* 97, no. 5 (1992): 1400-08.
Kenner, Hugh. *The Stoic Comedians: Flaubert, Joyce, and Beckett*. Berkeley: U of California P, 1962.
Kernan, Ryan James. *New World Maker: Radical Poetics, Black Internationalism, and the Translations of Langston Hughes*. Evanston, IL: Northwestern UP, 2022.
Khlebnikov, Velimir. *The King of Time: Selected Writings of the Russian Futurian*. Translated by Paul Schmidt. Cambridge, MA: Harvard UP, 1985.
Kindley, Evan. *Poet-Critics and the Administration of Culture*. Cambridge, MA: Harvard UP, 2017.
Kipling, Rudyard. *Rudyard Kipling's Verse: 1885-1918*. Garden City, NY: Doubleday, Page, 1919.
Kittler, Friedrich. "Benn's Poetry: 'A Hit in the Charts'; Song under Conditions of Media Technologies." *SubStance* 19, no. 1 (1990): 5-20.
Kittler, Friedrich. *Discourse Networks 1800/1900*. Translated by Michael Metteer and Chris Cullens. Stanford, CA: Stanford UP, 1990.
Kittler, Friedrich. *Gramophone, Film, Typewriter*. Translated by Geoffrey Winthrop-Young and Michael Wutz. Stanford, CA: Stanford UP, 1999.
Kittler, Friedrich. "Number and Numeral." *Theory, Culture & Society* 23, nos. 7/8 (2006): 51-61.
Kittler, Friedrich. "Observations on Public Reception." In *Radio Rethink: Art, Sound,*

and Transmission, edited by Daina Augaitis and Dan Lander, 75-85. Banff: Walter Phillips Gallery, 1994.

Koegel, John. "Preserving the Sounds of the 'Old' Southwest: Charles Lummis and his Cylinder Collection of Mexican-American and Indian Music." *ARSC Journal* 29, no. 1 (Spring 1998): 1-29.

Kunitz, Stanley. "Process and Thing: A Year of Poetry." *Harper's Magazine*, September 1960, 96-104.

Lacey, Kate. *Listening Publics: The Politics and Experience of Listening in the Media Age*. Cambridge: Polity, 2013.

Lamothe, Daphne. *Inventing the New Negro: Narrative, Culture, and Ethnography*. Philadelphia: U of Pennsylvania P, 2008.

Lane, Cathy. *Playing with Words: The Spoken Word in Artistic Practice*. Devon: Uniformbooks, 2015.

Lane, Cathy. "Voices from the Past: Compositional Approaches to Using Recorded Speech." *Organised Sound* 11, no. 1 (April 2006): 3-11.

Latouche, John. "The Muse and the Mike." *Vogue* 97, no. 5 (1941): 64, 124-25.

Laughlin, James, ed. *New Directions in Prose & Poetry*. Norfolk: New Directions, 1939.

Leavell, Linda. *Holding On Upside Down: The Life and Work of Marianne Moore*. New York: Farrar, Straus and Giroux, 2013.

Lechlitner, Ruth. "Verse-Drama for Radio: A New Direction." In *New Directions in Prose & Poetry*, edited by James Laughlin 110-15. Norfolk: New Directions, 1937.

Lee, Albert Rudolph, Jr. "The Poetic Voice of Langston Hughes in American Art Song." DM treatise, Florida State University, 2012.

Lesser, Jerry. "Radio Talent." *Billboard*, January 28, 1939, 9.

Letter to Norman Corwin, Box 24, Norman Corwin Papers. Special Collections Research Center, Syracuse University Libraries, Syracuse, NY.

Lévi-Strauss, Claude. *Mythologiques: Le cru et le cuit*. Paris: Plon, 1964.

Levin, Harry. "A Note on French Aspect." *Festschrift for Marianne Moore's Seventy Seventh Birthday*, edited by Meary James Thurairajah Tambimuttu, 40-43. New York: Tambimuttu & Mass, 1964.

Levine, Philip. *My Lost Poets: A Life in Poetry*. Edited by Edward Hirsch. New York: Knopf, 2011.

Lewis, Cara. *Dynamic Form: How Intermediality Made Modernism*. Ithaca, NY: Cornell UP, 2020.

Lindberg, Ulf. "Popular Modernism? The 'Urban' Style of Interwar Tin Pan Alley." *Popular Music* 22, no. 3 (2003): 283-98.

Lindsay, Vachel. *The Chinese Nightingale and Other Poems*. New York: MacMillan, 1918.

Lindsay, Vachel. *The Congo and Other Poems*. New York: Macmillan, 1922.

Lindsay, Vachel. *The Golden Whales of California and Rhymes in the American Language*. New York: Macmillan, 1920.
Locke, Alain. "Common Clay and Poetry." In *The Works of Alain Locke*, edited by Charles Molesworth, 55-57. New York: Oxford UP, 2012.
Lord, Albert B. *The Singer of Tales*. 2nd ed. Edited by Stephen Mitchell and Gregory Nagy. Cambridge, MA: Harvard UP, 2000.
Lott, Eric. *Black Mirror: The Cultural Contradictions of American Racism*. Cambridge, MA: Belknap, 2017.
Loviglio, Jason. *Radio's Intimate Public: Network Broadcasting and Mass-Mediated Democracy*. Minneapolis: U of Minnesota P, 2005.
Lowe, Ramona. "More Negroes in Radio Urged by Norman Corwin." *Chicago Defender*, February 17, 1945.
Lowell, Robert. "Acceptance Speech: National Book Award in Poetry (1960)." *National Book Foundation*. www.nationalbook.org/robert-lowells-accepts-the-1960-national-book-awards-in-poetry-for-life-studies/.
Lucey, Michael, and Tom McEnaney. "Language-in-Use and the Literary Artifact." *Representations* 137, no. 1 (Winter 2017): 1-173.
Lui, Alan. *Friending the Past: The Sense of History in the Digital Age*. Chicago: U of Chicago P, 2018.
Lyons, Louis M. "After Leap of 300 Miles Boston's W1XAL Covers World." *Daily Boston Globe*, May 22, 1938.
MacArthur, Marit. "Monotony, the Churches of Poetry Reading, and Sound Studies." *PMLA* 131, no. 1 (2016): 38-63.
Mackey, Nathaniel. *Blue Fasa*. New York: New Directions, 2015.
MacLeish, Archibald. *Collected Poems: 1917-1982*. Boston: Houghton Mifflin, 1985.
MacLeish, Archibald. *The Fall of the City: A Verse Play for Radio*. New York: Farrar and Rinehart, 1937.
MacLeish, Archibald. "A Letter." *Furioso* 1 (1939): 1-2.
Malanga, Gerard. "Some Thoughts on *Bottom* and *After I's*." *Poetry* 107, no. 1 (1965): 60-64.
Mao, Douglas. "The New Critics and the Text-Object." *ELH* 63, no. 1 (1996): 227-54.
Mao, Douglas. *Solid Objects: Modernism and the Test of Production*. Princeton, NJ: Princeton UP, 1998.
Mao, Douglas, and Rebecca L. Walkowitz. "The New Modernist Studies," *PMLA* 123, no. 3 (2008): 737-48.
Marinetti, F. T., and Pino Masnata. "La Radia." Translated by Stephen Sartarelli. In *Wireless Imagination*, edited by Douglas Kahn and Gregory Whitehead, 265-68. Cambridge, MA: MIT Press, 1992.
Martin, Lerone A. *Preaching on Wax: The Phonograph and the Shaping of Modern African American Religion*. New York: New York UP (2014).

Marvell, Andrew. *The Complete Poems*. Edited by Elizabeth Story Donno. Middlesex: Penguin, 1972.
Marx, Karl. *Der achtzehnte Brumaire des Louis Bonaparte*. Hamburg: Otto Meisner, 1869.
Marx, Karl. *Karl Marx: A Reader*. Edited by Jon Elster. Cambridge: Cambridge UP, 1986.
Mathiesen, Thomas J. *Apollo's Lyre: Greek Music and Music Theory in Antiquity and the Middle Ages*. Lincoln: U of Nebraska P, 1999.
Maud, Ralph. Introduction to *On the Air with Dylan Thomas: The Broadcasts*, edited by Ralph Maud, v-xiv. New York: New Directions, 1991.
McCabe, Susan. *Cinematic Modernism: Modernist Poetry and Film*. Cambridge: Cambridge UP, 2005.
McCaffery, Steve, and Abigail Lang. "The Poem as Score and the Genealogy of Sono-poetic Notation." *Revue française d'études américaines* 153, no. 4 (2017): 17-33.
McGann, Jerome. *Black Riders: The Visible Language of Modernism*. Princeton, NJ: Princeton UP, 1993.
McGann, Jerome. *The Romantic Ideology: A Critical Investigation*. Chicago: U of Chicago P, 1983.
McGann, Jerome. "The Text, the Poem, and the Problem of Historical Method." *New Literary History* 12, no. 2 (1981): 269-88.
McGeary, Thomas. *The Music of Harry Partch: A Descriptive Catalog*. New York: Institute for Studies in American Music, 1991.
McGill, Meredith L. "What Is a Ballad? Reading for Genre, Format, and Medium." *Nineteenth-Century Literature* 71, no. 2 (2016): 156-75.
McLane, Maureen N. *Balladeering, Minstrelsy, and the Making of British Romantic Poetry*. Cambridge: Cambridge UP, 2008.
McLane, Maureen N. "On the Use and Abuse of 'Orality' for Art: Reflections on Romantic and Late Twentieth-Century Poiesis." *Oral Tradition* 17, no. 1 (2002): 134-64.
McLuhan, Marshall. *Counterblast*. New York: Harcourt, Brace and World, 1969.
McLuhan, Marshall. *The Gutenberg Galaxy: The Making of Typographical Man*. Toronto: U of Toronto P, 1962.
McLuhan, Marshall. *Understanding Media: The Extensions of Man*. New York: McGraw-Hill, 1964.
Melanga, Gerard, and Mark Van Doren. Jacket copy for *Bottom: On Shakespeare*, Box 9, Folder 4, Louis Zukofsky Collection 1910-1985. Harry Ransom Center, University of Texas at Austin, Austin, TX.
Melillo, John. "The Politics of Noise in Henri Chopin's Audiopoems." *Revue TIES* 3 (2019): 91-101.
Melnick, David. "The 'Ought' of Seeing: Zukofsky's *Bottom*." *MAPS* 5 (1973): 55-65.
Meltzer, Milton. "Poetry on the Air." *New Masses* 31, no. 1 (1939): 28-29.

Michlin, Monica. "Langston Hughes's Blues." In *Temples for Tomorrow: Looking Back at the Harlem Renaissance*, edited by Geneviève Fabre and Michel Feith, 236-53. Bloomington: Indiana UP, 2001.

Middleton, Peter. *Distant Reading: Performance, Readership, and Consumption in Contemporary Poetry*. Tuscaloosa: U of Alabama P, 2005.

Middleton, Peter. "Lorine Niedecker's 'Folk Base' and Her Challenge to the American Avant-Garde." In *The Objectivist Nexus: Essays in Cultural Poetics*, edited by Rachel Blau DuPlessis and Peter Quartermain, 160-88. Tuscaloosa: U of Alabama P, 1999.

Middleton, Peter. "Poetry Reading." In *Princeton Encyclopedia of Poetry and Poetics*, edited by Roland Greene, 1068-70. Princeton, NJ: Princeton UP, 2012.

Miles, Barry. *Paul McCartney: Many Years from Now*. New York: Henry Holt, 1997.

Miller, Carolyn R. "Genre as Social Action." *Quarterly Journal of Speech* 70, no. 2 (1984): 151-67.

Miller, Cristanne. *Marianne Moore: Questions of Authority*. Cambridge, MA: Harvard UP, 1995.

Miller, Dayton C. "Analytical Study of Photographs Taken with the Phonodeik." In *Northern Ute Muse*, by Frances Densmore, 206-10. Washington, DC: Government Printing Office, 1922.

Miller, Dayton C. *The Science of Musical Sounds*. New York: MacMillan, 1922.

Miller, J. Hillis. *Poets of Reality: Six Twentieth-Century Writers*. Cambridge, MA: Harvard UP, 1965.

Miller, Karl Hagstrom. *Segregating Sound: Inventing Folk and Pop Music in the Age of Jim Crow*. Durham, NC: Duke UP, 2010.

Monroe, Harriet. "The Radio and the Poets." *Poetry* 36, no. 1 (1930): 32-35.

Moore, David Chioni. "Langston Hughes and His World." In *The Cambridge History of American Poetry*, edited by Alfred Bendixen and Stephanie Burt, 701-27. New York: Cambridge UP, 2014.

Moore, Marianne. *Becoming Marianne Moore: The Early Poems, 1907-1924*. Edited by Robin G. Schulze. Berkeley: U of California P, 2002.

Moore, Marianne. *The Complete Poems of Marianne Moore*. New York: Macmillan/Viking, 1967.

Moore, Marianne. *The Complete Prose of Marianne Moore*. Edited by Patricia C. Willis. New York: Viking, 1986.

Moore, Marianne. "In Lieu of the Lyre." Series 1, Box 2, Folder 32. Marianne Moore Collection. The Rosenbach, Philadelphia, PA.

Moore, Marianne. "I've Been Thinking . . ." *New York Review of Books*, October 31, 1963, 19.

Moore, Marianne. Jacket copy for *Bottom: On Shakespeare*, Box 9, Folder 4, Louis Zukofsky Collection 1910-1985. Harry Ransom Center, University of Texas at Austin.

Moore, Marianne. Letter to Celia Zukofsky, n.d., 1952, Box 19, Folder 3, Louis Zukofsky Collection 1910–1985. Harry Ransom Center, University of Texas at Austin.

Moore, Marianne. Letter to Russell Atkins, April 11, 1953. Author's copy, courtesy of Kevin Prufer.

Moore, Marianne. "Marianne Moore." October 10, 1957. San Francisco Museum of Art. MPEG/sound tape reel, 40 min. The Poetry Center. https://diva.sfsu.edu/collections/poetrycenter/bundles/191206.

Moore, Marianne. "Marianne Moore." October 18, 1957. University of California, Berkeley. MPEG/sound tape reel, 59 min. The Poetry Center, https://diva.sfsu.edu/collections/poetrycenter/bundles/191187.

Moore, Marianne. *A Marianne Moore Reader*. New York: Viking, 1961.

Moore, Marianne. "Marianne Moore Reading Her Poems with Comment." October 21, 1963, Coolidge Auditorium, Washington DC. MPEG/sound tape reel, 1:03, Archive of Recorded Poetry and Literature, Library of Congress. www.loc.gov/item/94838421?loclr=blogpoe.

Moore, Marianne. *New Collected Poems*. Edited by Heather Cass White. New York: Farrar, Straus and Giroux, 2017.

Moore, Marianne. "Poetry Reading." May 11, 1965. Sanders Theatre, Cambridge, MA. MPEG/sound tape reel, 49 min. Woodberry Poetry Room, Harvard University, Cambridge, MA.

Moore, Marianne. "The World in Books: Friday 13 July 1951." Marianne Moore Collection. Rosenbach Library, Philadelphia, PA.

Morin, Emilie. "'I Beg Your Pardon?' W. B. Yeats, Audibility and Sound Transmission." *Yeats's Mask: Yeats Annual*, no. 19 (2013): 191–219.

Morris, Adalaide, ed. *Sound States: Innovative Poetics and Acoustical Technologies*. Chapel Hill: U of North Carolina P, 1998.

Morrison, Hobe. "Words without Music." *Variety*, December 14, 1938, 42.

Morrisson, Mark. *The Public Face of Modernism: Little Magazines, Audiences, and Reception*. Madison: U of Wisconsin P, 2001.

Morton, Timothy. "An Object-Oriented Defense of Poetry." *New Literary History* 43, no. 2 (2012): 205–24.

Morton, Timothy. *Realist Magic: Objects, Ontology, Causality*. Ann Arbor, MI: Open Humanities Press, 2013.

Moten, Fred. *In the Break: The Aesthetics of the Black Radical Tradition*. Minneapolis: U of Minnesota P, 2003.

Mowitt, John. *Radio: Essays in Bad Reception*. Berkeley: U of California P, 2011.

Muntu Poets of Cleveland. *The Muntu Poets*. Edited by Sababa Akili and Mutawaf Shaheed. New York: Uptown Media Join Ventures, 2016.

Murphet, Julian. *Multimedia Modernism: Literature and the Anglo-American Avant-Garde*. Cambridge: Cambridge UP, 2009.

Mustazza, Chris. "James Weldon Johnson and the *Speech Lab Recordings*." *Oral Tradition* 30, no. 1 (2016): 95-110.

Mustazza, Chris. "Vachel Lindsay and *The W. Cabell Greet Recordings*." *Chicago Review* 59/60, nos. 4/1 (2016): 98-117.

Myrsiades, Linda. *Splitting the Baby: The Culture of Abortion in Literature and Law, Rhetoric and Cartoons*. New York: Peter Lang, 2002.

Nachman, Gerald. *Raised on Radio*. Berkeley: U of California P, 1998.

Nardone, Michael. "Our Format: PennSound and the Articulation of an Interface for Literary Audio Recordings." *ESC: English Studies in Canada* 44, no. 2 (2018): 101-24.

Neal, Allison. "Marianne Moore's Tone Technologies: Elocution, Poetry, Phonograph." *Modernism/Modernity Print Plus* 6, no. 2 (Fall 2021). https://modernism modernity.org/articles/neal-moore-tone-technologies.

Newman, Ian. "Moderation in the *Lyrical Ballads*: Wordsworth and the Ballad Debates of the 1790s." *Studies in Romanticism* 55, no. 2 (2020): 185-210.

Nicholls, Peter. "Lorine Niedecker: Rural Surreal." In *Lorine Niedecker: Woman and Poet*, edited by Jenny Penberthy, 193-217. Orono, ME: National Poetry Foundation, 1996.

Niedecker, Lorine. *Collected Works*. Edited by Jenny Penberthy. Berkeley: U of California P, 2002.

Niedecker, Lorine. "Letters to *Poetry* Magazine, 1931-1937." In *Lorine Niedecker: Woman and Poet*, edited by Jenny Penberthy, 175-92. Orono, ME: National Poetry Foundation, 1996, 1996.

Niedecker, Lorine. "Local Letters." In *Lorine Niedecker: Woman and Poet*, edited by Jenny Penberthy, 87-107. Orono, ME: National Poetry Foundation, 1996, 1996.

Niedecker, Lorine. "Mother Geese." In *New Directions in Prose and Poetry*, edited by James Laughlin, n.p. Norfolk, VA: New Directions, 1936.

Nielsen, Aldon Lynn. *Black Chant: Languages of African-American Postmodernism*. Cambridge: Cambridge UP, 1997.

Nielsen, Aldon Lynn. "Black Deconstruction: Russell Atkins and the Reconstruction of African-American Criticism." *Diacritics* 26, nos. 3/4 (1996): 86-103.

Nielsen, Aldon Lynn. *Integral Music: Languages of African-American Innovation*. Tuscaloosa: U of Alabama P, 2004.

Nielsen, Aldon Lynn. "Objects: For Russell Atkins." In *Russell Atkins: On the Life and Work of an American Master*, edited by Kevin Prufer and Michael Dumanis, 136-59. Warrensburg, MO: Pleiades Press, 2013.

Nielsen, Aldon Lynn. "Russell Atkins: 'Heretofore.'" In *Russell Atkins: On the Life and Work of an American Master*, edited by Kevin Prufer and Michael Dumanis, 117-35. Warrensburg, MO: Pleiades Press, 2013.

Niles, John D. "The Myth of the Anglo-Saxon Oral Poet." *Western Folklore* 62, nos. 1/2 (2003): 7-61.

North, Michael. *The Dialect of Modernism: Race, Language, and Twentieth-Century Literature*. New York: Oxford UP, 1994.

Obadike, Mendi, and Keith Obadike. "Crosstalk: Blurred Boundaries in American Speech Music." Liner notes for *Crosstalk*, assembled by Mendi + Keith Obadike. Bridge Records 9285, 2008, CD.

O'Donnell, Daniel Paul. "Material Differences: The Place of Caedmon's Hymn in the History of Anglo-Saxon Vernacular Poetry." In *Caedmon's Hymn and Material Culture in the World of Bede: Six Essays*, edited by Allen J. Frantzen and John Hines, 15-50. Morgantown: West Virginia UP, 2007.

Odum, Howard W., and Guy B. Johnson, eds. *The Negro and His Songs: A Study of Typical Negro Songs in the South*. Chapel Hill: U of North Carolina P, 1925.

O'Hara, Frank. "A Wreath for John Wheelwright." *Poetry* 130, no. 2 (1977): 75-77.

O'Keeffe, Katherine O'Brien. "Orality and Literacy: The Case of Anglo-Saxon England." In *Medieval Oral Literature*, edited by Karl Reichl, 121-40. Berlin: De Gruyter Lexikon, 2012.

Ong, Walter J. *Orality and Literacy: The Technologizing of the Word*. Oxon: Routledge, 2002.

Ong, Walter J. *The Presence of the Word: Some Prolegomena for Cultural and Religious History*. New Haven, CT: Yale UP, 1967.

Ortiz, Joseph M. *Broken Harmony: Shakespeare and the Politics of Music*. Ithaca, NY: Cornell UP, 2011.

Owen, Maureen. "Helen Adam, 1909-1993." In *A Helen Adam Reader*, edited by Kristin Prevallet, 381-82. Orono, ME: National Poetry Foundation, 2007.

Parker, Patricia. *Literary Fat Ladies: Rhetoric, Gender, Property*. London: Methuen, 1987.

Parry, Sarah. "The Inaudibility of 'Good' Sound Editing: The Case of Caedmon Records." *Performance Research* 7, no. 1 (2002): 24-33.

Partch, Harry. "Author's Preface." Box 8, Folder 2. Music and Performing Arts Library, Harry Partch Collection, 1914-2007. Sousa Archives and Center for American Music. University of Illinois, Urbana-Champaign, IL.

Partch, Harry. *Bitter Music: Collected Journals, Essays, Introductions, and Librettos*. Edited by Thomas McGeary. Urbana: U of Illinois P, 1991.

Partch, Harry. *Exposition of Monophony*. 1933. Box 24, Folder 6. Music and Performing Arts Library, Harry Partch Collection, 1914-2007. Sousa Archives and Center for American Music. University of Illinois, Urbana-Champaign, IL.

Partch, Harry. *The Genesis of a Music*. 2nd ed. New York: Da Capo, 1974.

Partch, Harry. Interview with Vivian Perlis. 1974. Box 24, Folder 4. Music and Performing Arts Library, Harry Partch Collection, 1914-2007. University of Illinois Archives, Urbana-Champaign, IL.

Partch, Harry. Letter to Betty Freeman, July 25, 1968. Box 4, Folder 6. Music and

Performing Arts Library, Harry Partch Collection, 1914-2007. University of Illinois Archives, Urbana-Champaign, IL.

Partch, Harry. Letter to Elizabeth S. Coolidge, January 25, 1932. Box 20, Folder 25. Music and Performing Arts Library, Harry Partch Collection, 1914-2007. University of Illinois Archives, Urbana-Champaign, IL.

Partch, Harry. Letter to Henry Allen Moe, December 28, 1933. Box 8, Folder 2. Music and Performing Arts Library, Harry Partch Collection, 1914-2007. University of Illinois Archives, Urbana-Champaign, IL.

Partch, Harry. Letter to Henry Allen Moe, December 24, 1934. Box 8, Folder 2. Music and Performing Arts Library, Harry Partch Collection, 1914-2007. University of Illinois Archives, Urbana-Champaign, IL.

Partch, Harry. Letter to Henry Allen Moe, January 27, 1934. Box 8, Folder 2. Music and Performing Arts Library, Harry Partch Collection. 1914-2007. University of Illinois Archives, Urbana-Champaign, IL.

Partch, Harry. Letter to Henry Allen Moe, January 29, 1934. Box 8, Folder 2. Music and Performing Arts Library, Harry Partch Collection. University of Illinois Archives, Urbana-Champaign, IL.

Partch, Harry. Promotional pamphlet (Chicago 1942) in Scrapbook. Box 18, Folder 1, Harry Partch Estate Archive 1918-1991. Sousa Archives and Center for American Music. University of Illinois, Urbana-Champaign, IL.

Partch, Harry. Southwest transcriptions. Box 15, Folder 9, Harry Partch Estate Archive 1918-1991. Sousa Archives and Center for American Music. University of Illinois, Urbana-Champaign, IL.

Paz, Octavio. *El arco y la lira: El poema, la revelación poética, poesía e historia*. México D.F.: Fondo de Cultural Económica. 1956.

Peirce, C. S. *The Philosophy of Peirce: Selected Writings*. Edited by Justus Buchler. London: Routledge, 1940.

Penberthy, Jenny, ed. *Lorine Niedecker: Woman and Poet*. Orono, ME: National Poetry Foundation, 1996.

Penberthy, Jenny, ed. *Niedecker and the Correspondence with Zukofsky: 1931-1970*. Cambridge: Cambridge UP, 1993.

Pence, Charlotte. Introduction to *The Poetics of American Song Lyrics*, edited by Charlotte Pence, xi-xx. Jackson: U of Mississippi P, 2012.

Perloff, Marjorie. *The Dance of the Intellect: Studies in the Poetry of the Pound Tradition*. Cambridge: Cambridge UP, 1985.

Perloff, Marjorie, and Craig Dworkin, eds. *The Sound of Poetry/The Poetry of Sound*. Chicago: U of Chicago P, 2009.

Perlow, Seth. *The Poem Electric: Technology and the American Lyric*. Minneapolis: U of Minnesota P, 2018.

Peters, John Durham. "Assessing Kittler's *Musik und Mathematik*." In *Kittler Now:*

Current Perspectives in Kittler Studies, edited by Stephen Sale and Laura Salisbury, 22-43. Cambridge: Polity, 2015.

Peters, John Durham. "Helmholtz, Edison, and Sound History." In *Memory Bytes: History, Technology, and Digital Culture*, edited by Lauren Rabinovitz, 177-98. Durham, NC: Duke UP, 2004.

Peters, John Durham. *The Marvelous Clouds: Toward a Philosophy of Elemental Media*. Chicago: U of Chicago P, 2015.

Peters, Margot. *Lorine Niedecker: A Poet's Life*. Madison: U of Wisconsin P, 2011.

Phillips, Ivan. "I Sing the Bard Electric: Dylan Thomas, a Poet for the Age of Mass Media." *Times Literary Supplement*, September 19, 2003, 14-15.

Phillips, Stephen. *Poems*. London: John Lane, 1898.

Pichaske, David R. *Beowulf to Beatles: Approaches to Poetry*. Edited by David R. Pichaske. New York: Free Press, 1972.

Picker, John M. *Victorian Soundscapes*. Oxford: Oxford UP, 2003.

Pinsky, Robert. *The Sounds of Poetry: A Brief Guide*. New York: Farrar, Straus and Giroux, 1999.

Plath, Sylvia. *Ariel: The Restored Edition*. New York: HarperCollins, 2005.

Ponce, Martin Joseph. "Langston Hughes's Queer Blues." *Modern Language Quarterly* 66, no. 4 (December 2005): 505-38.

Ponge, Francis. *Le parti pris des choses*. Paris: Éditions Gallimard, 1942.

Poovey, Mary. "The Model System of Contemporary Literary Criticism." *Critical Inquiry* 27, no. 3 (2001): 408-38.

Posmentier, Sonya. "Blueprints for Negro Reading: Sterling Brown's Study Guides." In *A Companion to the Harlem Renaissance*, edited by Cherene Sherrard-Johnson, 119-35. Malden, MA: Wiley-Blackwell, 2015.

Posmentier, Sonya. *Cultivation and Catastrophe: The Lyric Ecology of Modern Black Literature*. Baltimore: Johns Hopkins UP, 2017.

Pound, Ezra. *ABC of Reading*. New York: New Directions, 1960.

Pound, Ezra. *The Letters of Ezra Pound: 1907-1941*. Edited by D. D. Paige. London: Faber and Faber, 1951.

Pound, Ezra. *Literary Essays of Ezra Pound*. Edited by T. S. Eliot. New York: New Directions, 1968.

Pound, Ezra. *New Selected Poems and Translations*. Edited by Richard Sieburth. New York: New Directions, 2010.

Pound, Ezra. "A Retrospect." In *Literary Essays of Ezra Pound*, edited by T. S. Eliot, 3-14. New York: New Directions, 1968.

Preston, Carrie J. *Modernism's Mythic Pose: Gender, Genre, Solo Performance*. Oxford: Oxford UP, 2011.

Prevallet, Kristin. Introduction to *A Helen Adam Reader*, edited by Kristin Prevallet, 3-62. Orono, ME: National Poetry Foundation, 2007.

Prins, Yopie. "Historical Poetics, Dysprosody, and *The Science of English Verse*." *PMLA* 123, no. 1 (2008): 229-34.
Prins, Yopie. "'What Is Historical Poetics?,'" *Modern Language Quarterly* 77, no. 1 (2016): 13-40.
"Propaganda from the U.S.: The Nazis Hate and Fear Boston's Station WRUL." *Life*, December 15, 1941, 43-46.
Propertius, Sextus. *The Complete Elegies of Sextus Propertius*. Translated by Vincent Katz. Princeton, NJ: Princeton UP, 2004.
Prufer, Kevin, and Michael Dumanis, eds. *Russell Atkins: On the Life & Work of an American Master*. Warrensburg, MO: Pleiades Press, 2013.
Pryer, Anthony. "Graphic Notation." In *The Oxford Companion to Music*, edited by Alison Latham. Oxford: Oxford UP, 2011. www.oxfordreference.com/view/10.1093/acref/9780199579037.001.0001/acref-9780199579037-e-3008.
Quartermain, Peter. "Reading Niedecker." *Lorine Niedecker: Woman and Poet*, edited by Jenny Penberthy, 219-27. Orono, ME: National Poetry Foundation, 1996.
Ramazani, Jahan. *Poetry and Its Others: News, Prayer, Song, and the Dialogue of Genres*. Chicago: U of Chicago P, 2014.
Ramazani, Jahan. "Traveling Poetry." *Modern Language Quarterly* 68, no. 2 (2007): 281-303.
Rampersad, Arnold. "Langston Hughes's *Fine Clothes to the Jew*." *Callaloo*, no. 26 (1986): 144-58.
Rampersad, Arnold. *The Life of Langston Hughes*. Vol. 1, *1902-1941: I, Too, Sing America*. 2nd ed. Oxford: Oxford UP, 2002.
Rampersad, Arnold. *The Life of Langston Hughes*. Vol. 2, *1941-1967: I Dream a World*. 2nd ed. Oxford: Oxford UP, 2002.
Rasula, Jed. *The American Poetry Wax Museum: Reality Effects, 1940-1990*. Urbana, IL: National Council of Teachers of English, 1996.
Raulerson, Graham. "'A Fountainhead of Pure Musical Americana': Hobo Philosophy in Harry Partch's *Bitter Music*." *Journal of the Society of American Music* 11, no. 4 (2017): 452-69.
Reagan, Leslie J. *When Abortion Was a Crime: Women, Medicine, and Law in the United States, 1867-1973*. Berkeley: U of California P, 1997.
Redmond, Eugene B. *Drumvoices: The Mission of Afro-American Poetry: A Critical History*. Garden City, NY: Anchor Books, 1976.
Reed, Anthony. *Freedom Time: The Poetics and Politics of Black Experimental Writing*. Baltimore: Johns Hopkins UP, 2014.
Reed, Anthony. *Soundworks: Race, Sound, and Poetry in Production*. Durham, NC: Duke UP, 2020.
Reed, Brian. "Visual Experiment and Oral Performance." In *The Sound of Poetry / The*

Poetry of Sound, edited by Marjorie Perloff and Craig Dworkin, 270-84. Chicago: U of Chicago P, 2009.

Reich, Steve. *Writings on Music: 1965-2000*. Edited by Paul Hillier. Oxford: Oxford UP, 2004.

Rheinberger, Hans-Jörg. *Toward a History of Epistemic Things: Synthesizing Proteins in the Test Tube*. Stanford, CA: Stanford UP, 1997.

Rifkin, Libbie. *Career Moves: Olson, Creeley, Zukofsky, Berrigan, and the American Avant-Garde*. Madison: U of Wisconsin P, 2000.

Roach, Helen. "The Two Women of Caedmon." *Association for Recorded Sound Collections Journal* 19, no. 1 (1987): 21-24.

Robbins, Michael. *Equipment for Living: On Poetry and Pop Music*. New York: Simon & Schuster, 2017.

Robertson, Lisa. "In Phonographic Deep Song: Sounding Niedecker." In *Radical Vernacular: Lorine Niedecker and the Poetics of Place*, edited by Elizabeth Willis, 83-90. Iowa City: U of Iowa P, 2008.

Rockefeller Foundation. *Annual Report*. New York: Rockefeller Foundation, 1938.

Rosemont, Franklin, and Robin D. G. Kelley. *Black, Brown and Beige: Surrealist Writings from Africa and the Diaspora*. Austin: U of Texas P, 2009.

Rosenfeld, Alvin H. "John Wheelwright, Gorham Munson, and the 'Wars of Secession.'" *Michigan Quarterly Review* 14, no. 1 (Winter 1975): 13-40.

Rubery, Matthew. "Thomas Edison's Poetry Machine." *19: Interdisciplinary Studies in the Long Nineteenth Century* 18 (2014). https://19.bbk.ac.uk/article/id/1447/.

Rubery, Matthew. *The Untold Story of the Talking Book*. Cambridge, MA: Harvard UP, 2016.

Rukeyser, Muriel. *The Life of Poetry*. Ashfield, MA: Paris Press, 1996.

Said, Edward. "Yeats and Decolonization." In *Culture and Imperialism*, 220-38. New York: Random House, 1994.

Salvato, Nick. *Uncloseting Drama: American Modernism and Queer Performance*. New Haven, CT: Yale UP, 2010.

Sanders, Mark A. *Afro-Modernist Aesthetics and the Poetry of Sterling Brown*. Athens: U of Georgia P, 1999.

Sanders, Mark A. Foreword to *A Son's Return: Selected Essays of Sterling A. Brown*, edited by Mark A. Sanders, ix-xxi. Boston: Northeastern UP, 1996.

Sannazaro, Jacopo. *Arcadia and Piscatorial Eclogues*. Translated by Ralph Nash. Detroit, MI: Wayne State UP, 1966.

Sappho. *If Not, Winter: Fragments of Sappho*. Translated by Anne Carson. New York: Knopf, 2003.

Sargeant, Winthrop. "Humiliation, Concentration, and Gusto." *New Yorker*, February 16, 1957, 38-73.

Sarton, May. *Collected Poems: 1930-1993*. Digital ed. New York: Open Road, 2014.

Sartre, Jean-Paul. *Critique of Dialectical Reason*. Vol. 1. Translated by Alan Sheridan-Smith. London: Verso, 2004.
Saussure, Ferdinand de. *Course in General Linguistics*. Edited by Charles Bally and Albert Sechehaye. Translated by Wade Baskin. New York: Philosophical Library, 1959.
Savage, Barbara Dianne. *Broadcasting Freedom: Radio, War, and the Politics of Race, 1938-1948*. Chapel Hill: U of North Carolina P, 1999.
Schaeffer, Pierre. *Treatise on Musical Objects: An Essay across Disciplines*. Translated by Christine North and John Dack. Berkeley: U of California P, 2017.
Schuchard, Ronald. *The Last Minstrels: Yeats and the Revival of the Bardic Arts*. Oxford: Oxford UP, 2008.
Schulze, Robin G. *"Others."* In *Becoming Marianne Moore: The Early Poems, 1907-1924*, edited by Robin G. Schulze, 466-78. Berkeley: U of California P, 2002.
Schulze, Robin G. "Textual Darwinism: Marianne Moore, the Text of Evolution, and the Evolving Text." *Text* 11 (1998): 270-305.
Scott, Jonathan. *Socialist Joy in the Writing of Langston Hughes*. Columbia: U of Missouri P, 2006.
Scott, Walter. *Minstrelsy of the Scottish Border; Consisting of Historical and Romantic Ballads*. Vol. 2. Kelso, UK: Ballantyne, 1802.
Scott, Winfield Townley. "John Wheelwright and His Poetry." *New Mexico Quarterly* 24, no. 2 (1954): 178-96.
Scroggins, Mark. *Louis Zukofsky and the Poetry of Knowledge*. Tuscaloosa: U of Alabama P, 1998.
Scroggins, Mark. *The Poem of a Life: A Biography of Louis Zukofsky*. New York: Shoemaker Hoard, 2007.
Seiler, Claire. *Midcentury Suspension: Literature and Feeling in the Wake of World War II*. New York: Columbia UP, 2020.
Selch, Andrea. "Engineering Democracy: Commercial Radio's Use of Poetry, 1920-1946." PhD diss., Duke University, 1999.
Serres, Michel. *The Parasite*. Translated by Lawrence R. Schehr. Minneapolis: U of Minnesota P, 2007.
Shakespeare, William. *Pericles, Prince of Tyre*. In *The Complete Works of Shakespeare*, edited by W. G. Clark and W. Aldis Wright, 985-1010. Chicago: Morrill, Higgins, and Co., 1892.
Shakespeare, William. *The Sonnets*. New York: Signet, 1964.
Shakespeare, William. *Troilus and Cressida*. Edited by Barbara A. Mowat and Paul Werstine. New York: Washington Square Press, 2007.
Shand-Tucci, Douglass. *The Crimson Letter: Harvard, Homosexuality, and the Shaping of American Culture*. New York: St. Martin's Griffin, 2003.
Shaw, Lytle. *Narrowcast: Poetry and Audio Research*. Stanford, CA: Stanford UP, 2018.

Shelley, Percy Bysshe. "A Defence of Poetry." In *Shelley's Poetry and Prose*, edited by Donald H. Reiman and Neil Fraistat, 509-35. New York: Norton, 2002.
Shelley, Percy Bysshe. *Queen Mab: A Philosophical Poem*. New York: Baldwin, 1821.
Sheppard, W. Anthony. *Revealing Masks: Exotic Influence and Ritualized Performance in Modernist Music Theater*. Berkeley: U of California P, 2001.
Shockley, Evie. "Death Is Only Natural." In *Russell Atkins: On the Life and Work of an American Master*, edited by Kevin Prufer and Michael Dumanis, 108-16. Warrensburg, MO: Pleiades Press, 2013.
Siegert, Bernhard. *Cultural Techniques: Grids, Filters, Doors, and Other Articulations of the Real*. Translated by Geoffrey Winthrop-Young. New York: Fordham UP, 2015.
Silliman, Ron. "Louis Zukofsky." *Paideuma: Modern and Contemporary Poetry and Poetics* 7, no. 3 (1978): 405-6.
Silverstein, Michael. "Shifters, Linguistic Categories and Cultural Description." In *Meaning in Anthropology*, edited by Keith H. Basso and Henry A. Selby, 11-55. Albuquerque: U of New Mexico P, 1976.
Silverstein, Michael, and Greg Urban. "The Natural History of Discourse." In *Natural Histories of Discourse*, edited by Michael Silverstein and Greg Urban, 1-20. Chicago: U of Chicago P, 1996.
Skeele, David. "*Pericles* in Criticism and Production: A Brief History." In *Pericles: Critical Essays*, edited by David Skeele, 1-33. New York: Garland, 2000.
Skeele, David. *Thwarting the Wayward Seas: A Critical and Theatrical History of Shakespeare's Pericles in the Nineteenth and Twentieth Centuries*. Newark: U of Delaware P, 1998.
Skinner, Beverly Lanier. "Sterling Brown: An Ethnographic Odyssey." In *After Winter: The Art and Life of Sterling Brown*, edited by John Edgar Tidwell and Steven C. Tracy, 186-92. Oxford: Oxford UP, 2009.
Slatin, John M. *The Savage's Romance: The Poetry of Marianne Moore*. University Park: Pennsylvania State UP, 1986.
Smethurst, James. *The Black Arts Movement: Literary Nationalism in the 1960s and 1970s*. Chapel Hill: U of North Carolina P, 2005.
Smethurst, James. "'Don't Say Goodbye to the Porkpie Hat': Langston Hughes, the Left, and the Black Arts Movement." *Callaloo* 25, no. 4 (2002): 1225-36.
Smith, Charles Forster. "Stephen Phillips," *Sewanee Review* 9, no. 4 (1901): 385-97.
Smith, David Nowell. *On Voice in Poetry: The Work of Animation*. New York: Palgrave Macmillan, 2015.
Smith, Jacob. *Spoken Word: Postwar American Phonograph Cultures*. Berkeley: U of California P, 2011.
Smith, Judith E. "Literary Radicals in Radio's Public Sphere." In *American Literature in Transition: 1940-1950*, edited by Christopher Vials, 309-33. Cambridge: Cambridge UP, 2018.

Smith, Kathleen E. R. *God Bless America: Tin Pan Alley Goes to War*. Lexington: UP of Kentucky, 2003.
Spenser, Edmund. *The Faerie Queene*. London: Penguin, 1978.
Stapleton, Laurence. *Marianne Moore: The Poet's Advance*. Princeton, NJ: Princeton UP, 1978.
Staten, Henry. "The Origin of the Work of Art in Material Practice." *New Literary History* 43, no. 1 (2012): 43-64.
Steege, Benjamin. *Helmholtz and the Modern Listener*. Cambridge: Cambridge UP, 2012.
Steintrager, James A., and Rey Chow. "Sound Objects: An Introduction." In *Sound Objects*, edited by James A. Steintrager and Rey Chow, 1-19. Durham, NC: Duke UP, 2019.
Stephenson, William, ed. *British Security Coordination: The Secret History of British Intelligence in the Americas, 1940-1945*. Introduction by Nigel West. New York: Fromm International, 1999.
Sterling, Christopher H., and John M. Kittross. *Stay Tuned: A Concise History of American Broadcasting*. Belmont, CA: Wadsworth, 1978.
Sterne, Jonathan. *The Audible Past: Cultural Origins of Sound Reproduction*. Durham, NC: Duke UP, 2003.
Sterne, Jonathan. "Communication as Technē." In *Communication As . . . : Perspectives on Theory*, edited by Gregory J. Shepherd, Jeffrey St. John, and Ted Striphas, 91-98. Thousand Oaks, CA: Sage, 2006.
Stewart, Garrett. *Reading Voices: Literature and the Phonotext*. Berkeley: U of California P, 1990.
Stewart, Susan. *Crimes of Writing: Problems in the Containment of Representation*. New York: Oxford UP, 1991.
Stewart, Susan. *Poetry and the Fate of the Senses*. Chicago: U of Chicago P, 2002.
Stoever, Jennifer Lynn. *The Sonic Color Line: Race and the Cultural Politics of Listening*. New York: New York UP, 2016.
Strachey, Lytton. *Books and Characters: French and English*. London: Chatto and Windus, 1922.
Stuckenschmidt, H. H. "Contemporary Techniques in Music." *Musical Quarterly* 49, no. 1 (1963): 1-16.
Stuckenschmidt, H. H. *Twentieth Century Music*. New York: McGraw Hill, 1969.
Sturken, Marita, Douglas Thomas, and Sandra Ball-Rokeach. *Technological Visions: The Hopes and Fears that Shape New Technologies*. Philadelphia: Temple UP, 2004.
Taylor, Marjorie A. "The Folk Imagination of Vachel Lindsay." PhD diss., Wayne State University, 1976.
Teague, Jessica E. *Sound Recording Technology and American Literature: From the Phonograph to the Remix*. Cambridge: Cambridge UP, 2021.
Thomas, Dylan. *Collected Poems*. New York: New Directions, 2010.

Thomas, Dylan. "Poetic Manifesto." *Texas Quarterly* 4, no. 4 (1961): 44-53.
Thomas, Katherine Elwes. *The Real Personages of Mother Goose*. Boston: Lothrop, Lee & Shepard, 1930.
Thompson, Stith. "Transformation Flight." In *Motif-Index of Folk Literature: A Classification of Narrative Elements in Folktales, Ballads, Myths, Fables, Mediaeval Romances, Exempla, Fabliaux, Jest-Books, and Local Legends*. Vol. 2, edited by Stith Thompson, 77. Bloomington: Indiana UP, 1975.
Tidwell, John Edgar. "'Greatest Is the Song': Blues as Poetic Communication in Early Langston Hughes and Sterling A. Brown." In *Black Music, Black Poetry: Blues and Jazz's Impact on African American Versification*, edited by Gordon E. Thompson, 55-66. London: Routledge, 2014.
Tidwell, John Edgar, and Cheryl Ragar. "Langston Hughes Revisited and Revised: An Introduction." In *Montage of a Dream: The Art and Life of Langston Hughes*, edited by John Edgar Tidwell and Cheryl Ragar, 1-15. Columbia: U of Missouri P, 2007.
Tiffany, Daniel. *My Silver Planet: A Secret History of Poetry and Kitsch*. Baltimore: Johns Hopkins UP, 2014.
Tiffany, Daniel. *Radio Corpse: Imagism and the Cryptaesthetic of Ezra Pound*. Cambridge, MA: Harvard UP, 1995.
Time. "Quicker Fox." 32, no. 2 (July 11, 1938): 46-47.
Times. "Clifford's Inn-Hall." May 7, 1903.
Tkweme, W. S. "Blues in Stereo: The Texts of Langston Hughes in Jazz Music." *African American Review* 42, nos. 3/4 (2008): 503-12.
Trachtenberg, Alan. *Brooklyn Bridge: Fact and Symbol*. Chicago: U of Chicago P, 1965.
Tracy, Steven C. *Langston Hughes and the Blues*. Urbana: U of Illinois P, 2001.
Tracy, Steven C. "Vernacular Recordings Relevant to the Study of Sterling A. Brown's *Southern Road*." In *After Winter: The Art and Life of Sterling Brown*, edited by John Edgar Tidwell and Steven C. Tracy, 449-555. Oxford: Oxford UP, 2009.
Trotsky, Leon. *Writings of Leon Trotsky, 1933-34*. Edited by George Breitman and Bev Scott. New York: Pathfinder Press, 1975.
Trotter, David. *Literature in the First Media Age: Britain Between the Wars*. Cambridge, MA: Harvard UP, 2013.
Trudell, Scott A. *Unwritten Poetry: Song, Performance, and Media in Early Modern England*. Oxford: Oxford UP, 2019.
Twitchell-Waas, Jeffrey. "Keep Your Eyes on the Page: Zukofsky's *Bottom: On Shakespeare*." *Paideuma* 39 (2012): 209-47.
Twitchell-Waas, Jeffrey. "What Were the 'Objectivist' Poets?" *Modernism/modernity* 22, no. 2 (2015): 315-41.
Twitchell-Waas, Jeffrey. *Z-Site: A Companion to the Works of Louis Zukofsky*. www.z-site.net/.

Valéry, Paul. *Collected Works of Paul Valery*. Vol. 14, *Analects*. Translated by Stuart Gilbert. Princeton, NJ: Princeton UP, 1970.

Van Doren, Carl. Preface to *Thirteen by Corwin: Radio Dramas*, by Norman Corwin, vii. New York: H. Holt, 1942.

Van Loon, Hendrik. "WRUL: This Unique Short Wave Radio Station Sells Nothing and Builds International Good Will." *Current History* 52, no. 12 (1941): 22-23, 56.

Various Artists. *That's Why We're Marching: World War II and the American Folk Song Movement*. Smithsonian Folkways Recordings SFW20021, 1996.

Vendler, Helen. *Poems, Poets, Poetry: An Introduction and Anthology*. Boston: Bedford, 1997.

Verma, Neil. *Theater of the Mind: Imagination, Aesthetics, and American Radio Drama*. Chicago: U of Chicago P, 2012.

Vincent, John Emil. *Queer Lyrics: Difficulty and Closure in American Poetry*. New York: Palgrave Macmillan, 2002.

Wagner, Bryan. *Disturbing the Peace: Black Culture and the Police Power after Slavery*. Cambridge, MA: Harvard UP, 2009.

Wagner, Jean. *Black Poets of the United States: From Paul Laurence Dunbar to Langston Hughes*. Translated by Kenneth Douglas. Urbana: U of Illinois P, 1973.

Wald, Alan M. *The Revolutionary Imagination: The Poetry and Politics of John Wheelwright and Sherry Mangan*. Chapel Hill: U of North Carolina P, 1983.

Wald, Alan M. "Wheelwright and His Kind." *Spoke* 3 (2005): 194-208.

Walker, Jesse H. "Langston Hughes Lyrics in Revival of 'Street Scene.'" Clippings, Box 570, Folder 13600, Langston Hughes Papers, James Weldon Johnson Memorial Collection. Beinecke Library, Yale University. New Haven, CT.

Weheliye, Alexander G. *Phonographies: Grooves in Sonic Afro-Modernity*. Durham, NC: Duke UP, 2005.

Wheeler, Lesley. *Voicing American Poetry: Sound and Performance from the 1920s to the Present*. Ithaca, NY: Cornell UP, 2008.

Wheelwright, John. *Collected Poems of John Wheelwright*. Edited by Alvin H. Rosenfeld. New York: New Directions, 1972.

Wheelwright, John. Letter to unknown recipient, January 4, 1933. Box 7, Folder 1, John Wheelwright Papers 1920-1940. John Hay Library, Brown University, Providence, RI.

Wheelwright, John. Letter to unknown recipient, January 9, 1933. Box 7, Folder 1, John Wheelwright Papers 1920-1940. John Hay Library, Brown University, Providence, RI.

Wheelwright, John. "'Poetry Noon Hour,' Introductions for Readers." Box 33, Folder 5, John Wheelwright Papers 1920-1940. John Hay Library, Brown University, Providence, RI.

Wheelwright, John. Radio script for Kenneth Patchen broadcast, Box 33, Folder 5,

John Wheelwright Papers 1920-1940. John Hay Library, Brown University, Providence, RI.
Wheelwright, John. "Talking Machines." Box 36, Folder 5, John Wheelwright Papers 1920-1940. John Hay Library, Brown University, Providence, RI.
Wheelwright, John. "Toward the Recovery of Speech." *Poetry* 54, no. 3 (1939): 164-67.
Wheelwright, John. "U.S. 1." *Partisan Review* 4, no. 4 (1938): 54-56.
Wheelwright, John. "Verse + Radio = Poetry." Box 37, Folder 8, John Wheelwright Papers 1920-1940. John Hay Library, Brown University, Providence, RI.
White, Heather Cass. Introduction to *New Collected Poems*, by Marianne Moore, xv-xxvi. New York: Farrar, Straus and Giroux, 2017.
White, Newman Ivey, ed. *American Negro Folk-Songs*. Cambridge, MA: Harvard UP, 1928.
Williams, Cameron. "'Really, Music Was the Cause of It.'" Interview with Russell Atkins, June 2, 2016, at the Grand Pavilion, Cleveland, OH. *Jacket2*. https://jacket2.org/interviews/really-music-was-cause-it.
Williams, Raymond. *Marxism and Literature*. Oxford: Oxford UP, 1977.
Williams, Raymond. *Television: Technology and Cultural Form*. Abingdon: Routledge, 2003.
Williams, William Carlos. *The Collected Poems of William Carlos Williams*. Vol. 2, *1939-1962*. Edited by Christopher MacGowan. New York: New Directions, 2001.
Williams, William Carlos. *I Wanted to Write a Poem*. Edited by Edith Heal. Boston: Beacon Hill Press, 1958.
Williams, William Carlos. Letter to Louis Zukofsky, March 21, 1943, Box 29, Folder 10, Zukofsky Collection 1910-1985. Harry Ransom Center, University of Texas at Austin.
Williams, William Carlos. "Two Letters to Robert Lawrence Beum." In *Something to Say: William Carlos Williams on Younger Poets*, edited by James E. B. Breslin, 198-201. New York: New Directions, 1985.
Wimsatt, W. K. *The Verbal Icon: Studies in the Meaning of Poetry*. Lexington: U of Kentucky P, 1954.
Winn, James Anderson. *Unsuspected Eloquence: A History of the Relations between Poetry and Music*. New Haven, CT: Yale UP, 1981.
Wittke, Carl. *Tambo and Bones: A History of the American Minstrel Stage*. Durham, NC: Duke UP, 1930.
WNYC. "Blues Symposium." Recorded June 9, 1938. The New York Public Radio Archive Collections, courtesy of the NYPL Dance Division. Audio, ca. 39 min. www.wnyc.org/story/blues-symposium/.
Wollaeger, Mark. *Modernism, Media, and Propaganda: British Narrative from 1900 to 1945*. Princeton, NJ: Princeton UP, 2006.

Woods, Tim. *The Poetics of the Limit: Ethics and Politics in Modern and Contemporary American Poetry.* New York: Palgrave Macmillan, 2002.
Woolf, Virginia. *A Writer's Diary: Being Extracts from the Diary of Virginia Woolf.* London: Hogarth Press, 1953.
Wordsworth, William. *The Major Works Including "The Prelude."* Edited by Stephen Gill. Oxford: Oxford UP, 2008.
Yeats, W. B. *Ah, Sweet Dancer: W. B. Yeats, Margot Ruddock, a Correspondence.* Edited by Roger McHugh. New York: MacMillan, 1971.
Yeats, W. B. Letter to Edmund Dulac, November 21, 1934. *The Collected Letters of W. B. Yeats: Electronic Edition*, edited by John Kelly, InteLex 6130.
Yeats, W. B. Letters to Harry Partch, n.d. Box 8, Folder 2, Harry Partch Estate Archive, 1918-1991. The Sousa Archives and Center for American Music, Urbana-Champaign, IL.
Yeats, W. B. "Literature and the Living Voice." *Samhain* 6, no. 2 (December 1906): 4-14.
Yeats, W. B. "Music for the Plays." In *The Collected Works of W. B. Yeats*, vol. 2, *The Plays*, edited David R. Clark and Rosalind E. Clark, appendix B, 757-818. New York: Scribner, 2001.
Yeats, W. B. *The Poetical Works of William B. Yeats.* Vol. 2, *Dramatical Poems.* New York: Macmillan, 1907.
Yeats, W. B. "Speaking to Musical Notes." *Academy and Literature*, June 7, 1902, 590-91.
Yeats, W. B. "Speaking to the Psaltery." In *The Collected Early Works of W. B. Yeats*, vol. 4, edited Richard J. Finneran and George Bornstein, 12-24. New York: Scribner, 2007.
Zukofsky, Celia. *Pericles* in 1943 piano setting. Box 9, Folder 1-3, Louis Zukofsky Collection 1910-1985. Harry Ransom Center, University of Texas at Austin, Austin, TX.
Zukofsky, Celia, and William Carlos Williams. "Choral: The Pink Church." *Briarcliff Quarterly* 3, no. 3 (1946): 165-68.
Zukofsky, Celia, and William Carlos Williams. "Turkey in the Straw," notes and musical score. Box 35, Folder 6, Louis Zukofsky Collection 1910-1985. Harry Ransom Center, University of Texas at Austin.
Zukofsky, Louis. *"A."* New York: New Directions, 2011.
Zukofsky, Louis. *Anew: Complete Shorter Poetry.* New York: New Directions, 2011.
Zukofsky, Louis. "Bottom, a Weaver." In *Prepositions +: The Collected Critical Essays*, edited by Mark Scroggins, 167. Middleton: Wesleyan UP, 2000.
Zukofsky, Louis. Letter to Hugh Kenner, March 29, 1963. Box 50, Folder 4, Hugh Kenner Papers 1916-1994. Harry Ransom Center, University of Texas at Austin.
Zukofsky, Louis. Letter to Lorine Niedecker, n.d., 1952. Box 19, Folder 3, Louis

Zukofsky Collection 1910-1985. Harry Ransom Center, University of Texas at Austin.

Zukofsky, Louis, ed. An *"Objectivist" Anthology*. Le Beausset, Var, France: TO, 1932.

Zukofsky, Louis. *Prepositions +: The Collected Critical Essays*. Edited by Mark Scroggins. Middletown, CT: Wesleyan UP, 2000.

Zukofsky, Louis. "Sincerity and Objectification." *Poetry: A Magazine of Verse* 32, no. 5 (1931): 272-85.

Zukofsky, Louis, and Celia Thaew Zukofsky. *Bottom: On Shakespeare*. Middletown, CT: Wesleyan UP, 2002.

Index

Figures are indicated with an italic *f*.

"A" (Louis Zukofsky), 79, 81
"Abortionist, The" (Atkins), 202-3
Adam, Helen: emigration to United States by, 220; *New American Poetry* anthology's inclusion (1960) of, 221; reincarnation and, 221; revival of ballad tradition in twentieth century by, 220; Romanticism and, 222, 224; slippage between music and war in poetry of, 25; time-lapsing ballads of, 20-21, 221-22; von Praunheim's films and, 211. See also specific works
Adams, Henry, 75
Adapted Viola. See under Partch, Harry
Adorno, Theodor: on low and high art, 20; on modern poem as "philosophical sundial," 45; on the phonograph and music's relationship to writing, 59; pop music critiqued by, 157-58, 160; on "switching off" and radio, 13-14
aeolian harp, 27, 130, 228
Agha, Asif, 69
"Alastor, or The Spirit of Solitude" (Shelley), 27
Albright, Daniel, 137
Alcaeus, 24-25
Allen, Donald, 170, 221
Allen, Edward, 177
Allport, Gordon, 12-13
American Negro Folk-Songs anthology (1928), 123
"American Negro National Defense Song" (Hughes), 156

American Society of Composers, Authors and Publishers (ASCAP), 150
"America's Young Black Joe" (Hughes), 153, 155
Amos 'n' Andy (radio program), 114-15
Anacreon, 24
Andrews, Loring, 99
Arcadia (Sannazaro), 223
Aristotle, 78
Ashbery, John, 105, 200
Ask Your Mama: 12 Moods for Jazz (Hughes), 141
Atkins, Russell: abortion in the poetry of, 201-3, 205-7; Cleveland as home of, 190, 192; conspicuous technique approach and, 199, 200, 205-6; "egocentrical phenomenalism" and, 190, 200, 201; *Free Lance* magazine, 188, 195, 198; Hughes and, 188; Moore and, 173-74, 187, 199-200; Muntu Poets of Cleveland and, 188; music's visual nature for, 189-90, 192-93; object-forms and, 192-99, 201, 208; poetic objecthood and, 20, 173, 183, 188-89; psychovisualism and, 20, 174, 188-89, 191, 193-95, 197-99; Smith's disagreements about music with, 191-92. See also specific works
Auden, W. H., 9, 96
"Avant-Garde and Kitsch" (Greenberg), 135

Bakhtin, Mikhail, 66
Baldwin, James, 272n87

"Ballad of the Girl Whose Name Is Mud" (Hughes), 149, 269n52
ballads: blues music and, 144-45; content and context linked in, 220; distressed ballads and, 222, 224; freedom vis-à-vis material culture of, 225; intermediality and, 38, 218-19; kitsch and, 160; live traditions of musical practice as source of study of, 217; lyric and, 33; paraliterary nature of, 140; romance of orality and, 143, 219; Romanticism and, 6, 33-34, 135, 143-45, 160, 222; songwriting and, 38; suspensive temporalities and, 20-21, 38; time-lapsing ballads and, 20-21, 221-22
Baraka, Amiri, 139, 188, 199
Bartók, Béla, 58, 249n39
the Beats, 173
Bede, 29-33, 41, 44
Benét, Stephen Vincent, 89
Benjamin, Walter, 13-14, 87, 103
Benn, Gottfried, 4, 267n22
Berman, Marshall, 13
Bernstein, Charles, 38, 186, 240n26
Bernstein, Leonard, 99
Between the Bookends (radio program), 91, 99
Bewick, Thomas, 181
Biers, Katherine, 120
Billera, Jau, 193
"Birkenshaw, The" (Adam), 226-29
Bishop, Elizabeth, 177
Bishop, John Peale, 99
Bitter Music (Partch): excerpts from, 67-70, 67*f*, 68*f*, 69*f*, 70*f*; hobo life in transient shelters of American West chronicled in, 50, 63-65, 67; phaticity of speech and, 66-67, 69; same-sex desire and, 63-64; social-indexical dimension of language and, 66, 69; speech-music and, 63, 64-69
Black dialect: Brown and, 112, 116, 118-24, 131; Corwin's adaptations of, 115; folk idiom and, 118-21, 123, 131; Hughes and, 116, 118; Johnson and, 116, 118, 122; minstrel stereotypes and, 114, 115-16, 118, 119, 124, 146; Negro sermons and, 119-20; phonography and, 120-21, 123; standards of authenticity and, 122
Black Mountain College, 170, 173

Blake, William, 212-13, 220-22
Bolter, Jay, 90
Bonds, Margaret, 153
Bontemps, Arna, 149, 188
Book of American Negro Poetry (Johnson), 116, 119
Book of Ayres (Campion), 26
"The Book of the Dead" (Rukeyser), 98
Booth, Mark, 137-38
Bornstein, George, 203
Bottom: On Shakespeare (Louis and Celia Zukofsky): frontispiece of, 86*f*; gendered division of labor and, 85; literary-media theory presented in, 76-80; Love's mind and, 78-79, 87; lyric intermediality and, 78, 85; Moore on, 82; *Pericles, Prince of Tyre* score included in, 72, 73-74, 82, 84*f*, 86*f*; phatic dimension of language and, 82; on printed word and love, 79-80; rested totality and, 83, 85; on scored language, 81
Boufflers, Marquise de, 181-83
The Bow and the Lyre (Paz), 25
Bradford, Roark, 119
Brecht, Bertolt, 13, 96, 103
Breman, Paul, 201
British Security Co-ordination (BSC), 101
Brooklyn Bridge: Fact and Symbol (Trachtenberg), 182
Brooks, Gwendolyn, 203, 206
Brooks, Peter Chardon, 94
Brown, Earle, 193
Brown, Sterling: Black dialect and, 112, 116, 118-24, 131; blues music and the poetry of, 122-28; Bradford and, 119; contextual reading practices and, 119; as Federal Writers' Project Office of Negro Affairs director, 121; Great Mississippi Flood (1927) and, 126; latent remediation and, 19, 93, 112, 123, 131; on literary stereotypes of Black characters, 128; Negro sermons and, 119-20; phonography and, 121, 123-24, 130-31; on Russell, 116; sonic color line and, 124; *Words Without Music* program and the poetry of, 115. *See also specific works*
Browning, Robert, 171
"Burden" (Hughes), 133
Burke, Kenneth, 97

Index

Burns, Robert, 36
"By the Rivers of Babylon" (Partch), 55, 56*f*

"Cabaret" (Brown), 126
Caedmon Audio record label, 32, 42-43, 171, 178
"Caedmon's Hymn," 29-33, 41, 43-44, 224
Caesar, Irving, 164
Caetani, Marguerite, 188
Camlot, Jason, 42
Campion, Thomas, 25-27
Cantos (Pound), 17, 160
Cantril, Hadley, 12-13
Carby, Hazel V., 150-51
Carmichael, Hoagy, 164
Carmody, Todd, 121
Carpenter, Elliot, 153
Celan, Paul, 217
Celtic Renaissance, 51-52, 54
A Child's Christmas in Wales and Five Poems (Thomas), 42-43
Chinitz, David E., 144
Chion, Michel, 198
Chopin, Henri, 69
"Choral: The Pink Church" (William Carlos Williams and Celia Zukofsky), 73
Christgau, Robert, 136, 138
Chun, Wendy Hui Kyong, 44
Cohan, George M., 153
Coleridge, Samuel Taylor, 27, 220-21
Collins, Helen, 188
"The Congo" (Lindsay), 113
Connelly, Marc, 119
Connor, Steven, 171
Corman, Cid, 72
Corwin, Norman: audience share commanded by, 111; Black actors hired for radio programs by, 114-15; Black dialect affected on radio by, 115; canonical and popular poems adapted for radio by, 111-12; "Negro poetry" abusively adapted for radio by, 93, 112-15, 123-24, 130; as "poetry jockey" on WQXR radio, 111; as radio dramatist, 89, 92, 99, 111. *See also Words without Music*
Coughlin, Charles, 10
Crane, Hart, 102, 103
"Crazy Blues" (Mamie Smith), 145
"The Creation" (Johnson), 115, 119-20

Creeley, Robert, 188
Culler, Jonathan, 7-8, 22

Dahlberg, Edward, 72
"The Daniel Jazz" (Lindsay), 112-15, 124
Dante, 77
Davidson, Michael, 220, 222
Davis, Angela Y., 150-51
Davis, Stuart, 174-75, 182
Deep River Boys, 115
"Defence of Poetry" (Shelley), 27
"De Fust Banjo" (Russell), 115-16, 119, 122, 130
Delaney, Samuel R., 140
De Lange, Edgar, 164, 165*f*, 166*f*
de Man, Paul, 7, 173
"Democracy, Negroes, and Writers" (Hughes), 155
De Vulgari Eloquentia (Dante), 77
Dickinson, Emily, 209-10
"Dixie Negro to Uncle Sam" (Hughes), 156
Dolmetsch, Arnold, 52, 62
Donne, John, 102
Double V campaign, 153, 155-56, 162, 167
Du Bois, W. E. B., 129
Dulac, Edmund, 56, 62
dummy lyrics, 20, 164, 168
Dunbar, Paul Laurence, 149, 269-70n54
Duncan, Robert, 214, 220-21
Dylan, Bob, 21, 136

Eberhart, Richard, 99
Ecclesiastical History of the English People (Bede), 29-33, 41, 44
Edison, Thomas, 12, 17, 138, 171
Edwards, Brent Hayes, 38, 146, 148
The Eighteenth Brumaire of Louis Bonaparte (Marx), 107
"Elegy to Hurt Bird That Died (buried in matchbox)" (Atkins), 200
The Elfin Pedlar and Tales Told by Pixy Pool (Adam), 220
Eliot, T. S., 17, 100, 130, 135
Ellington, Duke, 153
Elliot, Jean, 34*f*, 35*f*, 36, 37*f*, 38, 220-21
Ellis, Ron, 39
Ellison, Ralph, 127
Encounter in April (Sarton), 100
Engels, Friedrich, 95

"The Eolian Harp" (Coleridge), 27
Ernst, Wolfgang, 58, 249n40
"Étude aux chemins de fer" ("Railway Study," Schaeffer), 195

The Fables of La Fontaine (Moore), 179, 181
The Faerie Queene (Spenser), 223
The Fall of the City (MacLeish), 89-90
Fang, Achilles, 181-83
Farr, Florence: immigration to Ceylon by, 55; on melody of spoken poetry, 53-54; photograph of, 53*f*; the psaltery and, 52, 53*f*, 59; speech-music and, 51, 54-56, 60, 65, 247n13; Yeats and, 51, 56, 60
Federal Transient Bureau, 63-64
Feldman, Morton, 193
"Fern Hill" (Thomas), 43
Fields of Wonder (Hughes), 132
Fine Clothes to the Jew (Hughes), 142, 144-45, 148
Finkelstein, Norman, 221, 229
"The Fish" (Moore), 203-5
Fletcher, Harvey, 60-61
Flodden, battle (1513) of, 36
"The Flowers of the Forest" (Elliot), 34*f*, 35*f*, 36, 37*f*, 38, 217, 220-21
Ford, Charles Henri, 187
"Fragment 118" (Sappho), 22-23
"Fragments" (Hughes), 133
"Freedom Road" (Hughes), 161-62, 167
Free Lance magazine, 188, 195, 198
Fried, Michael, 170
Friedberg, Leavia, 157
Frith, Simon, 136
"From Men Who Died Deluded" (Sarton), 100
Frye, Northrop, 22

"The Garden" (Marvell), 222-23
Gates Jr., Henry Louis, 116, 122
Genesis of a Music (Partch), 61, 63
"Georgia on My Mind" (Carmichael and Gorrell), 164
Gibson, Corey, 36, 38
Gilmore, Bob, 59
Ginsberg, Allen, 139, 214
Gitelman, Lisa, 139-40
"Go-and-Get-the-Enemy Blues" (Hughes and Handy), 153, 161

Goble, Mark, 113
"Go Down, Death" (Johnson), 115, 262n104
God's Trombones: Seven Negro Sermons in Verse (Johnson), 118, 120
"Good Morning, Stalingrad" (Hughes), 156
Good Neighbor Policy (Roosevelt), 101
Gorrell, Stuart, 164
Granade, S. Andrew, 63
"A Grave" (Moore), 203-5
Great Depression, 19, 50, 62-63. *See also* New Deal
Great Mississippi Flood (1927), 126
Greenberg, Clement, 135
The Green Pastures (Connelly), 119
Grobe, Christopher, 176
Grossman, Allen, 30-31
Grundberg, Carl, 214, 215*f*
Grusin, Richard, 44, 90
Guillén, Nicolás, 4
Guillory, John, 2, 44
Gunn, Thom, 176

Hague, Eleanor, 57-58
Hall, Juanita, 150
Hall Johnson Choir, 119
Hamlet (Shakespeare), 225-26
Handy, W. C., 124-25, 146, 148, 153, 161
"Hard Luck" (Hughes): blues style and, 142-43, 146, 148; divergent strategies of composition in, 151; image of musical setting of, 147*f*; Kingsley's setting for radio of, 146, 151-52, 167-68
Harjo, Joy, 224
Harkness, Nicholas, 66
Harper, Emerson, 153, 161
Harper, Toy, 153
Harris, Joel Chandler, 115-16
Harry Potter and the Sorcerer's Stone (Rowling), 216
Hartman, Saidiya, 130
Havelock, Eric, 76
Hayles, N. Katherine, 5, 25
"Helen All Alone" (Kipling), 213
Helmholtz, Hermann von, 47, 59-61
Henderson, Stephen E., 124, 128
Herbert, George, 224
Hermogenes, 22
Holdridge, Barbara, 32, 42-43
Hollander, John, 25

Index

"Honolulu Yaka-Hula Dixie" (Hughes), 153, 154*f*
Horace, 7, 24-25, 181
Howard, Richard, 105
Howarth, Peter, 175-76
Howe, Susan, 169
Hughes, Langston: Atkins and, 188; Black dialect and, 116, 118; blues poetry of, 142-46, 150-51, 158; *Chicago Defender* weekly column by, 152, 158; Double V campaign and, 153, 155-56, 162, 167; Harlem Renaissance and, 152; lyric intermediality and, 132-33; lyric poetry and, 132-33, 168; *The Man Who Went to War* radio program and, 153; *Music and Ballet* series and, 146; Office of Civilian Defense work by, 153; political affiliation shifts of, 134, 152; songwork and songwriting by, 6, 20, 23, 134, 141, 142, 146, 149-50, 152-53, 154*f*, 155, 157-64, 168, 217; studies of popular music standards by, 164, 165*f*, 166*f*; World War II and, 153, 154*f*, 155-56, 161-62; Yaddo artist community and, 153, 156. See also specific works
Hurston, Zora Neale, 116, 144

Imagism, 18, 98, 170, 186
"In Lieu of the Lyre" (Moore): *The Advocate* magazine's publication (1965) of, 175, 179, 182; audio recording at Harvard University of, 179, 183-84; displacements of poetic objecthood in, 178-79, 182-83; on Harvard University's refusal to admit Moore or other women before 1969, 179, 181, 184; inclusion in *Tell Me, Tell Me* volume of, 179; Moore's reading at Sanders Theatre (1965) of, 174, 175, 176, 180*f*, 182-83; transcription of onstage performance (1965) of, 180*f*; uncertain generic status of, 181
"The Invalidity of Dominant-Group 'Education' Forms for 'Progress' for Non-dominant Ethnic Groups as Americans" (Atkins), 190
Ives, Burl, 149

Jackson, Virginia, 7, 173, 209-10, 216, 221
Jakobson, Roman, 1-3, 23

Jamieson, Robert, 143
Jim Crow's Last Stand (Hughes), 153, 155-56, 158
Johnson, Barbara, 206
Johnson, Guy, 123
Johnson, James Weldon: on "black and unknown bards," 129; Black dialect and, 116, 118, 122; "folk blues" and, 144; *Words Without Music* program and the poetry of, 115, 130. See also specific poems
Jones, Gavin, 122
Jordan, Casper L., 188
Josephson, Matthew, 97
Joyce, James, 44, 76
"The Judgment Day" (Johnson), 127

Kahn, Douglas, 170-71
Kane, Brian, 195
Kassel, Richard, 58
Keats, John, 23, 217
Kelley, Robin D. G., 124
Kenner, Hugh, 76
Khlebnikov, Velimir, 12
King Oedipus (Yeats translation), 48-49, 57, 62, 65, 70
Kingsley, Herbert, 146, 147*f*, 148, 151-52, 163, 167
Kipling, Rudyard, 213
Kittler, Friedrich: discourse networks and, 52; on "informatic density" of printed poem, 138; on the lyre and Greek alphabet, 28; on the lyre as "epistemic thing" (Rheinberger), 29; lyric intermediality and, 139; on modernist poetry and audio technology, 9; on phonography and memory, 138; on phonography and poetry, 140, 146; on radio's programming of day and night, 11; radio technology compared to tap water by, 14-15; on Romanticism and poetic materiality, 33
Köhler, Wolfgang, 192
Kreymborg, Alfred, 89

"Labors of Hercules" (Moore), 177
Lacey, Kate, 13
"Lakefront, Cleveland" (Atkins), 201-3, 205-7
Landau, Irving, 161

latent remediation: Brown and, 19, 93, 112, 123, 131; definition of, 19, 92; psychovisualism and, 197; radio and, 19, 92-93; Wheelwright and, 19, 93, 103
Lawrence, D. H., 61
"Lazy Bones" (Carmichael and Mercer), 164
League of American Writers, 153
Lemmon, Walter, 99-101
LeNoire, Rosetta, 157
Levin, Harry, 179
Liebknecht, Karl, 108, 110
Liebknecht, Wilhelm, 108
The Life of Poetry (Rukeyser), 209
"Limbo Gate" (Adam): Adam (first man) depicted in, 212-13; apocalypse in, 211-12, 222; audio recording (1976) of, 214; Eve (first woman) depicted in, 213, 223; Grundberg's scoring (1982) of, 214, 215*f*, 216; intermediality of, 216, 218, 224, 225; Limbo as an eternal paradise in, 223; lyre, 224, 225; Marvell's "The Garden" and, 222-23
Lindsay, Vachel, 111-15
Li Po, 47
Locke, Alain, 146
Lomax, Alan, 144
Lomax, John, 121, 144
Lord, Albert, 58, 76, 249n40
"The Love Song of J. Alfred Prufrock" (Eliot), 137
Lowell, Robert, 99, 173
Lowell House, 181
Lu Chi, 181
Luxemburg, Rosa, 108
lyre: Adapted Viola and, 8; the aeolian harp, 27; ancient Greek alphabet and, 28; ancient Greek sung poetry and, 21; the bow and, 25; as epistemic "thing" (Rheinberger), 29; intermediality and, 21; lyric and, 8; lyric phaticity and, 23-24; lyric poetry and, 24, 217; as medium for composition and performance, 24; Partch and, 19; poetic autonomy and, 8; prosopopoeia and, 22; radio as, 8; tropological dimensions of, 24
lyric: ballads and, 33; definitional qualities of, 6-7; intermediality of, 5, 7-8, 15-16, 18, 19, 21-22, 28, 39, 42, 50-51, 61, 167, 172, 174, 196, 198-99, 210, 219, 224, 226; lyre and, 5, 8, 50, 71; media history and, 8-9; as media problem, 45; minimal media condition of, 3-4, 7-8; objecthood and, 20, 170, 204-5; reading practices and, 7; social practice and, 169-70; song and, 6, 18, 20, 45, 135-41, 157-58, 160-61, 168; sonorousness and, 4, 6, 33, 169; technological developments of twentieth century and, 9, 18, 19; transcription and, 33, 71
Lyrical Ballads (Wordsworth), 145
lyric poetry: apostrophic address and, 22-23; Dickinson and, 210; Hughes and, 132-33, 168; intermediality and, 28, 50, 218; lyre and, 24, 217; media history and, 224; musicality and, 5; phonography and, 138; as problematically "super-sized" genre, 141; song lyrics contrasted with, 140

Mackey, Nathaniel, 24
MacLeish, Archibald, 16, 89-90, 92, 99, 101
Malanga, Gerard, 72
Mallarmé, Stéphane, 138
Malone, Ted, 91, 99
Mantell, Marianne, 32, 42-43
"Ma Rainey" (Brown), 126
Marinetti, F. T., 12
Marvell, Andrew, 222-23
Marx, Karl, 13, 107
Marxism, 12, 14, 95, 104
Masnata, Pino, 12
"Matt Henson" (Hughes), 155
McCartney, Paul, 164
McCullers, Carson, 161
McGann, Jerome, 172, 186
McGill, Meredith L., 34, 140
McKay, Claude, 38
McLane, Maureen N., 33, 143-45
McLuhan, Marshall, 2, 17, 43-44, 76, 198
"Memphis Blues" (Brown): anaphoric call-and-response in, 126-27; contemporaneity of biblical catastrophe and, 125-26; Corwin's abusive appropriation for radio of, 115, 124, 130; emancipation and, 129; Great Mississippi Flood of 1927 and, 126; latent remediation and, 131; lyre and, 130; preacherly interpolations in, 127-28; "sperrichals" and, 125-26, 128-31

Index

"Memphis Blues" (Handy), 124–25
Mercer, Johnny, 164
"Message to the President" (Hughes), 156
Middleton, Peter, 41
Mill, J. S., 22
Millay, Edna St. Vincent, 91
Miller, Cristanne, 179, 181–82
Miller, Dayton Clarence, 250n51
Miller, J. Hillis, 170
Miller, Karl Hagstrom, 123
Minstrelsy of the Scottish Border, 35–36
Modernism: decadent zenith of print culture and, 76; didactic modernist poetics and, 96; Eliot and, 130; primitivism and, 52; printed sound and, 45; radio and, 102; sound recording and, 4, 43; speech-music and, 50
"Monophony" (Partch): "Americana" compositions and, 70; *King Oedipus* and, 62, 70; lyric intermediality and, 59; microtonal scales and, 47, 58, 60, 87, 218, 250n47; Native American music and, 57–58; phonography and, 56–59; ratio notation and, 55, 56f, 60; relations between performers and audiences in, 66; sound-technological developments and, 50, 59–60, 87; speech-music and, 56, 66; transcription of texts-in-recitation and, 49; Yeats and, 49
Monroe, Harriet, 90–91
Montage of a Dream Deferred (Hughes), 141
Moore, Marianne: Atkins and, 173–74, 187, 199–200; *Bottom: On Shakespeare* and, 82; habit of re-writing poetry on stage by, 177; incorporation of extra-poetic matter in the poetry of, 182; literary celebrity of, 20, 173, 174, 177, 184; *New Yorker* magazine and, 174–75, 176, 179, 181; objecthood of poetry and, 173, 175–76, 179, 184–86, 203–5, 207–8; performance voice of, 177; *Pericles, Prince of Tyre* performed by Zukofskys (1952) for, 51, 72–73; phonography and, 177–78, 207; Sanders Theatre reading of "In Lieu of the Lyre" (1965) by, 174, 175, 176, 180f, 182–83; Sargeant's profile of, 177–78; ubiquitous recording and, 176; "The World in Books" radio appearance (1951) by, 186–87; Zukofsky family's re-
lationship with, 71–72. *See also* "In Lieu of the Lyre" (Moore)
Morton, Timothy, 27
"the mother" (Brooks), 203, 206
"Mother Geese" (Niedecker), 11–13
Mozart, Wolfgang Amadeus, 87, 88
Muntu Poets of Cleveland, 188
Music and Ballet series (radio program), 146, 148
The Music of Speech (Farr), 55
musique concrète, 195
"My Adventures as a Social Poet" (Hughes), 132
"My Cricket" (Dickinson), 210
"My Heart Is a Lone Ranger" (Hughes), 161
Myrsiades, Linda, 203
My Silver Planet (Tiffany), 159–60

Neal, Allison, 177
The Negro and His Songs (Odum and Johnson), 123
"Negro Characters as Seen by White Authors" (Brown), 128
Negro Poetry and Drama (Brown), 116
"The Negro Speaks of Rivers" (Hughes), 150
New American Poetry, The: 1945–1960 (Allen, ed.), 170, 221
New Criticism, 99, 163, 169–70, 172, 186
New Deal, 10, 63, 89, 121. *See also* Great Depression
New Goose (Niedecker), 11–13, 16–17, 39–41, 91
new lyric studies, 209–10
"A New Wind a-Blowin'" (Hughes), 150
Niedecker, Lorine: archival alternatives and, 40–41; folk idiom and, 39; Marxist media critique by, 12, 14; on radio as medium for poetry, 16; radio's impact on the poetry of, 91; silent print in the poetry of, 15, 16; strategic disruption of media channels in poetry of, 15; transcriptional poetics of everyday speech and, 11–12, 14–17, 39–41. *See also specific works*
Nielsen, Aldon Lynn, 188
Niles, John D., 32
Norton, George A., 124
"A Note on Blues" (Hughes), 144–45
"Note on Commercial Theatre" (Hughes), 152

object-forms, 20, 192-99, 201, 208
objecthood of poetry: Atkins and, 20, 173, 183, 188-89; lyric and, 20, 170, 204-5; Moore and, 173, 175-76, 179, 184-86, 203-5, 207-8; psychovisualism and, 174, 188-89, 191; ubiquitous recording and, 20, 171, 175-76, 185-86
Objectivism, 19, 51, 72, 74-80, 87, 189
Objects for Piano (Atkins), 193-94
objet sonore (Schaeffer), 196-98
Oboler, Arch, 89
Occasionem Cognosce (Moore), 181
Ode 1.32 (Horace), 24-25
Odum, Howard, 121, 123
O'Keeffe, Katherine O'Brien, 31-32
Olds, Sharon, 139
Ol' Man Adam an' His Chillun (Bradford), 119
Olson, Charles, 5, 199
"On Lyric Poetry and Society" (Adorno), 45
"On Restless River" (Brown), 126
On the Sensations of Tone (Helmholtz), 47, 59
On Types of Styles (Hermogenes), 22
Outline for the Study of the Poetry of American Negroes (Brown), 119
"Over There" (Cohan), 153
Owen, Maureen, 220

Page, Thomas Nelson, 115
Parker, Charlie, 170
Parker, Patricia, 223-24
Parry, Milman, 58
Partch, Harry: Adapted Viola of, 8, 47, 48*f*, 49, 55-59, 89; *Bitter Music* and, 50, 63-70; "the corporeal" and, 75, 246n7; as ethnographer, 19, 65; "hobo years" in transient shelters of American West of, 19, 62-65, 67; in London (1934-35), 62; lyre and, 19; lyric intermediality and, 51, 78, 92; "Monophony" and, 47, 49, 55-62, 87, 218, 245n6; Native American songs transcribed by, 57-58; phonography and, 57-59; photo of, 48*f*; Pound and, 251n60; saltery and, 56; sound-technological developments and, 55-57, 59-61; Southwest Museum and, 57-58, 65; speech-music and, 47, 49-51, 54-57, 60-61, 63, 64-71, 73, 134, 217-18; Yeats and, 19, 47-51, 55-57, 62-63. *See also specific compositions*

Patchen, Kenneth, 99
Patton, Charley, 38
Paz, Octavio, 25
Penberthy, Jenny, 39
Pence, Charlotte, 136
Pericles, Prince of Tyre (Celia Zukofsky): excerpts from, 73*f*, 84*f*; "heavenly music" in, 83, 85; lack of an audio recording of, 74; lack of public performance of, 51, 72; love and, 79-82; lyre and, 86; lyric intermediality and, 19; poetry of Louis Zukofsky and, 74, 79; publication in *Bottom: On Shakespeare* of, 72, 73-74, 82, 84*f*, 86*f*; rested totality and, 87; score of, 80-82, 84*f*, 85-87, 254n80; Shakespeare's play as source of, 19, 74, 79, 82-83, 255n115; sound-technological developments and, 87; as speech-music, 74; visuality of, 73-74; Williams on, 73, 80; words as tuned objects in, 80-81; Zukofskys' performance for Moore (1952) of, 51, 72-73
Peters, John Durham, 28
Peters, Phillis Wheatley, 224
phatic apostrophe, 23-24
Philip, M. NourbeSe, 5, 26
Phillips, Stephen, 106-8
Philolaus, 28
the phonodeik, 250-51n51
phonography: appropriations of Black culture via, 93; auratic appeal of the authentic voice and, 87; Black dialect and, 120-21, 123; Brown's poetry and, 121, 123-24, 130-31; lyric intermediality and, 175; memory and, 138; "Monophony," 56-59; Negro sermons and, 119-20; nursery rhymes and, 17; as poetic medium, 92, 171; poetry's responses to, 10, 18; recording and playback linked in, 41-42; Thomas and, 44; transcriptional poetics and, 41; ubiquitous recording and, 9, 42
Pindar, 24
Pinsky, Robert, 1
The Plot to Overthrow Christmas (Corwin), 111
"Poetry" (Moore poem), 184-85
Poetry and the Fate of the Senses (Stewart), 226
Poetry Noon Hour (radio program), 99

Index

Political Self-Portrait (Wheelwright), 94-96, 98, 102, 108, 109*f*
Ponge, Francis, 170
Popular Front, 97
Porter, Cole, 38
Porter, Kenneth, 99
Posmentier, Sonya, 119, 126
poststructuralism, 22, 169, 172
Pound, Ezra: didactic modernist poetics and, 96; Imagism and, 170, 248n30; on the medium of poetry, 1, 3; Partch and, 251n60; poetics of radio and, 17-18, 19; on poetry and music, 6; radio broadcasts of, 18, 91; schema of poetic modes developed by, 75, 77
Primus, Pearl, 149
Prins, Yopie, 7, 173
prosopopoeia, 22-23
"Psalm 137" (Farr), 55, 57*f*
"Psalm 137" (Old Testament), 223
Psalms, Book of, 47, 65
the psaltery: Farr and, 52, 53*f*, 59; Partch and, 56; photograph of, 53*f*; semitonic design of, 52, 56; sound-technological developments and, 53; speech-music and, 52, 54-56; Yeats and, 19, 23, 50, 52, 59
psychovisualism: historical actualities of art production and, 191; latent remediation and, 197; lyric intermediality and, 198-99; *musique concrète* and, 195; as "non-dominant aesthetic theory," 189, 190-91, 197; "object-complexity density" and, 197; the object-form and, 193, 195; objecthood of poetry and, 174, 188-89, 191; postwar avant-garde and, 195; redress of uneven distributions of cultural knowledge and, 194-95; sonic inscription and, 198
"A Psychovisual Perspective for 'Musical' Composition" (Atkins), 193, 194*f*
Pythagoras, 61

radio: audience and, 90; connectivity and, 12-13; dramatic productions staged for, 10; fascism and, 18, 89, 94, 101; isolation and, 13; latent remediation and, 19, 92-93; local stations and, 11; as lyre, 8; as poetic medium, 16, 19, 89-92, 94, 96, 98-99, 101-4; poetry's responsiveness to the medium of, 10, 18, 91, 103; political mobilization and, 10; racist protocols during "golden age" of, 93; "sonic color line" and, 114-15, 122-24, 146; switching off of, 13-14; widespread US adoption (1919-1939) of, 9-10, 89. See also *specific programs*
Rahn, Muriel, 150
Ramazani, Jahan, 137
Rampersad, Arnold, 132, 152, 156, 158-59
Rankin, John E., 155
"The Record" (Louis Zukofsky), 87-88
"Red Cross" (Hughes), 156
Reich, Steve, 69, 253n74
Rheinberger, Hans-Jörg, 29
"Rights of Democracy" (Hughes), 153
Rivers, Conrad Kent, 188
Robeson, Paul, 150, 157
Rock and Shell (Wheelwright), 98, 258n25
Romanticism: aeolian harp and, 27, 130; ballads and, 6, 33-34, 135, 143-45, 160, 222; kitsch and, 159-60; poetic cognition and, 28; poetic materiality and, 33; sensualities of poetry and, 138
Roosevelt, Franklin D., 10, 62, 101
Rorem, Ned, 136
Rosseter, Philip, 26
Rukeyser, Muriel, 98, 209
Russell, Irwin, 115-16, 119, 121-22, 130

"Salute to Soviet Armies" (Hughes), 156
Sanders, Mark A., 118, 127, 129
San Francisco Renaissance, 173, 211
Sannazaro, Jacopo, 223
Sappho, 22-24, 28, 38
Sargeant, Winthrop, 177-78
Sarton, May, 99-100
Sartre, Jean-Paul, 14-15
Satorsky, Cyril, 86*f*
Saussure, Ferdinand de, 81
Savage, Barbara D., 114-15
Schaeffer, Pierre, 195-98
Schuchard, Ronald, 51
Schulze, Robin, 184-85
Scott, Walter, 34*f*, 35-36, 143
Scott, Winfield Townley, 94
Selch, Andrea, 112
"Sentir avec ardeur" (Boufflers), 181-83

Serbo-Croatian Folk Songs (Bartók and Lord), 58
Shakespeare, William: *Hamlet*, 225-26;
  Objectivism and, 76, 80; *Pericles, Prince
  of Tyre*, 19, 74, 79, 82-83, 255n115;
  Troilus and Cressida, 77-78; Zukofsky's
  literary-media theory and, 75-80
Shaw, George Bernard, 247n13
Shelley, Percy, 23, 27-28
Siegert, Bernhard, 44
Siegmeister, Elie, 153, 156
Silliman, Ron, 87
Silverstein, Michael, 252n71
"Simon Lee" (Wordsworth), 144
Smith, Bessie, 150
Smith, Hale, 191-92
Smith, Mamie, 145
"The Solitary Reaper" (Wordsworth), 33, 144
Song of the Andoumboulou (Mackey), 137
Songs of Innocence and of Experience (Blake), 137, 220
Songs with Music (Grundberg), 214, 215*f*
songwork: antinomies of writing and sound
  in, 134; blues poetry and, 149; dummy
  lyrics and, 164, 167-68; generic bound-
  aries and, 141, 168, 217; kitsch and,
  159-60; song plugging and, 157-58
"sonic color line" (Stoever), 114-15, 122-24, 146
Sorel, Felicia, 146
Southern Road (Brown poetry collection),
  116, 118, 125-26. See also "Southern
  Road" (Brown poem)
"Southern Road" (Brown poem): communal
  work song and traditional blues struc-
  tures in, 122-23; Corwin's abusive
  appropriation for radio of, 93, 115,
  123-24; latent remediation and, 123-24
Speech and Hearing (Fletcher), 61
Spengler, Oswald, 61
Spenser, Edmund, 223
Spinoza, Baruch, 78
Stept, Sam H., 164, 165*f*, 166*f*
Sterne, Jonathan, 60
Stewart, Susan, 143, 219, 222, 224, 226
Still, William Grant, 153
Stoever, Jennifer Lynn, 114-15

Tagore, Rabindranath, 4
Talmadge, Eugene, 155

"Tam Lin" (Scottish ballad), 226
Taniya, Kyn, 4
"Tea for Two" (Youmans and Caesar), 164
telephony, 4, 9, 60
Tell Me, Tell Me (Moore), 179, 275n23
Tennyson, Alfred, Lord, 106, 171
*They Fly Through the Air with the Greatest of
  Ease* (Corwin), 111
"This Is Worth Fighting For" (De Lange and
  Stept), 164, 165*f*, 166*f*, 167
Thomas, Dylan, 42-45
Tiffany, Daniel: kitsch and, 20, 134, 159-60,
  163, 167; poetic diction and, 163; on
  radio and fascism, 18
Tin Pan Alley music, 113, 135, 143, 153, 160
Titon, Jeff, 145
"Total War" (Hughes), 158-59, 163
Trachtenberg, Alan, 182
"Train Ride" (Wheelwright), 104-8, 110
"Trainyard at Night" (Atkins), 187, 195
*Traité des objets musicaux (Treatise on Musical
  Objects*, Schaeffer*)*, 196
Troilus and Cressida (Shakespeare), 77-78
Trotsky, Leon, 108
Trotter, David, 9

ubiquitous recording: magnetic tape and,
  171; objecthood of poetry and, 20, 171,
  175-76, 185-86; phonography and, 9,
  42; technological development of
  twentieth century and, 9, 170-71
Under Milk Wood (Thomas), 43
U.S. 1 (Rukeyser), 98
U.S. Highball (Partch), 253n76

Valéry, Paul, 136
Van Doren, Carl, 111
Van Doren, Mark, 72
Verma, Neil, 90
"Verse + Radio = Poetry" (Wheelwright), 103
von Praunheim, Rosa, 211

W1XAL radio station (Boston), 93-94, 96,
  99-101, 104. See also WRUL radio station
  (Boston)
Wagner, Bryan, 120-21
The War of the Worlds (Welles radio
  program), 111

Index

"The Waste Land" (Eliot), 99-100, 135, 138-39
"The Weary Blues" (Hughes), 133, 144
Weheliye, Alexander G., 120
Weill, Kurt, 153
Welles, Orson, 10, 111, 225-26
Wên-Fu (Lu Chi), 181
Wheelwright, John Brooks: Anglicanism and, 94-95; anti-war views of, 99; Boston as home of, 94, 97; death of, 96; latent remediation of radio and, 19, 93, 103; magazines published by, 96; Marxism and socialist politics of, 95, 96, 98, 104, 108, 110; media theory of modern poetry and, 102-3; Phillips and palimpsestic poetry by, 106-8; poetry on the radio and, 98, 100-104; political horizons of radio and, 10-11; W1XAL radio program of, 93-94, 96-97, 99-100, 104; WORL radio program of, 99. *See also specific works*
"When to her lute Corrina sings" (Campion), 25-27
White, Heather Cass, 184
White, Josh, 157, 161-62
White, Newman Ivey, 123
Whitman, Walt, 23
Williams, Raymond, 95, 104, 169
Williams, William Carlos, 10, 39, 71-73, 80
Wimsatt, W. K., 163
Wittgenstein, Ludwig, 78
Wons, Tony, 99
Woolf, Virginia, 182
"The Word Is Deed" (Wheelwright), 95-96
Words without Music (Corwin radio program): Black actors hired to perform on, 115; canonical and popular poetry adapted for broadcast on, 111-12; excerpt from script of, 117*f*; "Memphis Blues" and, 124, 130; "Negro poetry" adapted for broadcast on, 93, 112-15, 118, 123-24, 130; original verse dramas broadcast on, 111; seventh episode (January 15, 1939) of, 111, 115; "Southern Road" and, 93, 115, 123-24
Wordsworth, William, 33, 144-45

World War II: Double V campaign and, 153, 155-56, 162, 167; patriotic music during, 20, 153, 154*f*, 155-56, 161-62; Soviet Union and, 156; Tin Pan Alley musicians and, 153
WORL radio station (Boston), 99-100
WQXR radio station (New York City), 111, 146, 148
WRUL radio station (Boston), 93-94, 101. *See also* W1XAL radio station (Boston)

Yeats, W. B.: bardic orality and, 39, 51, 65; Celtic Renaissance and, 51-52; Farr and, 51, 56, 60; folk idiom and, 39; *King Oedipus* translation by, 48-49, 57, 62, 65, 70; lyric intermediality and, 51; modernist primitivism and, 52; Monophony and, 49; Partch and, 19, 47-51, 55-57, 62-63; the psaltery and, 19, 23, 50, 52, 59; regenerative oral poetry and, 50, 52; speech-music and, 51-52, 54-57, 60, 65, 71, 73
"Yesterday" (McCartney), 164
Youmans, Vincent, 164

Zong! (Philip), 5
Zukofsky, Celia: *Pericles, Prince of Tyre* performed for Moore (1952) by, 72; *Pericles, Prince of Tyre* scored by, 72, 74; poetics of transcription and, 38; speech-music compositions of, 19; Williams and, 73
Zukofsky, Louis: "graph of culture" developed by, 75-76; latent remediation and, 74; literary-media theory developed by, 75-80, 85-86; lyric intermediality and, 74-75, 78; *New Goose* (Niedecker) and, 11; Objectivism and, 19, 51, 72, 74-75, 79, 87, 189; *Pericles, Prince of Tyre* performed for Moore (1952) by, 72; on *Pericles, Prince of Tyre*'s significance for his poetry, 74, 79; phonography and, 87; on poetry as "audibility in two-dimensional print," 3; Pound's schema of poetic modes and, 75, 77; on prosody, 78-79; Shakespeare and, 75-80
Zukofsky, Paul, 71-72, 87

www.ingramcontent.com/pod-product-compliance
Lightning Source LLC
Chambersburg PA
CBHW051207300426
44116CB00006B/462